Microscale Inorganic Chemistry

A Comprehensive Laboratory Experience

Zvi Szafran
Ronald M. Pike
Mono M. Singh
Department of Chemistry
Merrimack College

JOHN WILEY & SONS, INC.

Recognizing the importance of preserving what has been written, it is a policy of John Wiley & Sons, Inc. to have books of enduring value published in the United States printed on acid-free paper, and we exert our best efforts to that end.

Acquisition Editor / Dennis Sawicki
Managing Editor / Joan Kalkut
Designer / Kevin Murphy
Production Supervisor / Lucille Buonocore
Copy Editor / Jeannette Stiefel

Library of Congress Cataloging in Publication Data:

Szafran, Zvi.
Microscale inorganic chemistry: a comprehensive laboratory experience / Zvi Szafran, Ronald M. Pike, Mono M. Singh.
p. cm.
Includes index.
ISBN 0-471-61996-5
1. Chemistry, Inorganic—Laboratory manuals. I. Pike, Ronald M. II. Singh, Mono M. III. Title.
QD155.S96 1991
542—dc20 90-46328
CIP

To our families

Jill and Mark, and parents: Daniel and Simona

*Marilyn, Dana and Jane, Gretchen and John,
and grandchildren Benjamin, Daniel, Allison, Erik, Shane, Scott, and Beth*

Shashi, Yuvraj, Balraj, Pamela, and parents Basant Singh and Lal Kaur

Preface

All Science is formed necessarily of three things: the series of facts which constitute the science; the ideas which they call forth; the words which express them.

Lavoisier

Inorganic chemistry is the most mature branch of science, having been practiced from ancient times. It is also the "new kid on the block" because most of the theory and current practice of inorganic chemistry dates back only to the 1950s. It is an area chosen by relatively few chemists, when compared to organic and biochemistry, yet is is one of the fastest growing areas, dealing with research in room temperature superconductors, modern ceramics and materials science, and the huge field of organometallics. Even the controversies in inorganic chemistry are spectacular: Note the worldwide attention to the on-again, off-again, on-again nature of cold fusion in metal lattices. Note the potential applications of room temperature superconductivity. Note the worldwide protests against nuclear power generation.

The inorganic laboratory should be an equally spectacular place. The sad truth is that many institutions have no such course. In the past, students took one year of general chemistry, which largely consisted of descriptive inorganic chemistry and laboratory work. That course has evolved into introductory physical chemistry in most locations.

The American Chemical Society has recognized this problem and has made recommendations that chemistry degree programs radically increase the amount of exposure students get in the inorganic area, including the laboratory experience. Even so, this is a difficult endeavor. High material costs, recognition of chemical waste disposal difficulties, toxicity fears, and the long and tedious procedures needed to prepare many inorganic materials have all played their parts in retarding the implementation of a modern, vigorous laboratory experience.

We had several goals in mind when we set out to develop this laboratory textbook. Most importantly, we wanted to challenge and inspire students by emphasizing the sweep and beauty of the inorganic area. Too many previous approaches had been limited to the least expensive inorganic reagents, forcing the elimination of much that is of interest or of importance.

We want the laboratory to be a healthful and enjoyable place for the experimenter to work. We seek to incorporate modern instrumentation and analytical techniques into our procedures, so that students will be prepared for further professional growth. We want to accomplish these goals in an ecologically responsible manner. Finally, we have looked to the bottom line, and tried to accomplish this in an efficient and cost effective manner.

This laboratory textbook is a new approach to the teaching of the introductory and advanced inorganic chemistry laboratory. It offers a flexible strategy to assist in the revitalization of this important aspect of undergraduate and advanced chemical training. We recognize that inorganic chemistry, when presented at all, may appear at the sophomore, junior, senior, or graduate level. The experiments, therefore, are offered at a variety of different levels of sophistication.

There are four major divisions in this book: chemistry of the main group elements; chemistry of the transition metals; organometallic chemistry, and bioinorganic chemistry. Within each division, we have tried to incorporate the use of many different elements, some familiar and some less so. There are, of course, 105 different elements to choose from, and it would be impossible to include them all. We certainly did our best to try, though! Any slights to specific areas of the field are entirely unintentional.

The experiments follow one another in a (hopefully) logical sequence. For example, Experiments 20 and 21 involve the preparation of simple transition metal complexes. Experiment 22 prepares metal chelate complexes, one of which is studied chromatographically in Experiment 23 and spectroscopically in Experiment 29. Many experiments have multiple parts, which can be treated sequentially or individually. It may be desirable to assign different students to different options, and to have them collect and compare their results, and to draw conclusions. Chemistry is, after all, a collective enterprise, involving the work of thousands of scientists around the world!

We have adopted the microscale approach in this text. The inauguration of microscale chemistry to the organic area fomented a revolution in thinking about how instructional laboratories should be conducted. That, in turn, led us to consider its merits in the area of inorganic chemistry. Since 1986, we have offered a microscale inorganic laboratory at the sophomore level at Merrimack College. In many ways, the benefits in inorganic chemistry exceed those in organic chemistry. Look at each experiment, in turn, as you do them. Can you visualize the preparation of a rhodium carbonyl at the the traditional scale (5–25 g of reagent), with $RhCl_3 \cdot xH_2O$ "weighing in" at $60 per gram? Would you really want to work with a large amount of thiosemicarbazide, knowing its great toxicity? What quantity of $NI_3 \cdot NH_3$ or ICl_3 do you want to keep around the laboratory, when you only require a few milligrams to explore its chemistry and to characterize it? In short, why use or make more of anything than you need?

The down-sizing of the laboratory scale has also allowed us to down size the amount of time needed in the various preparations. The assembly, reaction, manipulation, workup, and characterization times needed for a given procedure are sharply reduced when they are carried out at the microscale level. Most experiments in this text, in marked contrast with earlier works, can be carried out in a single 3-h laboratory period, or less.

In summary, by adopting the microscale approach, we can accomplish a larger number of experiments having greater variety, at a lower cost, more efficiently, and with greater safety.

Each experiment is followed by a series of questions. Some of these are elementary and some are quite advanced. The last question in each experiment will require a search of the chemical literature in order to be answered.

As with any first edition, there will, no doubt, be some errors. We would appreciate any suggestions and feedback for improving this text that its users might wish to offer.

We acknowledge the assistance of the following students who helped in the development of several of the procedures in this text: Mark Gelinas, Mark Johnston, Kimberly Parthum, and Nancy Rogan. We further thank those students who, over the past four years, have accompanied us on our microscale inorganic adventure in our sophomore laboratory.

Many friends have contributed thoughts and suggestions about techniques, experiments, and procedures. They include John Woolcock (Indiana University of Pennsylvania, magnetic susceptibility), Stephen A. Leone (Merrimack College, phosphazines), John E. Frye (Northern Michigan University, ammonium hexachloroplumbate), and William Heuer (Franklin and Marshall College, copper glycine). We also thank the following individuals for their kind assistance: Michael Laing (University of Natal), Jerome D. Odom (University of South Carolina), and Herbert Beall (Worcester Polytechnic Institute).

We express our appreciation to our colleagues here at Merrimack College: J. David Davis, Stephen A. Leone, Irene McGravey, Diane Rigos, Carolyn Werman, Rita Fragala, Catherine Festa, and Charlene Mahoney.

We acknowledge the financial assistance in the preparation of this manuscript of John Wiley & Sons, Inc., and Merrimack College. We also appreciate the encouragement from our chemistry editor, Dennis Sawicki. We would also like to thank the reviewers of this text for their insightful comments. The following companies donated equipment that helped immeasurably: Ace Glass, Inc., Aldrich Chemical Co., Inc., Ballston, Inc., DuPont Instruments, Hewlett–Packard, Nicolet Instruments, and Varian Instruments.

Finally, and most importantly, we thank our wives, Jill, Marilyn, and Shashi for their patience, understanding, and love.

Zvi Szafran
Ronald M. Pike
Mono M. Singh

North Andover, MA

Contents

List of Tables

Introduction

A. A BRIEF HISTORY OF INORGANIC CHEMISTRY

The history of inorganic chemistry is as old as the history of humankind. The ancients knew how to smelt copper, and to produce bronze weapons and kitchen implements, golden jewelry, and to utilize clays as building materials. In fact, we name the ages of human progress in terms of inorganic chemistry: the Stone Age, the Bronze Age, the Iron Age, and the Atomic Age.

The Middle Ages were not scientifically barren, as many people think. Glassmaking, the refining of lead and zinc, the discovery of the mineral acids, and the crude utilization of chemicals in medicine began in this period. Most of the "chemical" attention was focused on the conversion of base metals into gold, but much chemical knowledge was gathered along the way.

In the 18th century, the world of chemistry was divided into two sweeping categories: organic chemistry and inorganic chemistry. It was believed that chemists could only isolate and prepare inorganic compounds—organic compounds being imbued with a vital essence, which was obtainable only from God. Mere human beings, it was believed, could not perform similar miracles.

These thoughts undoubtedly arose because the inorganic compounds known at the time were relatively simple materials, such as HCl, H_2O, and $NaBr$. Metals had been known and worked with for centuries, and in some instances, millennia. Organic compounds, on the other hand, were complex materials, not lending themselves readily to chemical analysis. As a result, inorganic chemistry predominated during this period. By 1820, about 50 elements were known, and their simple compounds had already been investigated.

Friedrich Wöhler, in 1824, accidentally carried out a seminal experiment, wherein he heated ammonium cyanate, an inorganic compound, and thereby obtained urea, an organic compound.

$$\underset{\text{Ammonium cyanate}}{NH_4OCN} \rightarrow \underset{\text{Urea}}{NH_2CONH_2}$$

Urea was known to be isolable from mammalian urine and to be a product of human metabolism. Wöhler had done the seemingly impossible and, in fact, did not initially believe his own accomplishment. This established organic chemistry as a definitive field, and was the first step in a long period of decline for inorganic chemistry. Although the chemical industry still relied on inorganic materials for the largest bulk of their manufacture, most chemical research from 1850–1950 was carried out in the area of organic chemistry.

This is not to say that inorganic chemistry was completely ignored over this time period. Werner, in the early 1900s, developed the elegant theory of primary and secondary valence, which led to today's coordination theory. The Curies were instrumental in the discovery of several radioactive elements and in the understanding of radioactivity itself. Dozens of new elements were discovered, the last nonradioactive example being rhenium, in 1925, which was discovered by Nodack, Tacke, and Berg.

In modern times, inorganic chemistry has enjoyed a huge resurgence. The growth of the electronics and computer industry has stirred an extensive interest in the chemistries of silicon, germanium, gallium, indium, and other main group elements. Metals have been found in biological systems as the active agents in oxygen and energy transport, as well as in the regulation of cells. Metal clusters have been extensively explored, because of their ability to act as and to mimic catalytic systems. Some of the formerly inert gases were found to react under a wide variety of conditions, the first compound being discovered in 1962 by Bartlett. The first superconductors able to function above liquid helium temperatures were discovered in 1986, consisting of oxides of copper–barium–lanthanide systems.

The immense field of organotransition metal chemistry blossomed in 1951 with the discovery of ferrocene by Pauson and Kealey and, separately, in 1952 by Miller, Tebboth, and Tremaine. Ferrocene was an unprecedented compound, consisting of a metal bonded only to carbon. It was later discovered that other organometallic compounds had been prepared earlier, although their true nature was not known. Examples include an ethylene compound of platinum (Zeise's salt, discovered in 1827) and bis(benzene)chromium (Hein, 1919). Metal carbonyls were known since 1890, when Mond discovered $Ni(CO)_4$ as a corrosive agent of nickel valves. One year later, $Fe(CO)_5$ was first synthesized by Berthelot, Mond, and Quinche. Organometallic chemistry is one of the fastest growing areas of chemistry today.

Inorganic chemistry was hampered in its development by a lack of coherent bonding theory for many known compounds and complexes. Oddities such as the various electron deficient boron hydrides, first discovered by Stock in 1912, had to wait for theoretical explanation until 50 years later by Lipscomb.

The central theories for the bonding in transition metal compounds were not proposed until 1929 by Bethe (crystal field theory) and 1935 by Van Vleck (ligand field theory). The current all-encompassing theory, molecular orbital theory, had its origins in the quantum mechanical revolution in the early 1900s, and is increasingly applied to inorganic systems today.

What does the future hold for inorganic chemistry? It is clear that with 105 different elements to work with (not to mention the limitless combinational possibilities), the surface has barely been scratched. We hope that you, as you undertake this laboratory experience, will be motivated to be a part of this exciting, ongoing enterprise.

B. THE MICROSCALE APPROACH

The microscale approach involves using small quantities of starting materials (typically 25–100 mg), rather than the "traditional scale" of 5–25 g. There has been a steady sequence of declines in scale over the past two centuries. Today, with the increasing emphasis placed on ecological and safety considerations, we seek to minimize the exposure and production of potentially dangerous products.

The purpose of a chemistry laboratory is to acquaint the experimenter with the various basic techniques and manipulations required to carry out procedures and processes for the preparation of new materials. Obviously, there is a lower practical limit to the amount of reagent that can be used. It would not be convenient, for example, to use an amount so small that one could only see it through a microscope! Until recently, the lowest practical amount of reagent was 5–25 g, because of the need to characterize the product. Most materials were investigated by various "wet chemical" tests and via the synthesis of derivatives.

Today, most products are characterized by instrumental techniques, including magnetic susceptibility (Section 5A); thermal analysis (Section 5B); visible, infrared, and nuclear magnetic resonance spectroscopy (Sections 6B–6D, respectively), and atomic absorption (Section 6E). These techniques are inherently microscale, in that they require far less than 1 g of product for complete analysis. Usually, ~50 mg of product will be sufficient to characterize a given material in several ways.

The chemistry laboratory must be responsive to changing times and technological advances. It is our belief that the microscale approach allows an up-to-date, environmentally sound, safe learning experience. During the development of this textbook, we have often been asked (even by our colleagues!) if something is not lost in working at this small scale. We obviously believe that the answer is "no." Industrial reactions, carried out in huge quantities, have problems associated with mixing and mass transfer, which are not present at the microsale level. They are also not present in traditional benchtop work. Whenever a reaction is "scaled up" from the microscale (or benchtop) to an industrial level, industry recognizes this fact by the necessity of having a pilot plant step in the sequence to investigate these problems. In this sense, the microscale level is equivalent, or superior, to the traditional benchtop scale. Of course, all experiments in this book can be scaled up, without difficulty, taking into consideration appropriate spot heating and mass transfer considerations.

C. A WORD TO THE STUDENT

You are embarking on an exciting voyage, using unfamiliar tools and methods to investigate new worlds in chemistry. At this point you may have doubts. It may seem to you that it is impossible to work efficiently at the small scale used in this text. The wide sweep of knowledge to be mastered in inorganic chemistry may seem overwhelming.

Unlike Columbus and his crew, however, this journey was made before, by other students. We found that the microscale approach to inorganic chemistry works well, and allows the student access to a broad variety of learning experiences that would not otherwise be possible. You will be amazed to find that your perspectives as to scale will change drastically. Believe it or not, 100 mg of product will seem like a ton! Keep in mind, however, that all journeys begin with single steps. We hope your journey is a pleasant and successful one!

Chapter 1

Safety in the Laboratory

1.A GENERAL SAFETY RULES

1.A.1 Introduction

Safety in the laboratory is a subject of the utmost importance. Since all chemicals are harmful to some degree, the best way to ensure safety in the laboratory is by minimizing contact with all chemicals. Thus, the main way in which we promote safety in the inorganic laboratory is by using the microscale technique. This technique lets us reduce the amounts of chemicals used by a factor of 100–1000 from the traditional multigram scale previously used in inorganic laboratories. There are several advantages to doing this, many of which are safety related:

- Less toxic waste is generated. This saves money on disposal costs, and helps protect both the chemist involved (you!), your instructor, and the environment.
- Using smaller amounts of flammable or potentially explosive compounds reduces the chances of fire or an explosion.
- The air quality in the laboratory is improved. Using smaller amounts of volatile compounds cuts down sharply on the amounts of chemicals present in the air, improving both the smell and the healthfulness of the laboratory.
- Exposure to any chemical should be minimized, because all chemicals should be viewed as being potentially dangerous. Cutting down on the amounts of chemicals used minimizes the exposure, especially if other safety measures are also adopted.

At this point, it may seem to you that using microscale techniques eliminates all the risks involved in a chemical laboratory. It is certainly true that working with microscale techniques will minimize the risks. However, even using these small amounts of material, some chemicals are still highly toxic, spills or splattering of a corrosive material still can occur, or a compound may still decompose to generate a noxious gas. Furthermore, in future work, a required scale-up of a reaction may be necessary. It may be that you need more than a micro-amount of a particular product. For these reasons, plus the fact that it just makes plain sense, it is prudent for each of us to be aware of several safety regulations concerning work in a chemical laboratory. It should be emphasized that, as an individual, you have an obligation to protect yourself and your fellow workers. To paraphrase Donne, no one is an island unto him- or herself in the chemical laboratory!

1.A.2 Before the Laboratory

Safety in the laboratory does not begin when you walk in the laboratory door. There are three initial steps that should be carried out before the experiment begins:

1. Read the directions of the experiment to be carried out with a critical eye, in advance of the laboratory.
2. Think about what you are reading and visualize the experimental sequence as to the chemicals used and the arrangement of equipment. Especially note any safety warnings such as "Use the hood" or "No flames allowed." The safety warnings in this laboratory text are there for your and your neighbors' protection. Many people find it useful to prepare a flow chart for each experiment, listing each step of the laboratory in sequence.
3. Check the toxicities of the chemicals involved. Toxicity data is given for each chemical used in the experiment, in a section called "Safety Recommendations." You should also check such safety sources as MSD sheets or the *Merck Index*, discussed in Sections 1.B.2 and 1.B.3. Toxicities of common solvents are given in Appendix A.

To enhance safety in the laboratory, there are several considerations with regard to what you wear:

1. Safety goggles or suitable protective glasses are crucial. They should be worn at all times in the laboratory. Any visitors to the laboratory should also be required to have suitable protection. Contact lenses are **not to be worn**, as corrosive fumes or chemicals may get underneath them and prevent effective breathing and flushing of the eyes, especially if an accident occurs.
2. Suitable clothing should be worn. It should be obvious to you that open-toed shoes or sandals offer no protection to the feet from chemical spills. They should never be worn in the laboratory. Long hair should be tied back, and if ties or similar loosely hanging clothing are worn, they should be tucked in in an appropriate manner. Clothing that offers protection against an accidental spill is most appropriate. Suitable laboratory aprons or coats are highly recommended. Similarly, clothing that leaves the midriff exposed (such as a cut-off T-shirt) should never be worn.

1.A.3 SAFETY RULES IN THE LABORATORY

Specific rules and regulations have been formulated based on the experience of those who have extensively studied the safety aspects of the laboratory. It is imperative that you learn these safety rules and follow them at all times. You will be expelled from the laboratory for failing to comply with these regulations. These rules are referred to in many laboratories as "the usual safety procedures."

1. Use your common sense.
2. Do not rush; do not take shortcuts. If you rush your work, at best you will get poor results. At worst, you will be dangerous to yourself and those around you.
3. Report any spill or accident *immediately* to your instructor.
4. Know the location and operation of safety equipment in the laboratory from the first meeting of the laboratory section. This includes the following

equipment:

- Eyewash fountains
- Safety showers
- Fire extinguishers
- Fire exits
- First aid kits
- Fire blankets

5. Smoking is absolutely forbidden in the laboratory. Volatile, flammable solvents can ignite easily and result in explosions or fire. Severe burns can result from carelessness due to smoking or the use of an open flame in the vicinity of flammable solvents. The microscale laboratory markedly reduces the possibility of this aspect of potential injury but we must always be on our guard nevertheless.

6. Never work alone in a chemical laboratory. In the rarest of periods when you might be working alone, have someone check on you at regular intervals. As in swimming, the "buddy" system is the safest way to go.

7. Minimize exposure to all chemicals, whether they are considered toxic or not. Handle all chemicals according to any specific directions on the container, or those given to you by your instructor. Never pick up spilled solids with your bare hands. Never directly smell a chemical. Never pipet by mouth.

8. Dispose of chemicals properly, in the containers provided, and according to the instructions given by the laboratory instructor. Do not simply pour waste chemicals down the sink. Recent governmental regulations have place stringent rules on industrial and academic laboratories for the proper disposal of chemicals. This area is a major factor in the cost of running a chemical operation. Severe penalties are levied on those who do not follow proper waste disposal procedures.

9. Do not put anything in your mouth under any circumstances while in the laboratory. This rule pertains to *food, drink, and pipets.* Do not taste any chemicals. Do not inhale vapors from volatile materials. Some specific chemicals can be absorbed through the skin. In these cases, wear protective gloves and/or clothing. Your instructor or the experimental instructions will provide information in such cases.

10. The majority of your equipment is made of glass, and thus proper procedures for assembling and dismantling should be followed. Your instructor will introduce you to the techniques required. Be particularly careful when inserting thermometers into rubber stoppers. *Be sure the stopper is lubricated with water or glycerin.* In any glass operation breakage may occur and a cut or laceration may result. Report any accident to your instructor immediately, no matter how insignificant it may seem. Quick treatment can often prevent infection or other complications from occurring.

11. Do not carry out unauthorized experiments. At this stage of your chemical development, it is imperative that you follow the procedures given in the laboratory manual for basic safety reasons. **Immediate expulsion from the laboratory and failure of the course is the penalty for disobeying this rule**.

12. Keep your laboratory space clean. This also pertains to the balance area and where the chemicals are dispensed. You or your fellow students unknowingly can be burned or exposed to toxic chemicals if you do not clean up a spill.

13. Replace caps on containers immediately after use. An open container is an invitation for a spill. Furthermore, some reagents are very sensitive to moisture, and may decompose if left open.

14. Never heat a closed system. Always provide a vent to avoid an explosion. Provide a suitable trap for any toxic gases generated such as sulfur dioxide, hydrogen chloride, and chlorine. Directions will be found in the experimental procedures.

15. Learn the correct use of gas cylinders. Even a small cylinder of gas can become a lethal bomb if not used properly. The use of such cylinders is discussed in Section 1.B.4.

16. When using a mercury bubbler, cover the surface of the mercury with a layer of mineral oil to prevent toxic vapors.

1.B PLANNING FOR CHEMICAL SAFETY

1.B.1 Introduction

Many chemicals that are familiar to you have certain dangers associated with their use. Hydrochloric acid, for example, is very corrosive, and can cause severe burns on the skin. Chloroform is a narcotic agent and is also carcinogenic. Lead compounds are toxic and can cause heavy metal poisoning. Carbon monoxide is a toxic gas, which binds to hemoglobin in the blood more strongly than oxygen. It could be deadly, by way of suffocation, if enough is present. How, then, does one go about designing a safe laboratory experiment?

In some cases, the dangerous material can be entirely avoided. We will be making a rhodium carbonyl compound in Experiment 42. The most common way of preparing such compounds is to have the metal (or a metal compound) react directly with carbon monoxide. This is a dangerous reaction, however, due to the above-mentioned toxicity of carbon monoxide. We use an alternate route, employing the much safer compound *N,N*-dimethylformamide (DMF) as the source of the CO group. We do this because the dangers associated with using DMF are much lower than for carbon monoxide. Keep in mind, however, that this does not mean that DMF is perfectly safe either.

When the metal itself is dangerous, as in the case of lead compounds, the safety problem is harder to solve. One solution is simply not to do experiments using lead. This may not be as bad a solution as it may immediately seem, because much of the chemistry of lead can also be shown using other, safer, metals such as tin. Not wishing to avoid lead altogether, we have provided a procedure (Experiment No. 10) in which lead is used in one of its safer forms, one not readily incorporated into the body. Furthermore, we protect the environment from the hazards associated with the disposal of lead compounds by recycling our product in the end, back to the starting material. In this way, we generate no toxic wastes.

1.B.2 Use of MSD Sheets

Many chemicals, as pointed out previously, have dangers associated with their use. One way of combating these dangers is by the use of Material Safety Data Sheets (MSDS), which are provided by each manufacturer or vendor as required by law for the chemicals purchased and used in your laboratory. The information given relates to the risks involved when using a specific chemical. These sheets are available to you as a laboratory worker. Your chemistry department should have these sheets on file. Material safety data sheets may also be available in book form or in CD-ROM form on compact discs for running on a personal computer. Check with your laboratory instructor.

A typical MSD sheet is shown in Figure 1.1 for sodium chloride. The sheet was obtained from the Sigma–Aldrich CD-ROM MSD sheet compact disc (July 1989 version). The sheet is divided into several sections: Identification, Toxicity Hazards, Health Hazard Data, Physical Data, Fire and Explosion Data, Reactivity Data, Spill or Leak Procedures, and Additional Precautions and Comments. The identification section provides additional names by which the compound is known (e.g., salt), the CAS (Chemical Abstract Service) number, and the Sigma or Aldrich catalog product number. The CAS number is especially useful, as

```
MATERIAL SAFETY DATA SHEET

Sigma-Aldrich Corporation
1001 West Saint Paul Ave, Milwaukee, WI  53233 USA

July 1989 version

            ------------------ IDENTIFICATION ------------------
    PRODUCT #: S9888          NAME: SODIUM CHLORIDE ACS REAGENT
    CAS #: 7647-14-5
    MF: CL1NA1
SYNONYMS
    COMMON SALT * DENDRITIS * EXTRA FINE 200 SALT * EXTRA FINE 325
      SALT * HALITE * H.G. BLENDING * NATRIUMCHLORID (GERMAN) *
      PUREX * ROCK SALT * SALINE * SALT * SEA SALT * STERLING *
      TABLE SALT * TOP FLAKE * USP
    SODIUM CHLORIDE * WHITE CRYSTAL *
            ----------------- TOXICITY HAZARDS -----------------
RTECS NO: VZ4725000
    SODIUM CHLORIDE
IRRITATION DATA
    SKN-RBT 50 MG/24H MLD                  BIOFX* 20-3/71
    SKN-RBT 500 MG/24H MLD                 28ZPAK -,7,72
    EYE-RBT 100 MG MLD                     BIOFX* 20-3/71
    EYE-RBT 100 MG/24H MOD                 28ZPAK -,7,72
    EYE-RBT 10 MG MOD                      TXAPA9 55,501,80
TOXICITY DATA
    ORL-RAT LD50:3000 MG/KG                TXAPA9 20,57,71
    ORL-MUS LD50:4000 MG/KG                FRPPA0 27,19,72
    IPR-MUS LD50:6614 MG/KG                COREAF 256,1043,63
    SCU-MUS LD50:3 GM/KG                   ARZNAD 7,445,57
    IVN-MUS LD50:645 MG/KG                 ARZNAD 7,445,57
    ICV-MUS LD50:131 MG/KG                 TYKNAQ 27,131,80
REVIEWS, STANDARDS, AND REGULATIONS
    EPA GENETOX PROGRAM 1988, NEGATIVE: IN VITRO CYTOGENETICS-
      NONHUMAN; SPERM MORPHOLOGY-MOUSE
    EPA GENETOX PROGRAM 1988, INCONCLUSIVE: MAMMALIAN
      MICRONUCLEUS
    EPA TSCA CHEMICAL INVENTORY, 1986
    EPA TSCA TEST SUBMISSION (TSCATS) DATA BASE, JANUARY 1989
    MEETS CRITERIA FOR PROPOSED OSHA MEDICAL RECORDS RULE
      FEREAC 47,30420, 82
  GET ORGAN DATA
    MATERNAL EFFECTS (OVARIES, FALLOPIAN TUBES)
    EFFECTS ON FERTILITY (PRE-IMPLANTATION MORTALITY)
    EFFECTS ON FERTILITY (POST-IMPLANTATION MORTALITY)
    EFFECTS ON FERTILITY (ABORTION)
    EFFECTS ON EMBRYO OR FETUS (FETOTOXICITY)
    EFFECTS ON EMBRYO OR FETUS (FETAL DEATH)
    SPECIFIC DEVELOPMENTAL ABNORMALITIES (MUSCULOSKELETAL
      SYSTEM)
            --------------- HEALTH HAZARD DATA ---------------
ACUTE EFFECTS
    MAY BE HARMFUL BY INHALATION, INGESTION, OR SKIN
      ABSORPTION.
    CAUSES EYE IRRITATION.
    CAUSES SKIN IRRITATION.
    MATERIAL IS IRRITATING TO MUCOUS MEMBRANES AND UPPER
    RESPIRATORY TRACT.
```

Figure 1.1 *Material safety data sheet: sodium chloride. (Reprinted with permission of Aldrich Chemical Co., Inc., Milwaukee, WI.)*

FIRST AID
IN CASE OF CONTACT, IMMEDIATELY FLUSH EYES WITH COPIOUS AMOUNTS OF WATER FOR AT LEAST 15 MINUTES.
IN CASE OF CONTACT, IMMEDIATELY WASH SKIN WITH SOAP AND COPIOUS AMOUNTS OF WATER.
IF INHALED, REMOVE TO FRESH AIR. IF NOT BREATHING GIVE ARTIFICIAL RESPIRATION. IF BREATHING IS DIFFICULT, GIVE OXYGEN. CALL A PHYSICIAN.

------------------ PHYSICAL DATA ------------------

MELTING PT: 801 C
SPECIFIC GRAVITY: 2.165

APPEARANCE AND ODOR
WHITE CRYSTALLINE POWDER

---------- FIRE AND EXPLOSION HAZARD DATA ----------

EXTINGUISHING MEDIA
NON-COMBUSTIBLE.
USE EXTINGUISHING MEDIA APPROPRIATE TO SURROUNDING FIRE CONDITIONS.

SPECIAL FIREFIGHTING PROCEDURES
WEAR SELF-CONTAINED BREATHING APPARATUS AND PROTECTIVE CLOTHING TO PREVENT CONTACT WITH SKIN AND EYES.

---------------- REACTIVITY DATA ----------------

INCOMPATIBILITIES
STRONG OXIDIZING AGENTS
STRONG ACIDS

HAZARDOUS COMBUSTION OR DECOMPOSITION PRODUCTS
NATURE OF DECOMPOSITION PRODUCTS NOT KNOWN

------------ SPILL OR LEAK PROCEDURES ------------

STEPS TO BE TAKEN IF MATERIAL IS RELEASED OR SPILLED
WEAR RESPIRATOR, CHEMICAL SAFETY GOGGLES, RUBBER BOOTS AND HEAVY RUBBER GLOVES.
SWEEP UP, PLACE IN A BAG AND HOLD FOR WASTE DISPOSAL.
AVOID RAISING DUST.
VENTILATE AREA AND WASH SPILL SITE AFTER MATERIAL PICKUP IS COMPLETE.

WASTE DISPOSAL METHOD
FOR SMALL QUANTITIES: CAUTIOUSLY ADD TO A LARGE STIRRED EXCESS OF WATER. ADJUST THE PH TO NEUTRAL, SEPARATE ANY INSOLUBLE SOLIDS OR LIQUIDS AND PACKAGE THEM FOR HAZARDOUS-WASTE DISPOSAL. FLUSH THE AQUEOUS SOLUTION DOWN THE DRAIN WITH PLENTY OF WATER. THE HYDROLYSIS AND NEUTRALIZATION REACTIONS MAY GENERATE HEAT AND FUMES WHICH CAN BE CONTROLLED BY THE RATE OF ADDITION.
OBSERVE ALL FEDERAL, STATE, AND LOCAL LAWS.

– PRECAUTIONS TO BE TAKEN IN HANDLING AND STORAGE –

CHEMICAL SAFETY GOGGLES.
USE PROTECTIVE CLOTHING, GLOVES AND MASK.
SAFETY SHOWER AND EYE BATH.
MECHANICAL EXHAUST REQUIRED.
DO NOT BREATHE DUST.
DO NOT GET IN EYES, ON SKIN, ON CLOTHING.
WASH THOROUGHLY AFTER HANDLING.
IRRITANT.
KEEP TIGHTLY CLOSED.
HYGROSCOPIC
STORE IN A COOL DRY PLACE.

Figure 1.1 *(Continued)*

------ ADDITIONAL PRECAUTIONS AND COMMENTS ------
SECTION 9 FOOTNOTES
REACTS VIOLENTLY WITH BROMINE TRIFLUORIDE AND LITHIUM.
THE ABOVE INFORMATION IS BELIEVED TO BE CORRECT BUT DOES NOT PURPORT TO BE ALL INCLUSIVE AND SHALL BE USED ONLY AS A GUIDE. SIGMA-ALDRICH SHALL NOT BE HELD LIABLE FOR ANY DAMAGE RESULTING FROM HANDLING OR FROM CONTACT WITH THE ABOVE PRODUCT. SEE REVERSE SIDE OF INVOICE OR PACKING SLIP FOR ADDITIONAL TERMS AND CONDITIONS OF SALE

Figure 1.1 *(Continued)*

one can access several data bases using this number to obtain listings of papers and books that use this compound.

The **toxicity hazards section** contains results of studies detailing the toxicity of the compound in various animal and inhalation tests. Sodium chloride is a well studied compound, so many such tests have been performed. There are several common abbreviations used.

HMN	Human
IPR	Intraperitoneál dose (inside the smooth membrane, lining the interior of the stomach)
IVN	Intravenous (in the blood stream)
LD50	The dose with which 50% of the test subjects will die
LDLo	Lowest lethal dose
MUS	Mouse
ORL	Oral dose
RBT	Rabbit
SKN	Skin

Other abbreviations may be found in the NIOSH 1980 Registry: Toxic Effects of Chemical Substances, U.S. Dept. of Health and Human Services. Thus, the listing ORL-RAT LD50: 3000 mg/kg indicates that when sodium chloride was given via oral dose to a test sampling of rats, the dose that would kill 50% of the rats was 3000 mg/kg of the rat's weight. If one could extrapolate directly from a rat to a human being, it would require an oral dose of 240 g of sodium chloride to kill one half of a random group of 80-kg humans. Needless to say, this is well above the amount of sodium chloride one would expect to inadvertently ingest in a laboratory. We can conclude that sodium chloride is not very risky in this regard.

The **Health Hazard Data** section indicates that inhalation, ingestion, or skin absorption may be harmful, and that the material is irritating to mucous membranes and the upper respiratory tract. This may be surprising for as "innocent" a material as salt, but it is certainly well known that salt will sting the eyes, and that prolonged exposure of the skin to salt water can be harmful. While spilling a small amount of sodium chloride on the skin would not be harmful, this warning illustrates the general principle of trying to minimize contact with any chemical. This section also gives the treatment for having contact with salt in the eyes: flushing with water for at least 15 min.

The **Fire and Explosion Hazard Data** and **Reactivity Data** sections provide information about chemical incompatibilities and other chemical reaction dangers. We are told that sodium chloride does not combust, and that it may react with strong oxidizing agents or strong acids. The steps to be taken if material is released or spilled generally refer to large, industrial amounts. Specific information will be provided in the experimental procedures for materials with unusual handling characteristics.

Waste disposal methods are also given. Again, they generally refer to quantities much larger than those used in these laboratory experiments. In addition, the methods are designed for materials that are less than pure. Pure sodium chloride can be quickly added to water, and the pH will, of course, be neutral. This is not necessarily true of various industrial grades of sodium chloride, so that care should be indicated. The precautions to be taken listed in the handling and storage section offers some practical advice on how to deal with the compound, as well as recommendations as to safety equipment that should be on hand (shower and eye bath).

Finally, the **Additional Precautions and Comments** section details specific dangers associated with this compound. Sodium chloride is known to react violently with lithium or bromine trifluoride under certain conditions. These materials must never be used in the same reaction step.

It may seem to you that the MSD sheet is too detailed. This is certainly true in the microscale usage of sodium chloride, but keep in mind that these sheets are designed for many different kinds of use. A judicious reading of the sheets will provide the chemist with much useful information, and you will quickly learn what aspects of safety to "focus in" on. It is much better to have the detailed information and not to need it than to be in the opposite predicament.

1.B.3 The *Merck Index*

Similar information in a more compact form can be found in the *Merck Index* (Merck). This basic reference work gives the "bottom line" on the toxicity of chemicals, and their incompatibilities. In the case of sodium chloride (see Fig. 1.2), the index lists under Human Toxicity: "Not generally considered poison-

---

8430. Sodium Chloride. Salt; common salt. ClNa; mol wt 58.45. Cl 60.66%, Na 39.34%. NaCl. The article of commerce is also known as *table salt, rock salt* or *sea salt.* Occurs in nature as the mineral *halite.* Produced by mining (rock salt), by evaporation of brine from underground salt deposits and from sea water by solar evaporation: Faith, Keyes & Clark's *Industrial Chemicals*, F. A. Lowenheim, M. K. Moran, Eds. (Wiley-Interscience, New York, 4th ed., 1975) pp 722–730. Comprehensive monograph: D. W. Kaufmann, *Sodium Chloride*, ACS monograph Series no. 145 (Reinhold, New York, 1960) 743 pp.

Cubic, white crystals, granules, or powder; colorless and transparent or translucent when in large crystals. d 2.17. The salt of commerce usually contains some calcium and magnesium chlorides which absorb moisture and make it cake. mp 804° and begins to volatilize at a little above this temp. One gram dissolves in 2.8 ml water at 25°, in 2.6 ml boiling water, in 10 ml glycerol; very slightly sol in alcohol. Its soly in water is decreased by HCl and it is almost insol in concd HCl. Its aq soln is neutral. pH: 6.7–7.3. d of satd aq soln at 25° is 1.202. A 23% aq soln of sodium chloride freezes at −20.5°C (5°F). LD_{50} orally in rats: 3.75 g/kg, Boyd, Shanas, *Arch. Int. Pharmacodyn. Ther.* **144,** 86 (1963).

Note: Blusalt, a brand of sodium chloride contg trace amounts of cobalt, iodine, iron, copper, manganese, zinc is used in farm animals.

Human Toxicity: Not generally considered poisonous. Accidental substitution of NaCl for lactose in baby formulas has caused fatal poisoning.

USE: Natural salt is the source of chlorine and of sodium as well as of all, or practically all, their compds, e.g., hydrochloric acid, chlorates, sodium carbonate, hydroxide, etc.; for preserving foods; manuf soap, dyes—to salt them out; in freezing mixtures; for dyeing and printing fabrics, glazing pottery, curing hides; metallurgy of tin and other metals.

THERAP CAT: Electrolyte replenisher, emetic; topical antiinflammatory.

THERAP CAT (VET): Essential nutrient factor. May be given orally as emetic, stomachic, laxative or to stimulate thirst (prevention of calculi). Intravenously as isotonic solution to raise blood volume, to combat dehydration. Locally as wound irrigant, rectal douche.

Figure 1.2. Merck Index: *sodium chloride. [Reprinted by permission (10th ed., 1983) Merck & Co., Inc. Rahway, NJ.]*

ous. Accidental substitution of NaCl for lactose in baby formulas has caused fatal poisoning." While the information in the *Merck Index* is not as complete as on the MSD sheets, it is generally sufficient for our purposes at the microscale laboratory level. The index also supplies some interesting information about the common usages of the chemicals listed, with a special emphasis on medical usages. References to the literature are also provided.

1.B.4 Compresed Gas Cylinders and Lecture Bottles

Some of the experiments described in this text require the use of gases obtained from compressed gas cylinders. The commonly used gases obtained from such cylinders are N_2, He, CO_2, NH_3, Cl_2, BCl_3, HF, SO_2, N_2O_4, gaseous HCl, CO, Ar, and so on. The various cylinder types and the methods to safely handle gas cylinders are described below.

Commercial compressed gas cylinders come in various sizes and shapes. The commonly used nitrogen cylinder usually measures 9 in. in diameter and 60 in. in height, inclusive of the valve and cap. The dimensions and internal volumes of other commonly used cylinders are given in Table 1.1. Note that as the cylinder size numerals increase, the cylinder dimensions decrease. Each gas cylinder is identified by gas name (stencil or label) and tagged for ready identification. Hazard information is also provided. Lecture bottles are a convenient gas source for a small sized class.

All manufacturers supply these cylinders with specific instructions. We strongly suggest that prior to the use of any gas cylinder, you consult the manufacturer's catalog and strictly adhere to the instructions that come along with it. If you have additional questions, do not hesitate to ask your instructor. Make sure that all pertinent data (MSD sheets) and instructions have been consulted, prior to using the cylinder.

Each cylinder (see Fig. 1.3) comes with a threaded cap on the cylinder head, which must be removed and stored in a safe place. This cap is replaced when the cylinder is returned for refilling. Most cylinders are fitted with standard valve outlet fittings as recommended by Compressed Gas Association (CGA). Table 1.2 contains the CGA recommended valve outlet and corresponding connection numbers.

A typical compressed gas tank comes with an on–off valve (main valve), a gas outlet device, and a safety seal or an emergency pressure release fitting that looks like a nut. Never tamper with this safety seal. The seal is designed so that if the pressure inside the tank exceeds the limit of the cylinder capacity, the seal will rupture, preventing explosion. Note that gas cylinders containing flammable gases such as H_2, CO, or methane come with special fittings having left-handed threads. The connector for these gases can be recognized by V-shaped markings on the edges of the hexagonal nut. These unusual features help the users of

Table 1.1 Cylinder Sizes, Tare Weights, and Internal Volumes[a]

Cylinder Size	Dimensions Diameter × Length (cm)	Tare Weight (kg)	Internal Volume (L)
1A	23 × 130	55	43.8
1B	31 × 97	80	60.9
1C	38 × 137		128.0
2	22 × 65	29	16.7
3	16 × 47	13	6.9
4	10 × 34	5	2.3
LB	5 × 30	1.6	0.44

[a] From Matheson Gas Products.

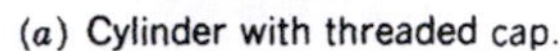

(*a*) Cylinder with threaded cap.

(*b*) Cylinder with outlet valve and gas leak detector.

Figure 1.3. *Gas cylinder. (Courtesy of Matheson Gas Products, Inc., Secaucus, NJ.)*

these gases recognize that they must adopt utmost caution in the use of these potentially dangerous materials.

It is preferable to use a lecture bottle of a gas (see Fig. 1.4) for microscale work. The bottle should be secured with the help of a strong three-prong clamp fastened to a stable ring stand or other device, capable of holding the cylinder. Alternatively, some manufacturers supply specialized holders for lecture bottles. Holders for lecture bottles can also be made economically using a wooden box with a lid, into which circular holes can be cut of proper size to hold the lecture bottles. The box can also be half-filled with sand to help stabilize the cylinders.

When using a large gas cylinder, always fasten the cylinder with a strap to a cylinder support, tightly secured to an edge of a working table. Never work with an improperly secured cylinder.

Figure 1.4. *Gas lecture bottle and safety stand. (Courtesy of Matheson Gas Products, Inc., Secaucus, NJ.)*

Table 1.2 Cylinder Valve Outlets and Connection Numbers[a]

Gas	CGA Valve Outlet and Connection Number	Gas	CGA Valve Outlet and Connection Number
Acetylene	510	Ammonia	705[b]
Argon	580[c]	Boron Trichloride	660[b]
Carbon Dioxide	320[c]	Carbon Monoxide	350[c]
Chlorine	660[d]	Ethylene	350[c]
Helium	580[c]	Hydrogen	350[c]
Hydrogen Chloride	330[c]	Nitric Oxide	660
Nitrogen	580[c]	Oxygen	540[c]
Propane	510[c]	Sulfur Dioxide	660[c]

[a] As standardized by CGA and accepted by American Standard Association. From Matheson Gas Products.
[b] Lecture bottle uses 180.
[c] Lecture bottle uses 170.
[d] Lecture bottle uses 110.

The main valve on a cylinder acts as an on–off valve, which when opened allows the gas to rush out from the cylinder. Note that the main valve does not provide a mechanism for the control of the pressure or of the flow rate of the gas. In order to control the pressure and flow rate of the gas, it is advisable to use some sort of gas control device, such as a pressure regulator. Depending on the nature of the work, one can use two kinds of regulators—a needle valve regulator or a single, two-stage pressure regulator.

A needle valve regulator is the least expensive type and provides for manual control of the gas flow. The needle valve is attached to the main cylinder valve. Such regulators are used only when the system to be flushed has an unobstructed outlet for the gas. Since needle valves do not control the pressure, their use in closed systems is not recommended. These valves can also conveniently be used for cylinders containing gases that liquify under low pressure (e.g., NH_3 and BCl_3). A constant flow can be easily maintained with a needle valve as long as the gas is liquified within the cylinder. In the case of a compressed gas cylinder (e.g., with CO, N_2, and Ar, which do not exist as liquids under high pressure), a needle valve cannot maintain a constant flow of the gas, because as the gas is used up, the pressure in the cylinder will gradually drop, resulting in a slow change of the outlet pressure of the gas. A continuous adjustment of the gas flow rate would be necessary.

In order to maintain N_2 or Ar flows, a good pressure regulator, which controls the pressure as well as the flow rate, must be used. There are two basic types of pressure regulators: two stage and single stage (see Fig. 5.8). Most pressure regulators incorporate two pressure gauges to monitor the pressures. If you stand facing the gauges on the regulator, the right-hand side gauge is for monitoring the cylinder gas pressure; the left-hand side gauge is used for setting and adjusting the outlet pressure. Note that neither the regulator nor the gauges control the flow rate of the gas; flow rate is controlled by a needle valve situated at the outlet end of the pressure regulator.

The two-stage pressure regulator reduces the pressure in two steps prior to delivery. In the first stage, the high-pressure gas is automatically adjusted to a preset intermediate pressure range. In the second stage of the control system, the desired pressure is manually adjusted. The two-stage regulators come with two diaphragms for two-step pressure control. Two-stage regulators are used when precise control of the pressure as well as of the flow rate is needed.

Single-stage regulators have the same functions as the two-stage regulators. They regulate the pressure of the gas as well as the flow rate in one step, using a single diaphragm. Periodic adjustment of the pressure must be made to compensate for the decreasing pressure in the cylinder as the gas is continuously used up.

Almost all of the inert atmosphere experiments described in this book use N_2 gas from the compressed gas cylinder, which can be procured from the supplier as a prepurified and moisture-free gas, without further purification. For example, Matheson's research purity N_2 gas contains less than 1 ppm (parts per million) impurities each of the gases oxygen, methane, carbon monoxide, carbon dioxide, and water. It contains less than 3 ppm of argon, which is an inert gas itself. Occasionally, further prepurification of a commercial gas is necessary, usually for creating a very dry and oxygen-free environment for an experiment. In such cases, the gas from the tank is passed through a simple purification train as described below. Supported copper or MnO are the most convenient solid oxygen scavengers used for removing oxygen from the commercial gas. Molecular sieves 4 or 5 Å (Linde type) are efficient desiccants for removing water. The inert gas from the needle valve outlet is passed through a desiccant column packed with the molecular sieves, followed by a column containing the oxygen scavenger (BTS catalyst or Ridox) and finally through another column of molecular sieves. The valve or stopcock is closed when the drying and purification towers are not in use, or when the compressed gas cylinder needs to be changed.

Ridox and BTS catalysts are available from Fisher Scientific and from Fluka Chemical Corp., respectively. The catalysts are supplied in the oxidized form and must be activated by reduction at elevated temperatures. The general procedure for regenerating the BTS catalyst is to pass 5% H_2 (diluted with N_2) through the catalyst bed, packed in a tower maintained at a temperature of 150 °C.

A cylinder should never be emptied completely. When the pressure regulator reads approximately 25 psig or 2-atm pressure, close the main valve, remove the pressure regulator, recap the head of the cylinder, and return it to the supplier for refilling. If the cylinder cannot be dispatched immediately for refilling, put an "empty" sign tag around the neck of the cylinder, and store it, belted securely until its return.

1.B.5 Fire Safety

Many solvents commonly used in the inorganic or organic chemistry laboratory are flammable. Some obvious examples are toluene, ethanol, ether, hexane, and acetone. (Appendix A gives safety data for common organic solvents, including fire safety data.) Chemical fires, due to rapid oxidation, can also occur through the use of strong oxidizing agents, or because of rapid reduction through the use of strong reducing agents. There is also a small risk (due to the use of electronic equipment) of an electrical fire caused by a short circuit, frayed electrical cord, or power surge.

If a fire should occur, the most important thing to remember is to **keep calm**. Call your instructor. Several methods for dealing with small fires are contained in most laboratories. The most obvious is the fire extinguisher. Laboratory extinguishers should weigh no more than 10 lbs, so as to be of convenient size to lift and employ rapidly. Ideally, there should be at least one fire extinguisher for every laboratory bench. Several types of extinguishers are available, the most common being of the dry chemical (bicarbonate powder under pressure) or compressed carbon dioxide type. Most small fire extinguishers are activated by pointing the nozzle toward the base of the fire and squeezing the handle. A jet of compressed powder or foam will then discharge from the nozzle, smothering the fire. Sometimes, it may be necessary to pull a pin from the handle before it can be squeezed. Some fire extinguishers only operate when turned upside down. It is imperative that each student be familiar with the proper use of the fire extinguishers located in the laboratory.

Alternate ways also exist for putting out small fires. Fires in small vessels can be extinguished by inverting a beaker or a similar container over the burning vessel, thereby excluding oxygen. A second way of putting out such fires is by covering the vessel with soaking wet towels. Never use dry towels for this purpose.

Liquid nitrogen is one of the best possible fire extinguishers. The liquid nitrogen can be poured directly over the fire.

CAUTION: *Oil fires can be spread by using this technique.*

Fires caused by reactive metals (Na, K, Al, Ca, and Mg) should not be extinguished with normal fire extinguishers, as a chemical reaction may occur. A better method for putting out such fires is to use powdered graphite, Pyrene G-1, or a special fire extinguisher designed for metal fires.

If a fire should occur, it is important to immediately remove any flammable material from the vicinity, especially bottles of flammable solvents and gas cylinders. If fire comes in contact with these items, an explosion can occur.

Whenever a fire occurs, there is also an associated danger caused by inhalation of smoke or toxic fumes. Inhalation of smoke and toxics is potentially more dangerous than the fire. Thus, if the air is not fit to breathe, the fire should be abandoned, and the fire department should be called. Any persons overcome by fumes should be removed to a well-ventilated area and health professionals should be called immediately. If the fire is too large to contain, the area should be evacuated, and the fire department should be called immediately. The fire department should be apprised of the specific nature of the laboratory fire, so that the proper equipment can be brought.

In the event that your clothing should catch on fire, all laboratories should be equipped with safety showers and fire blankets. To activate a safety shower, merely stand beneath it and pull down on the lever or chain. Remain under the shower until you are thoroughly soaked. To use a fire blanket, grasp the rope or material at the end of the blanket, and turn so that you are surrounded tightly by the blanket to smother the fire.

REFERENCES

—, *Hazards in the Chemical Laboratory*, D. Muir, Ed., 2nd ed., The Chemical Society: London, 1977.

—, *Informing Workers of Chemical Hazards: the OSHA Hazard Communication Standard*, American Chemical Society: Washington, DC, 1985.*

—, *Less is Better (Laboratory Chemical Management for Waste Reduction)*, American Chemical Society: Washington, DC, 1985.*

—, *Merck Index of Chemicals and Drugs*, 11th ed., Merck and Co.: Rahway, NJ, 1990.

—, *RCRA and Laboratories*, American Chemical Society: Washington, DC, 1986.*

—, *Safety in Academic Chemistry Laboratories*, 4th ed., American Chemical Society: Washington, DC, 1985.*

Szafran, Z.; Singh, M. M.; Pike, R. M. "The Microscale Inorganic Laboratory: Safety, Economy and Versatility." *J. Chem. Educ.* **1989**, *66*, A263.†

Craig, P. J., "Environmental Aspects of Organometallic Chemistry" in *Comprehensive Organometallic Chemistry*, G. Wilkinson, Ed., Pergamon: Oxford, 1982, Vol. 2, Chapter 18, p. 979.

Lenga, R. E., Ed., *The Sigma–Aldrich Library of Chemical Safety Data*, 2nd ed., Sigma–Aldrich: Milwaukee, WI, 1987.

* Available at minimal cost from the American Chemical Society.

† Available from the authors.

Chapter 2

Laboratory Equipment

2.A GLASSWARE

The majority of experiments in the inorganic microscale laboratory involve the use of glassware, much of which is expensive. Your responsibility is to keep this glassware clean and to give it proper care. As you study the following listings and discussion, you will note that many of the items used here are also used in the microscale organic laboratory.[1]

Figures 2.1 and 2.2 illustrate the glassware that you will use in the majority of experiments presented in this book. Figure 2.1 presents the more common glassware found in your laboratory locker. This includes, for example, beakers and Erlenmeyer flasks (10, 25, and 50 mL), a Hirsch funnel (plastic or porcelain), graduated cylinders (10 and 25 mL), suction filter flasks (25 and 50 mL), a separatory funnel, graduated glass or plastic pipets, and glass funnels and test tubes of various sizes. At this stage of your development it is logical to assume that you are familiar with this common glassware from your experience in prior course work in chemistry. If you are not, consult your instructor before you proceed in the use of that particular item.

Figure 2.2 illustrates the type of standard-taper glassware (ground-jointed glassware) that is used in many of the experiments you will perform. These items include round-bottom boiling flasks, conical reaction vials, reflux condensers, a Hickman–Hinkle distillation column (with or without a side arm), and various miscellaneous items. These items may be in a kit or as individual pieces depending on your laboratory setup.

Standard-taper glassware is preferred to the older corks or rubber stopper approach because it has the advantage of ease of assembly and safety. Various reagents often reacted with the cork stoppers, thus leading to contamination of the reaction system. Figure 2.3 depicts a typical standard-taper joint. These glass-taper joints come in various sizes and are designated by a $\mathfrak{T}$ symbol followed by two numbers. For example, $\mathfrak{T}$ 14/10, $\mathfrak{T}$ 14/20, and $\mathfrak{T}$24/40 are common ground-glass joint sizes. The first number is the outer diameter of the male joint in millimeters (mm) at its widest point (or the inner diameter of the female joint) and the second number is the length of the joint in millimeters.

When working at the microscale level, the $\mathfrak{T}$ 14/10 joint size is convenient to use. These joints, when connected in the assembly of a particular apparatus,

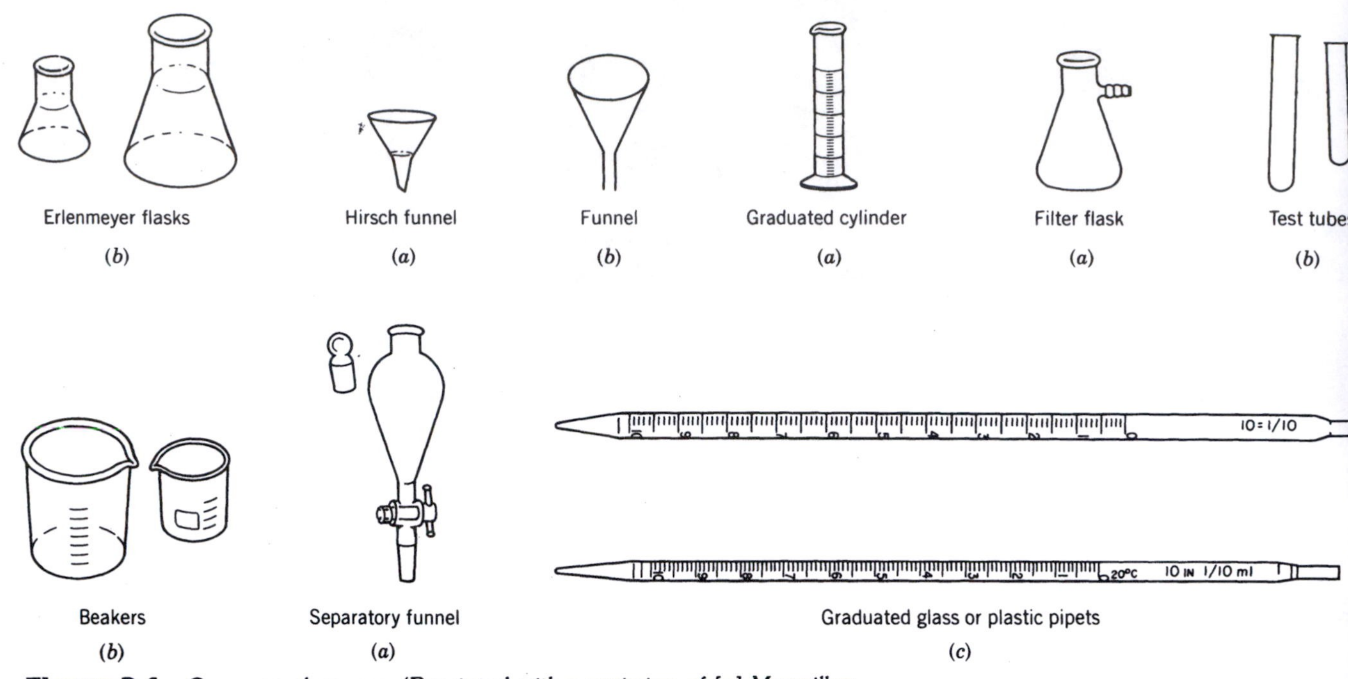

Figure 2.1. *Common glassware. (Reprinted with permission of [a] Macmillan Publishing Co., from* Experimental Organic Chemistry *by C. F. Wilcox, Jr. Copyright © 1985 by Macmillan Publishing Company. [b]* From *H. D. Durst; G. W. Gokel,* Experimental Organic Chemistry, *copyright © 1987 by McGraw-Hill Book Co., NY. [c] Courtesy of Sargent–Welch Scientific Co., a VWR Company, Skokie, IL.]*

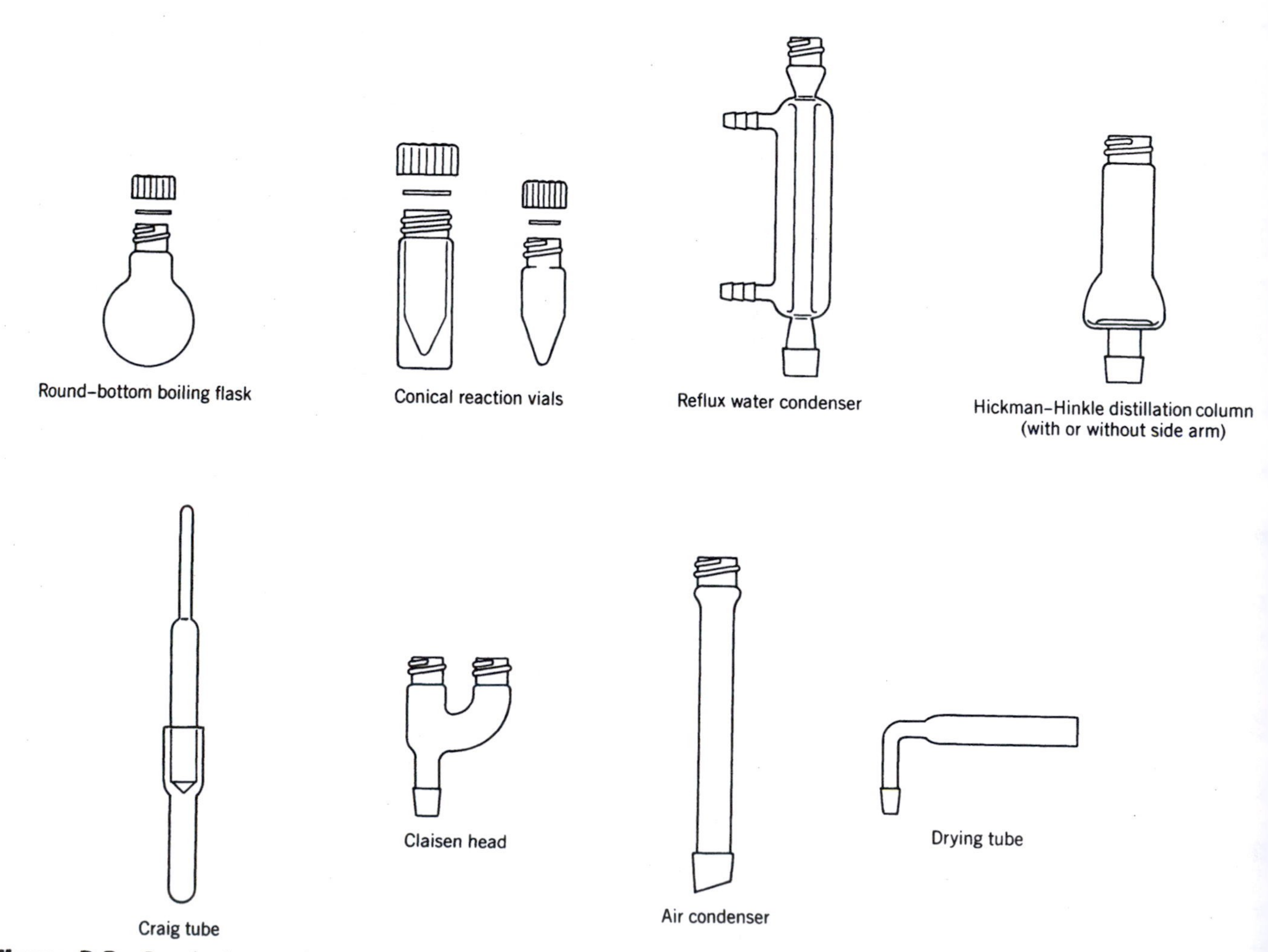

Figure 2.2. *Standard-taper glassware. (Courtesy of Ace Glass, Inc., Vineland, NJ.)*

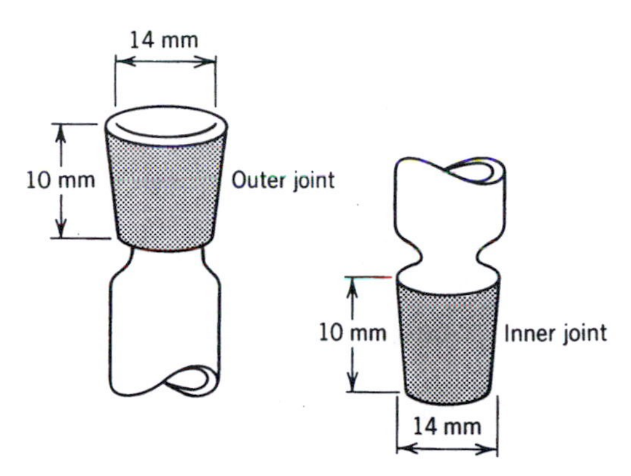

Figure 2.3. *Typical standard-taper Joint.* From *Zubick, J. W.,* The Organic Chem Lab Survival Manual, *2nd ed., Wiley: New York, 1988. (Reprinted by permission of John Wiley & Sons, New York.)*

do not require a lubricant. This is particularly advantageous when working at the microscale level, since any contaminant in the reaction system may well spell disaster! However, it is important to remember to disconnect any glassware joints *immediately* after use. This will prevent freezing or locking of the joints together. When using larger joints (such as the ₮ 24/40 size), a suitable lubricant such as a silicone or hydrocarbon based grease is required. In some cases, for example, when a strong alkaline solution is used, a lubricant is employed even with the ₮ 14/10 connecting joints. If your laboratory has a supply of the older ₮ 14/20 glassware kits, it is important to note that this size may be used interchangeably with the newer ₮ 14/10 glassware.

A Note of Caution: *Never apply strain or pressure when setting up or dismantling standard-taper glassware equipment.* It is also best to clean glassware *immediately* after use. Soap and water or a solvent recommended by your instructor may be used depending on the type of chemicals used in a particular experiment.

There are other special pieces of glassware equipment that you may use in the course of your inorganic laboratory experience, for example, working with a vacuum line in the handling of air-sensitive materials. This special equipment is described in Section 5.C.

2.B OTHER LOCKER EQUIPMENT

Your locker contains other pieces of equipment that are essential in carrying out the basic manipulations in an inorganic microscale laboratory. This may include items such as Pasteur pipets with bulbs, plastic disposable pipets, spatulas (micro and regular), a Hoffman screw clamp, a clay drying tile, assorted microclamps, an iron ring, a wire mesh, a gas striker, a Keck clip, a thermometer, and a glass plate approximately 12 × 13 in. with a half-white half-black background. A collection of these items is shown in Figure 2.4.

Special mention is made of the glass plate indicated above. This item is strongly recommended for use when working at the microscale level. Since a large portion of the equipment is small and various stoppers, vials, pipets, and so on, are used, placement of these items on the glass plate will assist you in keeping a neat and orderly approach to the operations at hand. The dark and white background gives a good contrast to various objects. These plates are easily made by taking two window-glass sheets, placing white and black paper between them ($\frac{1}{2}$ of each) and then a cardboard backing on one side. The edges

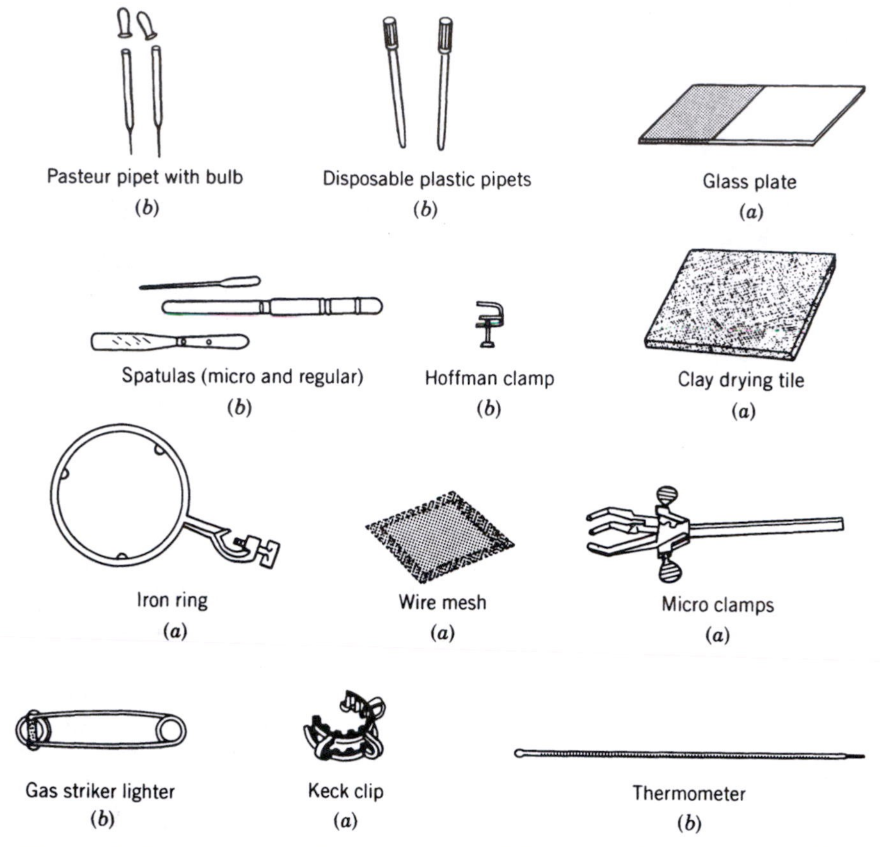

Figure 2.4. *Standard locker equipment. [a] Courtesy of Sargent–Welch Scientific Co., a VWR Company, Skokie, IL. [b]* From Durst, M. D.; Gokel, G. W., Experimental Organic Chemistry, *copyright © 1987 by McGraw-Hill Book Co., New York. (Reprinted by permission of McGraw-Hill Book Company, New York.)*

of the glass–paper–glass–cardboard sandwich are then taped to hold the sections together.

Caution: *Sharp edges!*

2.C MEASURING QUANTITIES OF CHEMICALS

2.C.1 Weighing

Weighing is one of the most critical measurements made at the microscale level. The use of single-pan electronic balances with automatic taring (zeroing) and digital readout[2] has made the move to the microscale laboratory a reality. It is essential that each worker take the responsibility of keeping the balance(s) clean, as well as the balance area. These are expensive and delicate instruments. *Chemicals should never be placed directly on the balance pan.*

Solid chemicals are weighed using aluminum or plastic boats, or various glass containers. Weighing paper or filter paper may also be used in certain situations.

Liquid chemicals are generally weighed directly into the reaction flask being used or in the product flask when isolated as a liquid. In any event, a glass container is usually preferred.

2.C.2 Liquid Volumes

Various techniques are used to measure small volumes of liquids at the microscale level. When 1–2 mL of a reagent, solvent, or solution is required to be accurately measured, it is convenient to use one of the measuring devices listed below (see Fig. 2.5).

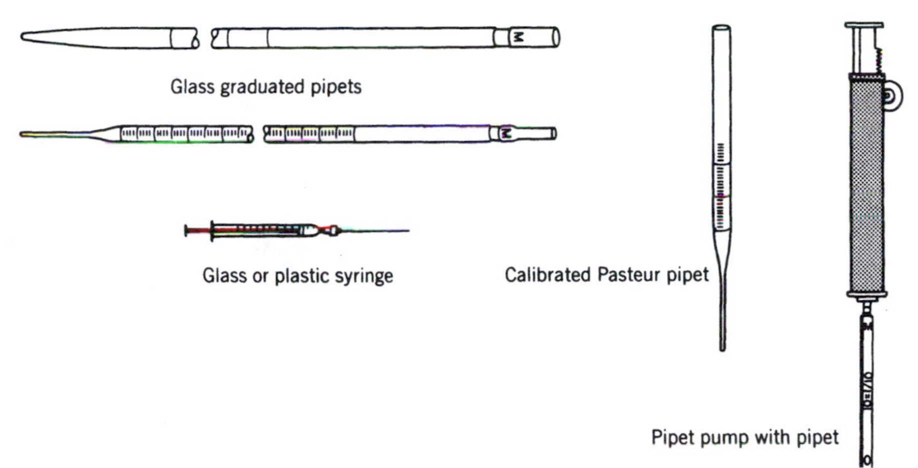

Figure 2.5. *Equipment to measure volume. (Courtesy of Ace Glass, Inc., Vineland, NJ: Thomas Scientific, Swedesboro, New Jersey.)*

1. A glass graduated pipet with a pipet pump.
2. A graduated 1- or 2-mL glass or plastic syringe.
3. A calibrated Pasteur pipet with pipet pump or bulb.

Several types of pipet pumps are commercially available, as are glass, polypropylene, and disposable syringes. Disposable syringes or pipets are cheap and often convenient to use, especially when aqueous solutions are dispensed.

The standard calibrated glass pipets (1, 2, and 5 mL) come in various designs. Pipets are calibrated in one of two ways: (a) to deliver a given amount or (b) to contain a given amount between the limiting calibration marks. It is important for you to recognize which pipet type you have. **NEVER** use mouth suction to draw *any liquid* into the pipet! Serious injury may result if this advice is not followed.

A glass Pasteur pipet may be easily calibrated by drawing the desired amount of liquid from a filled 10-mL graduated cylinder into the pipet using a pump or bulb, and marking the level of the liquid. It is recommended that you calibrate several pipets showing volumes of 0.5, 1.0, 1.5, and 2.0 mL. For permanent use, a light file mark may be scored on the pipet. Plastic pipets may be calibrated in a similar manner.

In many experiments, it is recommended that a Pateur *filter* pipet be used. To make one, a small cotton plug is inserted into the tip of a Pasteur pipet by use of a copper wire, as shown in Figure 2.6. It is important to use the correct amount of cotton, so that the plug is not so tight as to prevent easy flow of the liquid or so loose as to come out when a liquid is dispensed. The old saying holds "practice makes perfect!"

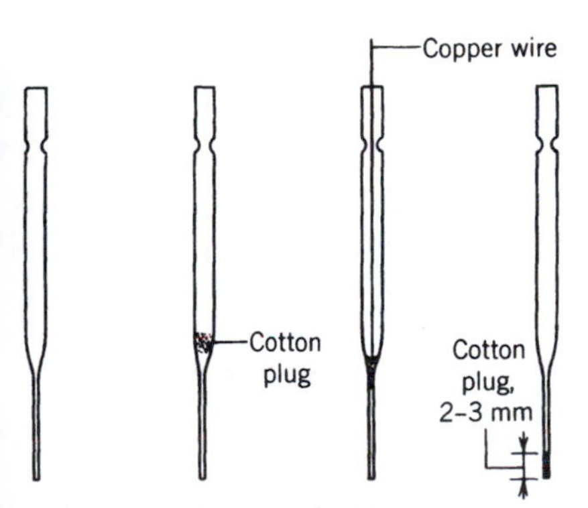

Figure 2.6. *Preparation of Pasteur filter pipet.* From *Mayo, D. W.; Pike, R. M.; Butcher, S. S.* Microscale Organic Laboratory, *2nd ed., Wiley: New York, 1989. (Reprinted by permission of John Wiley & Sons, New York.)*

The Pasteur filter pipet offers two distinct advantages over the regular Pasteur pipet. First, the solution is automatically filtered each time it is taken into the pipet. This is necessary, since it is important to remove dust or other suspended material in the solution at the microlevel, where purity of reagents and products is critical. Second, when a volatile solvent such as ether is used, a back pressure rapidly builds up in the bulb, which causes the solvent or solution to drip from the tip of the pipet. The small plug allows the operator sufficient time to make a transfer rapidly and completely from one container to another without loss of material.

An Important Suggestion: *When transferring a material from one container to another, hold the container with the material to be transferred as close to the*

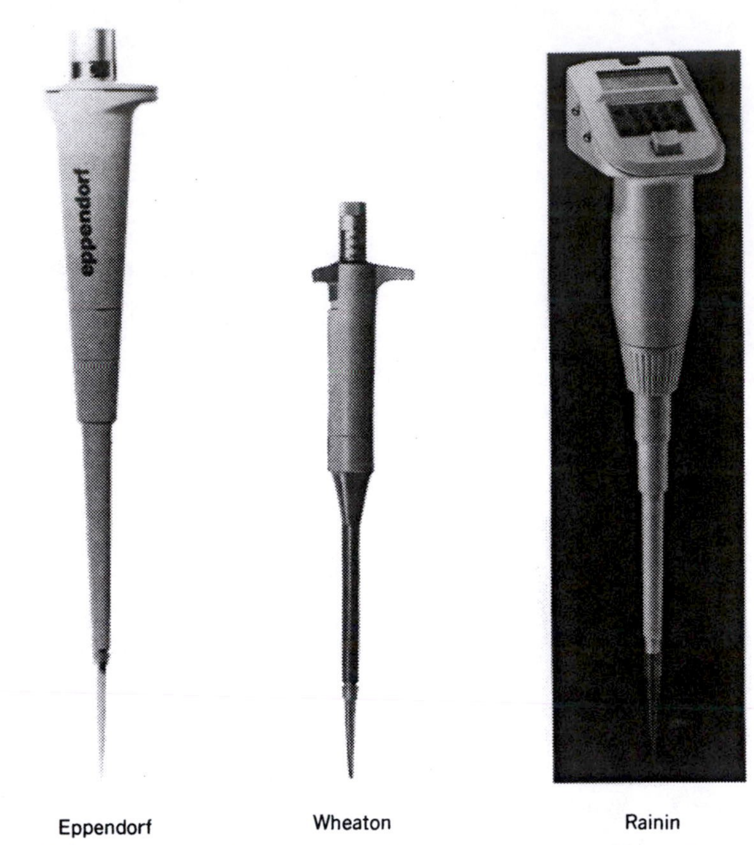

Figure 2.7. *Automatic delivery pipets. (Courtesy of Sargent–Welch Scientific Co., a VWR Company, Skokie, IL; Ace Glass, Inc., Vineland, NJ; Rainin Instrument Co., Woburn, MA.)*

receiving container as possible, preferably holding both in one hand. This arrangement allows for a smooth, quick transfer with minimal loss.

Automatic delivery pipets are very useful when the amount of liquid required is less than 1 mL. Several types are commercially available, three of which are shown in Figure 2.7. All have a control to adjust the pipet to deliver the desired volume. They are also designed so that the liquid being transferred comes in contact only with the plastic tip. This tip can be automatically ejected from the pipet, after delivery of the liquid is complete.

A few tips on the use of automatic delivery pipets:

1. They are relatively expensive, so treat them with respect.
2. Always store, hold, and use the pipet in a vertical position. Liquid running back up into the pipet mechanism can cause damage to the controls.
3. Never immerse the pipet in a liquid—only ~5 mm of the plastic tip should come in contact with the liquid to be transferred.
4. Develop a smooth depression and release technique of the control mechanism. Consistent results come with care and practice.

It is often necessary to convert volume measurements into weight measurements, and vice versa. For most calculations (unless extremely accurate data is needed), the familiar relationship

$$\text{density } (\text{g}\cdot\text{mL}^{-1}) = \frac{\text{mass (g)}}{\text{volume (mL)}}$$

using literature values[3] for density (or specific gravity) may be used at the ambient working temperature.

One final note on the measurement of liquid volumes: Glassware such as Erlenmeyer flasks, beakers, and test tubes *are not* calibrated for use as accurate measuring devices! Graduated cylinders are somewhat more accurate, burets if used properly are quite accurate, but we recommend the use of the calibrated types of pipets discussed previously for accurate measurement of small amounts of liquids at the microlevel.

2.D HEATING METHODS

2.D.1 The Microburner

In general practice, the open *free flame* of a burner is not recommended for routine heating use, even in the microscale laboratory. This is due to the presence of many individuals in a given laboratory space, and the possible presence of volatile, flammable solvents. However, if certain precautions are observed (mainly, no flammable solvents in use, and the gas being turned off immediately after use of the burner), the microburner (Fig. 2.8) does have a useful role to play in the laboratory.

Microburners may be used to prepare thin-layer chromatography (TLC) spotting capillaries (see Section 5.G.2), or as a heating device in the sublimation technique (see Section 5.H).

2.D.2 Steam Bath

Since most laboratories are equipped with steam lines, the steam bath (Fig. 2.8) is a cheap, convenient, and very safe method of heating when temperatures below 90 °C are required. Generally, baths with concentric rings are used so that various sizes of flasks, and so on, can be accommodated.

A general precaution to follow in the use of steam baths is to *avoid excess steam*. Not only can one be burned by hot steam, but the presence of excess moisture could introduce water into the reaction system by condensation. This could have a negative effect on a reaction system that is moisture sensitive, a common situation in inorganic chemistry.

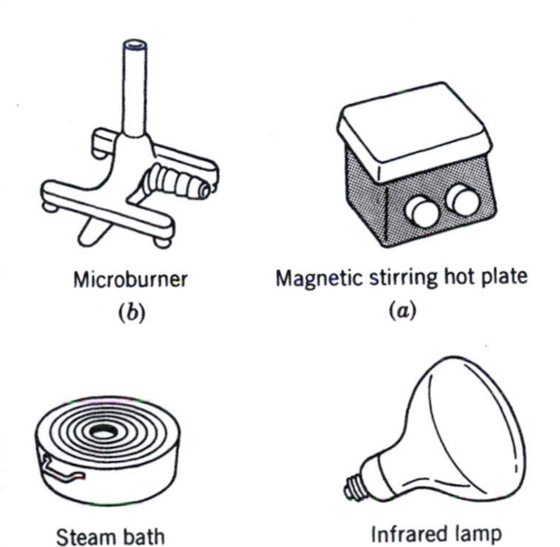

Figure 2.8. *Heating devices. [a] Ace Glass, Inc., Vineland, NJ. [b]* From *Durst, H. D.; Gokel, G. W.,* Experimental Organic Chemistry, *copyright © 1987 by McGraw-Hill Book Co., NY. (Reprinted by permission of McGraw–Hill Book Company, New York.)*

2.D.3 Oil Baths

The oil bath has largely been displaced by sand baths or by aluminum blocks (see Section 2.D.5), but is still used in many research laboratories. It is not highly recommended for the initial instructional laboratory because of several drawbacks:

1. Oil baths present a potential safety hazard, since a spill of hot oil can lead to severe burns. Also, depending on the type of oil used, a potential fire hazard may be present. For this reason, if an oil bath must be used, it is recommended that the relatively safe silicone oil be chosen.
2. Oil baths are slow to heat.
3. The flask retains an oily residue, which is messy at best.

The advantages of an oil bath are:

1. The temperature is easy to control and thus a steady heat source is maintained.
2. No flames are present to cause ignition of flammable vapors.
3. It is relatively inexpensive.

The oil bath may be heated by placing the container holding the oil on a hot plate. A more effective and efficient technique, which is also inexpensive, is to

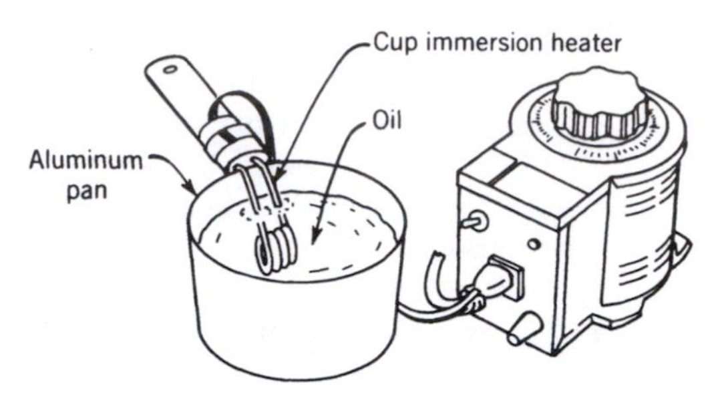

Figure 2.9. *Insulated immersion heater and Variac. From Durst, H. D.; Gokel, G. W.,* Experimental Organic Chemistry, *copyright © 1987 by McGraw-Hill Book Co., NY. (Reprinted by permission of McGraw–Hill Book Company, New York.)*

use an insulated immersion heater (designed and sold for use in coffee cups). The heater is set in a porcelain or metal flat-bottom container (see Fig. 2.9). The heater is taped to the pot handle as shown. A Variac or other controlling device must be used to adjust the temperature of the bath.

One Caution: *Be sure to place the oil bath on a firm support to prevent any possibility of tipping.*

A flat bottom container is highly recommended for this reason.

2.D.4 Infrared Lamp

The infrared heating lamp (Fig. 2.8), when connected to a Variac or other controlling device, is a clean, safe (nonflammable) heat source. The use of such a lamp is especially effective when low-boiling solvents or reaction mixtures are to be warmed to reflux temperature. It is also an effective heating source when crystalline products are to be dried under vacuum (see Section 5.F.3). Reflection of the heat by use of an aluminum foil reflector increases the efficiency of this technique.

2.D.5 Sand Bath or Aluminum Block with Magnetic Stirring Hot Plate

A sand bath placed on a magnetic stirring hot plate (Fig. 2.10*a*) makes a very efficient heating device. It has been used extensively in the microscale organic laboratory.[1] This arrangement allows for stirring (see Section 2.E) and heating to be performed simultaneously. This approach is very inexpensive, provides a nonflammable source of heat, and does not decompose. *Make sure that the hot plate used has sealed electrical contacts on the switches.* If sparks are produced as the heater is switched on or off, ignition of a flammable solvent could occur. The sand is usually contained in a glass crystallizing dish or a metal container.

Sand, being a poor conductor of heat, in effect acts as an insulator. Thus, the temperature of a container placed in the sand will vary depending on the depth to which it is immersed. It is therefore recommended that each individual be assigned their own equipment (sand bath and hot plate), and that a calibration of this heating source be made at the first session of the laboratory. This approach is also quite effective at temperatures below 100 °C and because of this the sand bath has replaced the steam bath in many laboratories.

An alternative to the sand bath is the aluminum block (Fig. 2.10*b*)[4] placed on the magnetic stirring hot plate. These blocks are now available from several supply houses (or can be easily manufactured in a metal shop). Users of this device cite two major advantages over the sand bath method: (a) better heat transfer and (b) no problem with breakage of a crystallization dish or spilled sand. In addition, they are relatively cheap and store easily. One disadvantage

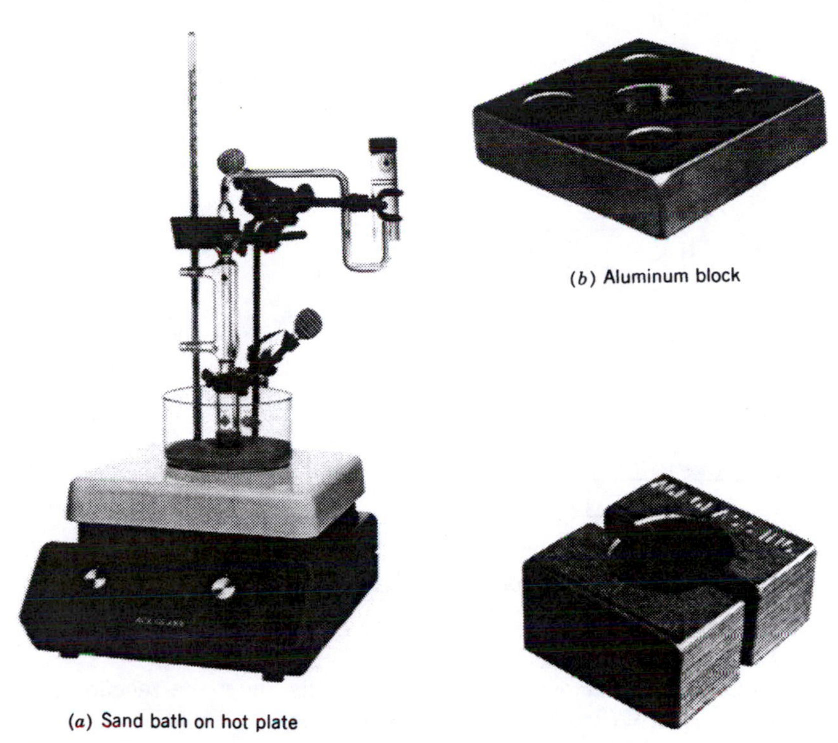
(b) Aluminum block

(a) Sand bath on hot plate

Figure 2.10. (a) *Sand bath on magnetic-stirring hot plate and* (b) *aluminum heating blocks. (Courtesy of Ace Glass, Inc., Vineland, NJ.)*

to this method is that burns can result from the impossibility of differentiating hot aluminum from cold aluminum.

2.E STIRRING

Stirring is often required in inorganic microscale reactions, and is generally carried out using magnetic stirring bars in flat or round-bottom containers or with magnetic spin vanes when conical vials are used (see Fig. 2.11).

Heating and stirring are conveniently carried out simultaneously by using a magnetic stirring hot plate as shown in Figure 2.8. You should be aware that when this type of arrangement is used, the container should be placed directly in the center of the hot plate so as to gain maximum efficiency of the magnetic flux. The spin rate of the bar or vane should be adjusted to obtain smooth mixing of the contents in the container. The insertion of a thermometer into the sand/oil bath or aluminum block easily allows you to monitor the temperature of the container and its contents. The bar or vane is easily removed from the container using forceps or a magnetic wand.

When the mixing of solutions or dissolution of a solid in a given solvent is desired and heating is not required, the use of a Vortex mixer (Fig. 2.11) is most convenient. These mixers are commercially available and are relatively inexpensive and durable. The mixer speed is adjustable and the mixing operation is fast and efficient. Vortex mixing is recommended over the techniques of *swirling* or *shaking* by hand, since much greater control is maintained during the mixing process.

2.F REFLUX AND DISTILLATION

Most reactions run in the laboratory proceed at a faster rate when carried out at a higher temperature. By increasing the temperature of a reaction system, we increase the average kinetic energy of the molecules and thus a larger portion of them have enough energy to react in a given instance. If the given reaction is run in a solvent system, the most convenient method of increasing the tem-

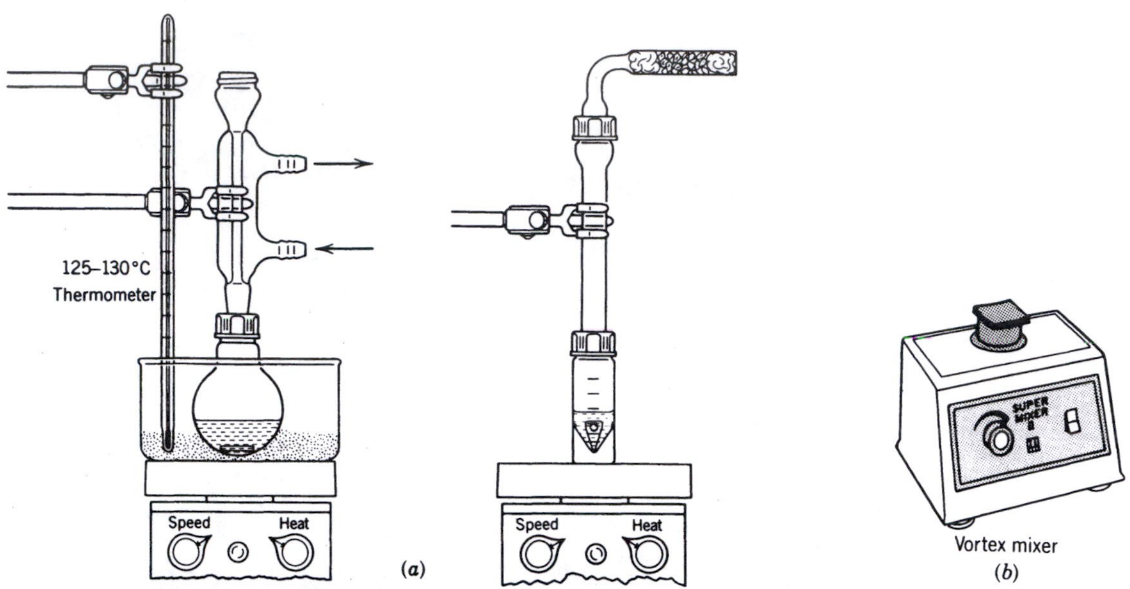

Figure 2.11. *Stirring arrangements. [a] From Mayo, D. W.; Pike, R. M.; Butcher, S. S.,* Microscale Organic Laboratory, *2nd ed., Wiley: New York, 1989. (Reprinted by permission of John Wiley & Sons, New York. [b] Courtesy of Sargent–Welch Scientific Co., A VWR Company, Skokie, IL.)*

perature is to "heat the reaction solution at *reflux* temperature" for a given period of time.

Heating at the *reflux* temperature means that when boiling of the liquid system occurs, the vapors produced will be condensed by an air-cooled or more often a water-cooled condenser, and continuously returned to the reaction pot as a liquid. Using this arrangement, the rate of heating is not important, as long as the system is at the boiling temperature. We should recognize that the reaction flask should not be heated at a temperature higher than that necessary to cause the reaction solution to boil. Also, it is important to understand that the top of the condenser is never stoppered—a closed system is never heated to reflux!

The temperature of the refluxing system is essentially the boiling temperature of the reaction solution. With a dilute solution, this reflux temperature is approximately that of the solvent used in the reaction. The reflux technique is the simplest and most convenient manner we have of running a chemical reaction at an elevated temperature.

The removal of solvent by distillation is straightforward, using standard techniques.[1] This method also allows for recovery of volatile solvents and often can be carried on outside a hood. Distillation should be used primarily for concentration of the solution followed by transfer of the concentrate with a Pasteur filter pipet to a vial for final concentration and isolation of the desired product.

Distillation is the process of heating a liquid to the boiling point, condensing the heated vapor by cooling it, and returning only a portion or none of the condensed vapors to the distillation flask. Distillation differs from the process of reflux only in that a fraction of the condensate is diverted from the boiling system. Distillations in which a fraction of the condensed vapors is returned to the boiler are often referred to as being under "partial reflux."

Distillation techniques often can be used for separating two or more components on the basis of their differences in vapor pressure. Separation can be accomplished by taking advantage of the fact that the vapor phase is generally richer in the more volatile component (lower boiling component) of the liquid mixture. Molecules in a liquid are in constant motion and possess a range of kinetic energies. Those with higher energies (a larger fraction of the lower boiling component) moving near the surface have a greater tendency to escape into the vapor (gas) phase. If a pure liquid (e.g., hexane) is in a closed container, eventually hexane molecules in the vapor phase will reach equilibrium with hexane molecules in the liquid phase. The pressure exerted by the hexane vapor molecules at a given temperature is called the *vapor pressure*, and given by the symbol $P°$, where the superscript ° indicates a pure component. For any pure component H, the vapor pressure would be P°_{H}.

Suppose a second component (e.g., toluene) is added to the hexane. The total vapor pressure (P_{total}), is then the sum of the individual component *partial vapor pressures* (P_H, P_T) as given by **Dalton's law**.

$$P_{total} = P_H + P_T$$

or in general, for N components,

$$P_{total} = P_A + P_B + P_C + \cdots + P_N$$

It is important to realize that the vapor pressure (P°_{A}) and the partial vapor pressure (P_A) are not equivalent, since the presence of a second component in the system has an effect on the vapor pressure of the first component. The partial vapor pressure of any one volatile component can be obtained using **Raoult's law**.

$$P_A = P^{\circ}_{A} X_A$$

where X_A is the mole fraction of component A in the liquid system.

Simple distillation involves the use of the distillation process to separate a liquid from minor components that are nonvolatile or that have boiling points substantially (>30–40 °C) above the major component. Apparatus were developed that achieve good separation of mixture samples smaller than 2.0 mL in volume. The Hickman–Hinkle still, shown in Figure 2.2, is one of the more useful still designs. This still may be employed in microscale experiments to purify solvents, carry out reactions, and concentrate solutions for recrystallization.

The most volatile component of the reaction mixture enters the gas phase, and condenses on the sides of the much cooler Hickman–Hinkle still, getting trapped in the still collar. The distillate may be removed using a Pasteur pipet. It is often advantageous to bend the tip of the pipet slightly (using a microburner) to make the collar more accessible.

REFERENCES

1. For example, see Mayo, D. W.; Pike, R. M.; Butcher, S. S., *Microscale Organic Laboratory*, 2nd ed., Wiley: New York, 1989.
2. These balances are available from several manufacturers. It is recommended that a balance having a capacity of 100 ± 0.001 g be used.
3. For example, see a current *Handbook of Chemistry and Physics*, CRC Press: Boca Raton, FL.
4. Lodwig, S. N. *J. Chem. Educ.* **1989**, *66*, 77.

Chapter 3

Writing Laboratory Reports

3.A INTRODUCTION

All scientists are called on to submit written reports of their work as part of their employment. These reports may include:

- Experimental procedures to workers in the laboratory group.
- Progress reports about experimental work submitted to supervisors.
- Papers submitted to a scientific journal for peer review.
- Undergraduate or graduate theses.

Although all scientists must write laboratory reports as part of their job descriptions, scientists also have a reputation (some of which is deserved) for poor writing. Part of this perception of poor writing comes about because the language of science is unfamiliar to general audiences. Another part comes from individuals trying to read scientific work in the same cursory manner as a science fiction novel. Unfortunately, a large part comes from just plain poor writing.

Like all types of good writing, scientific writing has certain rules associated with it, and a certain style that has been adopted by the scientific community.[1,2] This style is discussed in the following sections. The main point to keep in mind is that the primary purpose of any writing is to communicate information. If the intended audience does not clearly follow what is written, the attempt at communication has failed. For this reason, clarity is the hallmark of good scientific writing.

3.B MAINTENANCE OF THE LABORATORY NOTEBOOK

The laboratory notebook[3] has a number of purposes associated with its use. Foremost is the maintenance of a written record of experimental procedures, data, and results. The proper laboratory notebook entry (see Fig. 3.1) begins with the date and a title for the work being done at the time, for example,

April 5, 1989: Recrystallization of Iron(II) chloride

Note that the month is written out, instead of using the common notation 4/5/89. In the United States, the common order for dates is month/day/year. In Europe,

58 September 1989
Synthesis of Trans – $Rh(CO)Cl(PPh_3)_2$

In this experiment, Wilkinson's catalyst, $RhCl(PPh_3)_3$, reacts with n-heptaldehyde in toluene to produce trans - $Rh(CO)Cl(PPh_3)_2$ and hexane according to the following reaction:

$RhCl(PPh_3)_3 + C_6H_{13}CHO \rightarrow$ trans - $Rh(CO)Cl(PPh_3)_2 + C_6H_{14}$

The lemon yellow crystals of trans - $Rh(CO)Cl(PPh_3)_2$ are characterized by I.R. spectrum and melting point while hexane is detected by gas chromatography.

Reactants and Products

compound	F.W	amt/vol	mmol	mp (°C)	bp (°C)	density
n-heptaldehyde	114.19	.05 mL	.372	-43	153	.850
$RhCl(PPh_3)_3$	925.23	50 mg	.054	157-158		
toluene	92.14	2 mL	18.8	-95	110.6	.8669
trans - $Rh(CO)Cl$ - $(PPh_3)_2$	690.91	25 mg	.036	170 (decomp)		

Note: Ether is a highly flammable liquid; no open flame was used. This part of experiment was carried out in the HOOD. Also, due to its strong odor, n-heptaldehyde was used in the HOOD.

EXPERIMENTAL

Procedure

I A distillation apparatus was assembled using a 25-mL, round-bottom flask, a deep-well Hickman still, an air condenser, and a sand bath.

II To the flask was added:
- A. 50 mg Wilkinson's catalyst
- B. 2 mL toluene (using auto pipet)
- C. .05 mL n-heptaldehyde (using auto pipet)
- D. a magnetic stir bar

III The flask was placed on the sand bath and the mixture heated to a gentle reflux

IV While the distillation was under way, a known solution was prepared in a 25-mL Erlenmeyer flask using:
- A. 2 mL toluene
- B. .05 mL n-hexane
- C. The solution was covered to prevent evaporation of hexane

V The distillate was collected in a 25-mL Erlenmeyer flask and covered
- A. The gas liquid chromatography was run on both the distillate and the known solution (Gow-Mac instrument, model # 69-150) using injection port A at 125°C.

VI The trans - $Rh(CO)Cl(PPh_3)_2$ crystals were filtered by a Hirsch funnel after adding a few drops ethanol and cooling in an ice bath
- A. the crystals were washed with 2-3 drops ethanol
- B. washed with .5 mL ether
- C dried and weighed

VII An I.R. spectrum was run on: (Nicolet FTIR, model #
- A. Wilkinsons catalyst
- B trans - $Rh(CO)Cl(PPh_3)_2$

Calculations

$$50\ mg\ RhCl(PPh_3)_3 \times \frac{1\ mmol}{925.23\ mg} = .054\ mmol\ RhCl(PPh_3)_3 \quad \leftarrow \text{limiting reagent}$$

$$.05\ mL\ C_6H_{13}CHO \times \frac{.850\ g}{mL} \times \frac{1000\ mg}{g} \times \frac{1\ mmol}{114.19\ mg} = .372\ mmol\ C_6H_{13}CHO$$

$$.054\ mmol\ RhCl(PPh_3)_3 \times \frac{1\ mmol\ Rh(CO)Cl(PPh_3)_2}{1\ mmol\ RhCl(PPh_3)_3} \times \frac{690.91\ mg}{1\ mmol} = 37.3\ mg\ Rh(CO)Cl(PPh_3)_2$$

Figure 3.1. *Typical laboratory notebook entry.*

$$\% \text{ yield} = \frac{\text{actual}}{\text{theoretical}} \times 100 = \frac{25 \text{ mg}}{37.3 \text{ mg}} \times 100 = 67\% \text{ yield}$$

Conclusion

1. The trans-$Rh(CO)Cl(PPh_3)_2$ crystals weighed 25 mg for a 67% yield.
2. Heptaldehyde was decarbonylated as confirmed by the isolation of trans-$Rh(CO)Cl(PPh_3)_2$ and the detection of hexane in GC.
3. Trans $Rh(CO)Cl(PPh_3)_2$ was characterized by its melting point (170°C) and its I.R. spectrum in nujol mull. A strong carbonyl stretching band occurs at 1965 cm^{-1}, which is very close to the literature value.

Sharon L Colebaugh

Figure 3.1 *(Continued)*

however, the common order is day/month/year. Hence, a European would have interpreted the date given above as May 4, 1989. Writing out the month eliminates this problem. The use of the title allows those using the notebook to find quickly any particular experiment or step. It is also a good idea to leave several pages blank at the beginning of the notebook for later use as a table of contents after the laboratory notebook is full. Since the laboratory notebook is used as a record of experiments carried out, it may be submitted in a court of law as evidence for first discovery, and so on. To ensure that no material is added or removed following an entry, it is customary that the notebook be sewn-bound (not loose-leaf). If any space remains at the bottom of a page that will not be used, that space should be "x-ed" out.

The first item that a notebook entry should contain is information about the chemicals used, including any unusual dangers or toxicities, where unusual chemicals were purchased, and any prior purification of the chemicals that may be necessary for subsequent steps. If one were running a Grignard reaction, for example, it might be appropriate to write:

The solvent (ether or tetrahydrofuran) was dried over sodium and benzophenone, as Grignard reagents are extremely water sensitive.

This should be mentioned before any other experimental procedure, as the next scientist following the laboratory procedure outlined in the notebook may be unaware of these prior steps.

The experimental procedure should then be entered, but only as it is actually done. There is a great temptation to copy an experimental procedure into a notebook from some other source. In many cases, however, you will find that it will be necessary to modify given procedures, resulting from the unavailability of chemicals or specific glassware, or because of the discovery of an alternate procedure. The procedure written in the laboratory notebook should be the exact procedure that was actually followed. Any unusual assemblies of glassware should be sketched. Any numbers entered should have units associated with them. It is usually convenient to number each reaction step.

Any observations of chemical or physical change should be noted after the experimental step that produced them. For example, one might write:

3. The zinc–mercury amalgam is then added to the reaction mixture, which is gently swirled. The vanadium(V) solution changes color from yellow to blue at this point.

Physical data should be entered after the step that produces the data, for example:

7. The melting point is taken using a Thomas–Hoover melting point apparatus. mp = 125 °C, uncorrected.

When the experimental procedure is completed, any calculations made using the data obtained should be clearly performed, with an indication as to where any external numbers came from. Enough information should be provided to allow the next scientist using the notebook to perform another calculation on a new set of data.

If any conclusions are drawn following an experimental procedure, it is useful to include them in the laboratory notebook at the end of that procedure. Finally, you should sign the entry at the very end, and date the work once again. For a legally binding document, a witness should also sign the work.

3.C THE LABORATORY REPORT

The full laboratory report consists of several parts, which will be individually discussed. These parts include the abstract, introduction, experimental section, data section, discussion, conclusions, acknowledgments, and references. It may be that a shorter version of the laboratory report may be acceptable for a particular course. It is always wise to ask the laboratory instructor as to the proper format. Most scientific reports are written in the passive voice, for example, "The solution was mixed for 10 min" rather than "I mixed the solution for 10 min."

The **abstract** is a short (one or two paragraph) summary of the experimental procedure and results. It is found at the beginning of a laboratory report so that an individual can get a quick overview of the experiment and decide if they wish to read any further.

The abstract is followed by the **introduction**, where a brief background as to the technique(s) used in the experiment is outlined. The theory behind any unusual calculations should be presented in the introduction as well. The introduction should not go into great detail about material that is readily found elsewhere; sources for this information should be listed in the reference section. "Common knowledge" such as might be found in an introductory textbook need not be referenced, but one should reference more specific material.

The **experimental section** should begin with a listing of where the chemicals used were obtained, what the purity was, and any prior purification of the starting materials. The manufacturer and model number of all major equipment should also be listed. The manner in which spectra were obtained should be included, for example:

All solid infrared spectra were obtained on a Nicolet 730 FT IR as KBr pellets.

The experimental equipment and glassware should be described, with a diagram provided if necessary.

All steps performed in the experimental procedure should be listed in the order that they were performed, in exactly the manner in which you performed them. Do not list a step as written in some experimental procedure that was followed unless you actually performed that step in an identical fashion. The source of the experimental procedure (if one exists) should be referenced. Observations as to physical and chemical changes should be included in this section.

The **data section** should list all data obtained, in raw form, with information provided as to how the data was obtained, as well as the experimental accuracy of all measurements. The data should be compiled into tables, if appropriate. Calculations should be made here (percent yield, heat of reactions, etc.) with a sample calculation provided.

The data should be discussed and evaluated, both positively and negatively, in the **discussion section**. Do not try to twist the data to fit the results you think should be obtained. Let the data "speak for itself," and evaluate the data fairly, even if the data seem to contradict theory you may have been expecting the data to follow. Many times in history, great scientific opportunities were missed because the experimenter skewed the data to fit his/her preconceptions. A discussion of possible sources of error should be included in this section.

The results should be summarized and conclusions drawn in the **conclusions section**. In situations where more than one explanation of the data is possible, all explanations should be presented, compared, and contrasted. Any help that was provided to you over the course of the experiment should be mentioned in the **acknowledgment section**.

3.D PROPER CITATION AND PLAGIARISM

All materials that were used in writing the laboratory report or to gather background material should be listed in the **reference section**, the last section of a laboratory report. There are two common types of reference to printed work; that for a journal article and that for a book. Journal articles are referenced by listing the authors (last name first), the title of the journal (usually abbreviated, in italics), the year of publication (boldface), the volume number (italics), and the page number. For example:

Smith, R. A.; Jones, B. C. *J. Am. Chem. Soc.* **1965,** *80*, 295.

Books are referenced by listing the authors (last name first), the title of the book (italics), the edition (if other than first), the publisher (followed by a colon), city (and state if the city is small) of publication, and the year of publication. For example:

Smith, R. A.; Jones, B. C., *The Chemistry of Manganese*, Acme Press: New York, 1955.

Other styles of referencing were used in the past and are currently used in other countries.

References should be consecutively numbered, as encountered in the text of the laboratory report. The reference number should either be superscripted[x] or appear underlined in parentheses (X) following the phrase or idea that is being referenced.

Failure to reference material obtained elsewhere constitutes plagiarism and is grounds for immediate failure. Even if an idea from elsewhere is "put into your own words" rather than being directly copied or quoted, it must be referenced.

REFERENCES

1. Ebel, H. F., Bliefert; C., Rusey, W. E., *The Art of Scientific Writing*, VCH: Weinheim: Federal Republic of Germany, 1987.
2. Schoenfeld, R., *The Chemist's English*, 2nd ed., VCH: Weinheim: Federal Republic of Germany, 1986.
3. Kanare, H. M., *Writing the Laboratory Notebook*, American Chemical Society: Washington, DC, 1985.

Chapter 4

Literature Searching and the Inorganic Literature

4.A LITERATURE SEARCHING

To write a proper introduction to a laboratory report, or to properly investigate previous experimental work that could be helpful on a research project, it is necessary to do a literature search. We have included one question involving literature searching in each of the laboratory experiments in this text. In chemistry, there are two different common methods by which this can be efficiently done: the Chemical Abstracts method[1,2] and the Citation Index method.

4.A.1 Chemical Abstracts Method

Chemical Abstracts (CA) is a journal that is published weekly by the American Chemical Society. It lists and cross references the abstracts from all papers published in all major (and most minor) chemical journals. The abstracts are placed into one of 80 sections within CA, depending on the subject matter contained in the referenced paper. The sections cover the following material:

Sections 1–20	Biochemistry
Sections 21–34	Organic Chemistry
Sections 35–46	Macromolecules
Sections 47–64	Applied Chemistry and Chemical Engineering
Sections 65–80	Physical, Inorganic, and Analytical Chemistry

The sections that are of the most direct interest to inorganic chemists are:

Section 29	Organometallic Chemistry
Section 67	Catalysis, Reaction Kinetics, and Inorganic Reaction Mechanisms
Section 78	Inorganic Chemicals and Reactions
Section 79	Inorganic Analytical Chemistry

Articles of inorganic interest often appear in other sections, as well.

The abstracts are cross referenced in the index of each **weekly issue** of CA by author, patent number, and subject keyword. The weekly indexes are collected into six **annual indexes**, which cross reference the abstracts in various ways:

Author Index. Lists all papers in alphabetical order according to the author's name.

Chemical Substance Index. Lists all papers according to the proper name of all chemicals used in the paper.

Formula Index. Lists all papers according to the formula of the chemical compounds used in the paper, in the following order: carbon first, hydrogen second, and then all other elements in alphabetical order.

General Subject Index. Lists all papers according to subject keywords found in the paper.

Index of Ring Systems. Useful mainly for organic chemistry. Lists papers involving ring systems according to the type of ring system found.

Patent Index. Lists all chemical patents in numerical order, as well as subsidiary patents.

Every 5 years, a **Collective Index** is published (in the past, it was every 10 years, and called a Decennial Index). The 11th Collective Index is the current one,[3] with the 12th Collective Index now in press. These replace the Annual Indexes for the 5-year coverage period. Additionally, every 18 months, an **Index Guide** is published, listing all current keywords and chemical names used in the indexes.

Searching a Topic in *Chemical Abstracts*

There are several alternate schemes for searching a particular topic in *Chemical Abstracts*. Two of the more direct methods are described next.

Use of the Formula Index: Suppose, after doing Experiment 42, "Synthesis and Reactions of *trans*-Chlorocarbonylbis(triphenylphosphine)rhodium(I)," you wished to see if the starting material, $RhCl_3{\cdot}3H_2O$, had been used in the syntheses of any other rhodium complexes. The easiest way to proceed would be to look up the formula in the Formula Index. The Formula Index lists formulas in the following order: carbon first, hydrogen next, and then all other elements in alphabetical order. Hydrates are listed under the parent compound formula. In this case, therefore, we would look under the heading Cl_3Rh.

Generally, one begins a search in the most current volume of *Chemical Abstracts*, Volume 109 as of this writing. Two references are listed under the subheading "trihydrate," those being

110627m and 149765e

The listing is shown in Figure 4.1.

These two references tell us to look up abstracts numbers 110,627 and 149,765. These abstracts are shown in Figure 4.2. In older volumes (before 1966), a different listing style was used. A typical abstract number for this period would be

9135d

where the 9135 is the column number (two columns to the page), and the d

Cl_3Rh
Rhodium chloride ($RhCl_3$) *[10049-07-7]*. See
Chemical Substance Index
trihydrate *[13569-65-8]*. 110627m. 149765e

Figure 4.1. Chemical Abstracts *formula index listing. (Copyright © 1988 by the American Chemical Society. Reprinted by permission.)*

109:110627m Ortho-chelated arylrhodium(I) complexes. X-ray structure of $Rh^{I}[C_6H_3(CH_2NMe_2)_2\text{-}o,o'\text{-}C,N](COD)$. Van der Zeijden, Adolphus A. H.; Van Koten, Gerard; Nordemann, Richard A.; Kojic-Prodic, Biserka; Spek, Anthony L. (Anorg. Chem. Lab., Univ. Amsterdam, 1018 WV Amsterdam, Neth.). *Organometallics* 1988, 7(9), 1957–66 (Eng). The reaction $Li_n[C_6H_3(CH_2NMe_2)\text{-}o\text{-}R\text{-}o']_n$ with $[RhCl(diene)]_2$ yields the ortho-chelated arylrhodium(I) complexes $Rh[C_6H_3(CH_2NMe_2)\text{-}o\text{-}R\text{-}o'](diene)$ [R = CH_2NMe_2, diene = cyclooctadiene (COD) (**I**), norbornadiene; diene = COD, R = Me, H]. The solid-state structure of **I** was detd. by a single-crystal x-ray diffraction study. **I** consists of a Rh(I) center that has a square-planar coordination comprising the two double bonds of COD and a C atom and one of the N atoms of the monoanionic aryl ligand. In soln., the chelated arylrhodium(I) complexes exhibit dynamic behavior which involves a reversible dissocn. of the Rh-N bond and rotation of the aryl moiety around Rh-C. This process, which generates a highly unsatd. T-shaped 14 electron species, is accompanied by the relief of steric repulsions within the complex. **I** reacts with a range of electrophilic reagents leading to Rh-C bond breakage [HX, X = acetylacetonato, Cl, Br, OAc, OH, OMe, L-alanyl; MX_nL_m, $SnMe_2Br_2$, $NiBr_2(PBu_3)_2$, $ZrCl_4$, $PdCl_2(NCPh)_2$, $HgCl_2$, $PtBr_2(COD)$, and $\{IrCl(COD)\}_2$]. A redox reaction of **I** with AgX (X = OAc, NO_3) forms $RhX_2[C_6H_3(CH_2NMe_2)_2\text{-}o,o'](H_2O)$.

109: 149765e Tetrathiometalate complexes of rhodium, iridium, palladium, and platinum. Structures of $[(C_5Me_5)RhCl]_2WS_4$ and $[(C_3H_5)Pd]_2WS_4$. Howard, Kevin E.; Rauchfuss, Thomas B.; Wilson, Scott R. (Sch. Chem. Sci., Univ. Illinois, Urbana-Champaign, Urbana, IL 61801 USA). *Inorg. Chem.* 1988, 27(20), 3561-7 (Eng). Synthetic routes to tetrathiometalate complexes of rhodium, iridium, palladium, and platinum are described. Acetonitrile solns. of WS_4^{2-} reacted with $[Rh(diene)Cl]_2$ [diene = 1,5-cyclooctadiene (COD) or norbornadiene (NBD)], $[Ir(COD)Cl]_2$, $[(\eta^5\text{-}C_5Me_5)RhCl_2]_2$, and $[Pd(C_3H_5)Cl]_2$ to give good yields of $\mu\text{-}WS_4$ complexes $[Rh(diene)]_2WS_4$, $[Ir(COD)]_2WS_4$, $[(\eta^5\text{-}C_5Me_5)RhCl]_2WS_4$, and $[(C_3H_5)Pd]_2WS_4$, resp. $(COD)PtCl_2$ reacted with MS_4^{2-} to give $(COD)PtMS_4$, (M = Mo, W). The latter complexes reacted further with PPh_3 or WS_4^{2-} to give $(PPh_3)_2PtMS_4$ and $Pt(WS_4)_2^{2-}$, resp. $[Rh(COD)]_2WS_4$ reacted with Me_3CNC to give $[Rh(Me_3CNC)_2]_2WS_4$. $RhCl_3 \cdot 3H_2O$ and $IrCl_3 \cdot 3H_2O$ reacted with WS_4^{2-} to give the octahedral complexes $M(WS_4)_3^{3-}$ isolated as their Et_4N^+ salts. The ^{183}W NMR spectrum of $(Et_4N)_3[Rh(WS_4)_3]$ showed a doublet with $J(^{183}W,^{103}Rh)$ = 4.8 Hz. Thermal gravimetric analyses of $[Rh(COD)]_2WS_4$ and $(COD)PtWS_4$ indicated that loss of org. coligands and stoichiometric amts. of sulfur occur at moderate temps. The compd. $[(\eta^5\text{-}C_5Mes)RhCl]_2WS_4 \cdot CHCl_3$ and $[(\eta^3\text{-}C_3H_5)Pd]_2WS_4$ were characterized by x-ray crystallog.

Figure 4.2. Chemical Abstracts *abstracts listing. (Copyright © 1988 by the American Chemical Society. Reprinted by permission.)*

indicates the position within the column. In the collective volumes, the abstract number will appear as

109: 110627m

where the 109: indicates that the abstract is in volume 109, necessary as the Collective Index covers more than one volume.

Note that the compound we are searching for, $RhCl_3 \cdot 3H_2O$, does not actually appear in abstract 110,627m (*Organometallics* **1988,** 7, 1957–1966), although it does in abstract 149,765e (*Inorg. Chem.* **1988,** *27*, 3561–3567). The reason that 110,627m turned up in our search was that $RhCl_3 \cdot 3H_2O$ was listed in the Registry Number listing at the end of the paper (p. 1966 of the reference). Any compound listed in the Registry Number listing in a paper will be cross-referenced in the annual and collective indexes of *Chemical Abstracts*. This is a major advantage of the use of *Chemical Abstracts*—the compound does not need appear in the abstract to be cross-referenced.

The abstracts provide a synopsis of the highlights of the papers, and one can then decide whether to look up the articles in the journals themselves. Since

CA covers many minor journals, one frequently obtains listings in journals that are not available at a particular library. In such cases, a copy of the paper can be obtained through the Chemical Abstracts Document Delivery Service (write to CAS Customer Service) or by interlibrary loan. The Institute for Scientific Information also provides a document copy service.

The references found in the papers obtained in the search provide additional sources of information on the subject being searched. These references are not available in *Chemical Abstracts*, only in the paper itself. This is in contrast with the *Citation Index* (see Section 4.A.2), which provides the references but not the abstract. Searching earlier volumes of *Chemical Abstracts* would provide additional papers detailing the use of $RhCl_3{\cdot}3H_2O$.

Use of the General Subject Index: Suppose, for example, after reading Sections 5.A (Magnetic Susceptibility) and 5.B (Thermal Analysis), you wanted to see if it were possible to use thermogravimetric analysis to make magnetic susceptibility measurements. Since this is a question about general techniques rather than about a specific reaction or compound, the General Subject Index would be the logical place to start.

Consulting a recent General Subject Index, we could look up either "magnetic susceptibility" or "thermogravimetric analysis." If either of these terms did not appear in the General Subject Index, it might be necessary to consult the Index Guide to see the closest term to the above that does appear. A number of subheadings appear under the major heading "thermogravimetric analysis," one of which is "—use of in magnetic susceptibility detn." The reference that follows is the number

203402z

This number tells you to look up abstract number 203,402 in volume 109 of CA. Looking up these keywords in volumes 106–108 show no additional references. One could carry the search as far back into the past as desired.

Looking up the reference yields a paper entitled "Magnetic Susceptibility Measurement with a DuPont Thermogravimetric Analyser," authored by L. M. Razo and R. Gomez, which appeared in 1988 in the *Journal of Thermal Analysis*, volume 34, pages 89–92. The references cited within that paper would provide additional information about the given topic.

CAS Online: *Chemical Abstracts* may also be accessed by computer data base searching. The search may be conducted using various types of input, such as structures, molecular formula, CAS registry numbers (provided in this text for all chemicals used), or keywords. More complete information about the use of CAS Online may be found in Refs. 4 and 5. Tutorial programs are available from STN^{SM} International.

Advantages and Disadvantages of *Chemical Abstracts*: The *Chemical Abstracts* method of literature searching is the most common method for several reasons.

- Most colleges with chemistry departments subscribe to CA, so the resource is readily available.
- *Chemical Abstracts* covers a huge variety of journals, so any work published in the desired area is almost certain to be abstracted.
- The abstract is provided to help determine if the article will be of interest.
- *Chemical Abstracts* has been published since 1907, thereby covering most literature (CAS Online is only available since 1967).

There are several disadvantages as well:

- *Chemical Abstracts* is quite expensive, so many colleges that subscribe to the journal do not subscribe to the Collective Indexes. Looking up a keyword or formula over a long stretch of time is very tedious.
- Finding a reference is a four step process: determine the keyword or formula, look it up in the index, look up the abstract, and look up the paper.
- Many times, a paper that would be useful does not contain your keyword in its title or abstract, and would not be located. This is not a problem with formulas, as the formula only need be listed in the Registry.

4.A.2 Use of the *Science Citation Index*

The *Science Citation Index* (SCI)[6,7] is published every 3 months by the Institute for Scientific Information. Papers from all major and most minor journals are cross-referenced by use of authors, keywords, journal, location, and most importantly, by their references. The SCI is now available on compact disc, which can be accessed using a personal computer.

The printed SCI consists of three major parts:

Citation Index. Lists all papers as a function of the references they cite.

Source Index. Lists all papers alphabetically, by the first author.

Permuterm Subject Index. Lists all papers according to subject keywords.

The computer version allows searches by the following criteria:

Source Author. The author of the article.

Cited Author. The author of the reference cited by the article.

Address. The address(es) of the authors of the paper.

Journal. The journal that the paper appeared in.

Title Word. By subject keywords appearing in the title of the paper.

Searching Using the Printed *Science Citation Index*

Suppose, for example, you were interested in the use of thermal analysis to investigate Rh(III) complexes. The first step would be to enter the search using the subject keywords "thermal analysis" or "Rh(III)" in the latest issue of the **Permuterm Subject Index**. For the first 3 months of 1988, there were 16 papers listed under the keyword "Rh(III)." Various subheadings also appear, including "thermal." The first author's name, "Poston S," for example, appears under this subheading. The name would then be looked up in the **Source Index**, yielding the reference

Poston S

Reisman A—Physical, Thermal and Optical Characterization of Rhodium(III) Acetylacetonate

J. Electronic Materials 17(1):57–61 88 13R

as well as the address of the author. In this listing, A. Reisman is the coauthor of the paper, which appeared in 1988 in the *Journal of Electronic Materials*, volume 17, issue 1, on pages 57–61. The notation 13R indicates that the paper has 13 references. A code number for ordering the article would also appear after the 13R notation. This bibliographic information is followed by a list of the references cited by the paper, the first of which in this instance is

Barnum DW 61 J Inorg Nucl Chem 21 221

This indicates that the reference is by D. W. Barnum, and appeared in 1961 in the *Journal of Inorganic and Nuclear Chemistry*, volume 21, on page 221.

Searching Using the Compact Disc *Science Citation Index*

Searching for a particular reference is much simpler when the CD version of the SCI is available. One would simply insert the CD into a disc reader, turn on the computer, and follow the menu to perform the search.

In the previous example, using the title word "Rh(III)," the computer responds with 16 listings. These listings can be shown by title only, or the complete listing can be shown. More than one Title Word can be searched at a time, using logical commands such as "and" and "or."

Searching Using the *Citation Index*

The true power of the SCI is seen when we search using a previous reference or author, using the ***Citation Index***. In most cases, there is a well-known (or not so well-known) original paper on a given subject, which all subsequent papers are likely to cite. For example, if we were interested in the topic of "isomerism in transition metal complexes," the key papers were written by the chemist who pioneered this area, Alfred Werner.

We would look up the name "Werner, A" in the *Citation Index*, and under his name would come a series of subheadings, each being a particular paper that Werner published. Under each subheading would appear a series of references to papers who listed Werner's paper as one of their references. In this way, we could quickly obtain a list of all papers on a particular subject area.

The CD computer version is especially powerful. Each listed paper also has a list of references and related papers. In the case of the paper by Poston and Reisman discussed previously, there are 13 references and 4 related records. The listing for the paper is shown in Figure 4.3. We can then list the 13 references found in this paper, shown in Figure 4.4. If desired, each of the referenced authors listed can then be searched to see if any paper abstracted in the given issue of the CD *Citation Index* cites any of their work. We can also list the four related records, along with their references, an example of which is shown in Figure 4.5. A related record is a paper that has a reference in common with (in this case) the paper by Poston and Reisman. In the case of the first related record, by Hassan, Abubakr, Ahmed, and Seleim, the reference in common is

Holm-RH 1958 J Am Chem Soc V80 P5658

The paper by Hassan et al. itself has 21 references and 15 related records, each of which could be listed. This would constitute a **second-order search**. Each of those 15 related records could then be searched in a **third-order search**, and so on. By using the references and related records, a huge bibliography of work on any particular subject can quickly and efficiently be built up.

POSTON-S REISMAN-A

PHYSICAL, THERMAL AND OPTICAL CHARACTERIZATION OF RHODIUM(III) ACETYLACETONATE (English) → Article

JOURNAL OF ELECTRONIC MATERIALS
Vol 17 No 1 pp 57–61 1989 (L9568)
References: 13 Related Records: 4

Figure 4.3. *Typical* Science Citation Index *listing. From Social SCISEARCH®. Copied with permission of the Institute of Scientific Information®, Philadelphia, PA.)*

POSTON-S REISMAN-A

PHYSICAL, THERMAL AND OPTICAL CHARACTERIZATION OF RHODIUM(III) ACETYLACETONATE (English) → Article

JOURNAL OF ELECTRONIC MATERIALS
Vol 17 No 1 pp 57–61 1988 (L9568)
References: 13 Related Records: 4

BARNUM-DW 1961 J-INORG-NUCL-CHEM V21 P221
BEECH-G 1971 THERMOCHIM-ACTA V3 P97
CHARLES-RG 1958 J-PHYS-CHEM-US V62 P440
DEARMOND-K 1968 J-CHEM-PHYS V49 P466
FACKLER-JP 1963 INORG-CHEM V2 P97
HOLM-RH 1958 J-AM-CHEM-SOC V20 P5658
LINTVEDT-RL 1975 CONCEPTS-INORGANIC-P
OHRBACH-KH 1983 THERMOCHIM-ACTA V67 P189
OPRYSKO-MM 1986 SEMICONDUCTOR-INT V1 P92
OSBURN-C 1987 J-ELECTRON-MATER V16 P223
REISMAN-A 1975 J-ELECTRON-MATER V4 P721
SIEVERS-RE 1967 J-INORG-NUCL-CHEM V29 P1931
YOSHIDA-I 1973 J-INORG-NUCL-CHEM V35 P4061

Figure 4.4. Science Citation Index *reference listing*. From *Social SCISEARCH®. Copied with permission of the Institute of Scientific Information®, Philadelphia, PA.*

HASSAN-MK ABUBAKR-MS AHMED-MA SELEIM-MM

SYNTHESIS AND STRUCTURAL STUDIES ON MONONUCLEAR, HOMOBINUCLEAR, AND HETEROBINUCLEAR COMPLEXES OF CU(II), CD(II) AND HG(II) WITH N,N'-ETHYLENEBIS (SALICYLIDENEIMINS) (English) → Article

ANNALI DI CHIMICA
Vol 78 No 1–2 pp 107–116 1988 (M5096)
References: 21 Related Records: 15

ASTECTWIESER INTRO-ORGANIC-CHEM P627
BAKER-AW 1959 J-AM-CHEM-SOC V81 P1223
BELLAMY-LJ 1958 INFRARED-SPECTRA-COM
COLEMAN-WM 1979 J-INORG-NUCL-CHEM V41 P95
FENTON-DE 1982 INORG-CHIM-ACTA V62 P57
FREEDMAN-HH 1962 J-AM-CHEM-SOC V83 P2900
FUJITA-J 1956 J-AM-CHEM-SOC V78 P3963
GEARY-WJ 1971 COORD-CHEM-REV V7 P91
HOLM-RH 1958 J-AM-CHEM-SOC V80 P5658
ISSA-IM 1973 J-CHEM V16 P18
JOSH-KC 1977 AGR-BIOL-CHEM-TOKYO V41 P543
KOVACIC-JE 1967 SPECTROCHIM-ACTA-A V23 P183
MELNIK-M 1982 COORD-CHEM-REV V42 P250
PRINCE-RH 1974 INORG-CHIM-ACTA V10 P89
SCOTT-W 1952 STANDARD-METHODS-CHE
SIMMONS-MG 1980 J-CHEM-SOC-DA P1827
TAJMIRRIAHI-HA 1983 CAN-J-SPECTROSC V28 P129
TAYIM-HA 1975 INORG-NUCL V11 P395
UENO-K 1956 J-PHYS-CHEM-US V60 P1370
VIDALI-M 1980 INORG-CHIM-ACTA V38 P58
WELCHER-RJ 1941 ANAL-USES-ETHYLENE-D

Figure 4.5 Science Citation Index *related reference listing*. From *Social SCISEARCH®. Copied with permission of the Institute of Scientific Information®, Philadelphia, PA.*

Advantages and Disadvantages of the *Science Citation Index*

Advantages of the SCI method include:

- The method is very fast, especially in the CD version.
- If the compact discs are available, there is no need to visit a library to perform the search.
- Papers on a given subject are related by virtue of their references. Thus, once you have a key paper on a subject (one all subsequent papers are likely to cite), searches can be accomplished without using keywords at all.
- Related papers without your keyword in their title are much easier to find, as they will turn up as related references.
- Since the SCI is computerized, more than one keyword can be used in a given search, narrowing down a list of papers very quickly.
- Second and higher order searches are easily done.

Disadvantages include:

- The SCI only goes back to 1980 on compact disc and to 1961 in print. Older papers do, however, appear in the *Citation Index*.
- No abstracts are provided.
- The *Citation Index* is found in fewer libraries than *Chemical Abstracts*.

4.A.3 Comparison of the Two Methods

With two different searching methods, it is natural to ask the question "Which one is best?" As is usually the case in such matters, the authors of this book will evade the question by answering that it depends on what you are searching for.

If the goal of the search is to obtain information on a given subject or chemical, CA will generally be easier to use. Since CA provides the abstract of the paper, the searcher can quickly determine if the paper obtained in the search actually is likely to contain the desired data. If the goal, however, is to build up a "library" of papers on a particular topic, especially after having a key paper in hand, the SCI will be much more convenient. The two methods complement each other nicely, with CA generally being used for the initial search, and the SCI being used to obtain a broad coverage of the subject.

4.B THE INORGANIC CHEMICAL LITERATURE

4.B.1 Introduction

There are literally thousands of journals published in the area of chemistry, most of which have papers within them that would be of interest to the inorganic chemist. In addition to the journals, there are thousands of serial monographs and books published annually on the subject. Needless to say, it would be impossible to keep up with all of them! Selective literature searching solves part of the problem (see Section 4.A). The most important information appears, however, in a relative handful of major journals and serial monographs, listed and briefly described below. Coverage has been restricted to major journals that provide at least abstracts in English. To the average student, the most useful parts of each article will be the introduction (describing earlier work) and the results–discussion section.

4.B.2 Purely Inorganic Journals

Inorganic Chemistry (Every 2 Weeks, ACS)

The major American journal with general coverage of inorganic chemistry. English.

Inorganica Chimica Acta (Every 2 Weeks, Elsevier)
A major European inorganic journal. English.

Journal of Organometallic Chemistry. (Weekly, Elsevier)
A major journal in the area of organometallic chemistry. Disadvantage: a very expensive journal. Various issues have review articles. All abstracts in English. Most articles in English, ~20% German.

Journal of The Chemical Society Dalton Transactions (Monthly, Royal Society)
The major British journal with general coverage of inorganic chemistry. English.

Organometallics (Monthly, ACS)
A relatively new journal with a high impact, covering the area of organometallic chemistry. English.

Zeitschrift für Anorganische und Allgemeine Chemie (Monthly, Johann Ambrosius Barth Verlag)
The major German journal with general coverage of inorganic chemistry. Abstracts appear in English, some articles do as well.

4.B.3 General Coverage Journals

Angewandte Chemie International Edition in English (Monthly, VCH Publishers)
A major German chemical journal published in translation. Excellent review articles appear in most issues.

Journal of the American Chemical Society (Every 2 Weeks, ACS)
The premier chemistry journal in the world. English.

Chemische Berichte (Monthly, VCH Publishers)
The major German chemical journal covering all areas of chemistry. Part A covers inorganic and organometallic chemistry. Mainly in German, ~10% English. (All abstracts appear in English.)

Helvetica Chimica Acta (Monthly, Verlag Helvetica)
A major European chemical journal covering all areas of chemistry. Mainly English, ~10% German.

4.B.4 Review Journals and Monographs Covering Inorganic Chemistry

Accounts of Chemical Research. (Monthly, ACS)
A general coverage review journal, some inorganic coverage.

Advances in Inorganic Chemistry (Annual Monograph, Academic Press).
Formerly called *Advances in Inorganic Chemistry and Radiochemistry.* Published since 1959.

Angewandte Chemie International Edition in English (Monthly, VCH Publishers)
A general coverage journal, it usually contains one or two review articles per issue.

Chemical Reviews (Monthly, ACS)
A general coverage review journal, some inorganic coverage.

Chemical Society Reviews (Quarterly, Royal Society)
Formerly called *Quarterly Reviews of the Chemical Society.* A general coverage review journal, some inorganic coverage.

Coordination Chemistry Reviews (7 per year, Elsevier Science Publishers) Published since 1966.

Inorganic Syntheses (Annual Monograph, Wiley)
Published since 1939. This series is the place to find fully tested methods to synthesize inorganic compounds.

Progress in Inorganic Chemistry (Annual Monograph, Wiley)
Published since 1959.

4.B.5 Major Comprehensive Books on Inorganic Chemistry

Note should also be made of the following comprehensive surveys of inorganic chemistry, published by the Pergamon Press. These multivolume works are extremely comprehensive, with each section surveying the chemistry of an element and written by the "authorities" of the particular field. These surveys are extremely expensive (~\$2500 each), but well worth the expenditure.

Comprehensive Inorganic Chemistry, Eds. J. C. Bailor, H. J. Emeleus, R. S. Nyholm, and A. F. Trotman Dickenson. Published in 1974. A new edition of this should appear in the near future.

Comprehensive Organometallic Chemistry, Eds. G. Wilkinson, F. G. A. Stone and E. W. Abel. Published in 1982.

Comprehensive Coordination Chemistry, Eds. G. Wilkinson, R. D. Gillard and J. A. McCleverty. Published in 1987.

An even more comprehensive series of books on inorganic chemistry is the *Gmelin Handbuch der Anorganische Chemie*. First published in 1924, the series consists of hundreds of books, each about an individual element's chemistry. Most elements have multiple volumes devoted to them (some over 20 volumes!). Not all elements have been covered to this point. Earlier volumes (pre-1980 or so) are in German and are relatively inexpensive. More recent volumes are in a combination of German and English or in English, and are extremely expensive (~\$3000 for each volume!). Needless to say, the entire series will only be found in the largest libraries.

The main source for information on industrial chemistry (both inorganic and organic) is the Kirk–Othmer *Encyclopedia of Chemical Technology* (Wiley). This multivolume work covers a broad range of inorganic preparations, properties, and uses.

A frequently useful source, especially in the area of bioinorganic chemistry, is the *Merck Index* (Merck Chemical Co.), now in its 11th edition. Properties, preparations, and pharmacological data are given on a wide variety of compounds (mostly organic). Another useful source is the *Aldrich Catalog Handbook of Fine Chemicals* (available at no cost from Aldrich Chemical Co.), which provides physical properties, toxicological data, and references to spectra of many inorganic and organic compounds.

REFERENCES

1. —, *How to Search Printed CA*, American Chemical Society: Washington, DC, 1984. This 24 page pamphlet is available at no charge from the American Chemical Society.
2. Stowell, J. C., *Intermediate Organic Chemistry*, Wiley: New York, 1988. Chapter 2 provides a good overview of the use of Chemical Abstracts, including computer searching of CA.
3. Interestingly, the 10th Collective Index of CA is listed in the *1989 Guiness Book of world Records* as the "longest index" ever published, at 75 volumes and a weight of 380 lb. The number of entries is 23,948,253 in 131,445 pages. The Book of World Records is behind the times, however, as the

11th Collective Index is even larger, and the forthcoming 12th edition promises to be larger still.

4. Schulz, H., *From CA to CAS Online*, VCH: Weinheim, Federal Republic of Germany, 1988.

5. Maizell, R. E., *How to Find Chemical Information*, 2nd ed., Wiley: New York, 1987.

6. —, *Science Citation Index*, Institute for Scientific Information: Philadelphia, 1988. This eight page pamphlet is available at no charge from ISI.

7. Garfield, E., "How to Use Science Citation Index (SCI)," *Current Contents* **1983,** (9), 5. Reprints of this article are available at no charge from ISI.

General References

Gould, R. F., Ed., *Searching the Chemical Literature* Advances in Chemistry Series, No. 30, American Chemical Society: Washington, DC, 1961.

Schulz, H., *From CA to CAS Online* VCH: Weinheim, Federal Republic of Germany, 1988.

Maizell, R. E., *How to Find Chemical Information*, 2nd ed., Wiley: New York, 1987.

Chapter 5

Inorganic Microscale Laboratory Techniques

5.A MICROSCALE DETERMINATION OF MAGNETIC SUSCEPTIBILITY

5.A.1 Introduction

Transition metals, by definition, have at least one oxidation state with an incomplete *d* or *f* subshell. Since electrons spin and generate a magnetic field, the magnetic properties of transition metals are of great interest in determining the oxidation state, electronic configuration, and so on. Most organic compounds and main group element compounds have all their electrons paired. Such molecules are diamagnetic and have very small magnetic moments. Many transition metal compounds, however, have one or more unpaired electrons, and are termed paramagnetic. The number of unpaired electrons on a given metal ion determines the magnetic moment, μ, affecting it both by virtue of their spin and their orbital motion. The spin part is the more important, and a close estimate of the magnetic moment can be obtained using the equation

$$\mu_s = g\sqrt{S(S+1)} \tag{5.1}$$

where g is the gyromagnetic ratio for an electron (≈ 2) and S is the total spin of the unpaired electrons (at $\frac{1}{2}$ each).

For one unpaired electron (as in Ti^{3+})

$$\mu_s = 2\sqrt{\frac{1}{2}\left(\frac{1}{2}+1\right)} = 1.732 \tag{5.2}$$

The units of the magnetic moment are Bohr magnetons (BM). Actual magnetic moments are somewhat larger than the spin-only values obtained above, because of the orbital contribution.

Magnetic moments are not measured directly. Instead, the magnetic moment is calculated from the magnetic susceptibility, as described in Section 5.A.2. Experiments 24 and 25 directly utilize magnetic susceptibility measurements to help determine bonding within a complex. The technique could also be used to determine the electronic configuration of any of the transition metal complexes prepared or used in this book.

5.A.2 Measurement of Magnetic Susceptibility

There are a number of techniques that were used to determine the magnetic susceptibility of transition metal complexes. These include the Gouy method, the Faraday method, and the determination of magnetic susceptibility by nuclear magnetic resonance (NMR). Of these techniques, only the last two qualify as microtechniques, and can be carried out practically with 50 mg or less of sample. More recently, a new type of magnetic susceptibility balance, developed by D. F. Evans of Imperial College, London and Johnson Matthey/AESAR has appeared.[1] The balance is compact, lightweight, and self-contained. It does not require a separate magnet or power supply, and is therefore easily portable. The instrument has a digital readout that provides quick and accurate readings and whose sensitivity matches that of traditional methods. This balance can handle microscale quantities of solids as well as determine the magnetic susceptibility of liquids and solutions. As such, it is an ideal instrument for microscale inorganic work. The balance is shown in Figure 5.1.

In the Gouy method, a sample is suspended from a balance between the two poles of a magnet. The balance measures the apparent change in the mass of the sample, because it is repelled or attracted by the magnetic field. The attraction is due to the magnetic field generated by the unpaired electrons in a paramagnetic sample. Diamagnetic samples are repelled by the balance. The Evans balance has the same basic equipment configuration as in the Gouy method, but instead of measuring the force that the magnet exerts on the sample, it measures the equal and opposite force the sample exerts on a suspended permanent magnet. The Evans balance measures the change in current required to keep a set of suspended permanent magnets in balance after their magnetic fields interact with the sample. The magnets are on one end of a balance beam, and after interacting with the sample, change the position of the beam. This change is registered by a pair of photodiodes set on opposite sides of the balance beam's equilibrium position. The diodes send signals to an amplifier that in turn supplies current to a coil that will exactly cancel the interaction force. A digital voltmeter, connected across a precision resistor, in series with the coil, measures the current directly and this is displayed on the digital readout.

The general expression for the mass magnetic susceptibility, X_g, for the Evans

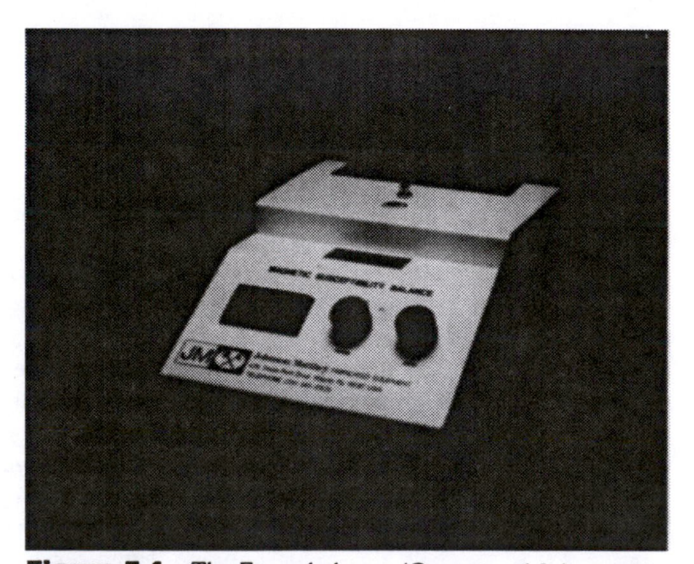

Figure 5.1. *The Evans balance. (Courtesy of Johnson Matthey, Catalytic Systems Division, Wayne, PA.)*

balance is

$$X_g = \frac{L}{m}\left\{C(R - R_0) + X'_v A\right\} \quad (5.3)$$

where L = sample length in centimeters
m = sample mass in grams
C = balance calibration constant (different for each balance; printed on the back of the instrument)
R = reading from the digital display when the sample (in the sample tube) is in place in the balance
R_0 = reading from the digital display when the empty sample tube is in place in the balance
X'_v = volume susceptibility of air (0.029×10^{-6} erg·G^{-2} cm^{-3})
A = cross-sectional area of the sample

The calibration standards usually employed in magnetic susceptibility measurements are $Hg[Co(SCN)_4]$ or $[Ni(en)_3]S_2O_3$, and have values of 1.644×10^{-5} and 1.104×10^{-5} erg·G^{-2} cm^{-3}, respectively. The volume susceptibility of air is usually ignored with solid samples, so that the mass magnetic susceptibility equation can be rewritten as follows:

$$X_g = CL(R - R_0)/(1 \times 10^9(m)) \quad (5.4)$$

where X_g is in centimeter-gram-second (cgs) units of erg·G^{-2} g^{-1}.

The magnetic susceptibility of liquids and solutions must include the volume susceptibility term, and, if the density of the liquid or the solution is known, the mass susceptibility of the solution, X_s, or of the solvent, X_0, may be calculated by

$$X_s \text{ or } X_0 = \frac{C(R - R_0)}{1 \times 10^9 A\, d_s} + \frac{X'_v}{d_s} \quad (5.5)$$

The terms C, A, R, R_0, and X'_v are as defined above and d_s is the density of the liquid or solution. The mass susceptibilities X_s and X_0 are in cgs units of erg·G^{-2} g^{-1}. The solution susceptibility, X_s, can be converted to the mass suceptibility, X_g, using the Weidmann additivity relationship

$$X_s = \frac{m_1}{m_1 + m_0} X_g + \frac{m_0}{m_1 + m_0} X_0 \quad (5.6)$$

where m_1 is the mass of the sample and m_0 is the mass of the solvent in grams.

The other method that can be used to determine the magnetic susceptibility of solutions is the NMR method developed by Evans.[2] The relationship between the mass susceptibility, X_g, and the change in frequency, $\Delta\nu$, of an indicator species in the solution is given by

$$X_g = \frac{3\Delta\nu}{2\pi\nu c} + X_0 + \frac{X_0(d_0 - d_s)}{c} \quad (5.7)$$

where the other terms given in the equation are defined as shown below:

ν = frequency of the nucleus under investigation in hertz (Hz)
c = concentration of the sample in grams per milliliter ($g \cdot mL^{-1}$)
X_0 = mass susceptibility of the solvent in cgs units
d_0 = density of the solvent in grams per milliliter
d_s = density of the solution in grams per milliliter

The third term is used to correct for the difference in density between the solvent and solution. This term usually makes only a small correction and is often ignored. Thus, the equation reduces to

$$X_g = \frac{3\,\Delta\nu}{2\,\pi\nu\,c} + X_0 \tag{5.8}$$

5.A.3 Calculation of Magnetic Moment from Magnetic Susceptibility

Once the mass susceptibility of the sample has been determined by one of the previous methods, the effective magnetic moment in Bohr magnetons (BM) can be calculated from the molar susceptibility (which has cgs units of $erg \cdot G^{-2}$ mol^{-1}). However, this is after diamagnetic corrections for the inner-core electrons, ligands, atoms, or ions in the compound have also been made. The molar susceptibility, X_M, is calculated as shown below.

$$X_M = X_g \cdot (\text{molecular weight in } g \cdot mol^{-1}) \tag{5.9}$$

Table 5.1 Diamagnetic Corrections for Ions and Molecules[a]

Cations	Correction	Anions[b]	Correction	Molecules[b]	Correction
Li^+	1	F^-	9	H_2O	13
Na^+	7	Cl^-	23	NH_3	16
K^+	15	Br^-	34	en	47
Rb^+	22	I^-	50	py	49
Cs^+	33	$CH_3CO_2^-$	29	PPh_3	167
NH_4^+	13	$C_6H_5CO_2^-$	71		
Mg^{2+}	4	CN^-	13		
Ca^{2+}	9	CNO^-	23		
Sr^{2+}	16	CNS^-	34		
Ba^{2+}	26	ClO_4^-	32		
Cu^+	15	CO_3^{2-}	28		
Ag^+	27	$C_2O_4^{2-}$	28		
Zn^{2+}	13	HCO_2^-	17		
Cd^{2+}	20	NO_3^-	19		
Hg^{2+}	36	O^{2-}	6		
Tl^+	36	OH^-	11		
Pb^{2+}	32	S^{2-}	28		
First-row transition metals[c]	13	SO_4^{2-}	38		
		$S_2O_3^{2-}$	46		
		$acac^-$	55		

[a] All values are $10^6\,X_M$ cgs, that is, $Li^+ = 1 \times 10^{-6}\ erg \cdot G^{-2}\ mol.^{-1}$
[b] acac = acetylacetonate; en = ethylenediamine; PPh_3 = tgriphenylphosphine; and py = pyridine.
[c] Inner-core electrons.

The molar susceptibility includes diamagnetic contributions from the other atoms in the molecule or compound, making the apparent molar susceptibility smaller than it really is. Thus, these contributions must be added to the value of X_M, so that the molar susceptibility of the paramagnetic atom, X_A, can be determined. There are a number of tables of diamagnetic corrections in the literature.[3,4] Table 5.1 lists some of these values. If a ligand or other group in the molecule does not appear in this table, the diamagnetic correction can be obtained by summing the values for each atom or type of atom in the group.

In order to use these corrections and to find the molar susceptibility of the paramagnetic atom, the following equation is used.

$$X_A = X_M + \text{sum of all diamagnetic corrections} \quad (5.10)$$

The relationship between X_A and μ_{eff} can be used to calculate the effective magnetic moment in Bohr magnetons.

$$\mu_{eff} = \left[\frac{3kTX_A}{N\beta^2}\right]^{1/2} \quad (5.11)$$

where k is Boltzmann's constant, N is Avogadro's number, and β is the Bohr magneton. This reduces to

$$\mu_{eff} = 2.828(X_A T)^{1/2} \quad (5.12)$$

The value of μ_{eff} can then be compared to the calculated value from the "spin-only" formula that assumes the ligand field "quenches" the orbital angular momentum.

$$\mu_s = g\sqrt{S(S+1)} \quad (5.13)$$

Values of μ_s for various numbers, n, of unpaired electrons are given next.

n	μ_s (BM)	n	μ_s (BM)
1	1.73	5	5.92
2	2.83	6	6.93
3	3.87	7	7.94
4	4.90		

The value of μ_{eff} may vary slightly from one compound to another. Table 5.2 has a list of transition metal ions with the typical range of μ_{eff} values for each. The effective magnetic moment for lanthanides and actinides cannot be calculated from the spin-only equation, since the orbital motion contribution cannot be ignored.

Once the effective magnetic moment is determined, it is a simple matter to find the number of unpaired electrons, and from that, the electron configuration of the metal. This can then lead to an understanding of the geometry and bonding in the molecule. For example, $Fe_2(CO)_9$ is known to have three terminal carbonyl groups on each iron and three bridging carbonyl groups between the two iron atoms.[3] The magnetic susceptibility of the complex shows it to be diamagnetic (no unpaired electrons). Therefore, there must be a metal–metal bond joining the two iron atoms with the two electrons paired. This is confirmed by a short Fe—Fe distance in the X-ray structure.

Table 5.2 Oxidation States and Magnetic Moments for Octahedral Complexes

Metal Ion	Configuration	Number of Unpaired Electrons	Magnetic Moment (BM)
Ti^{3+}	d^1	1	1.7–1.8
Ti^{4+}	d^0	0	0
V^{3+}	d^2	2	2.7–2.9
V^{4+}	d^1	1	1.7–1.8
V^{5+}	d^0	0	0
Cr^{2+}	d^4	4 (High spin)	4.8–5.0
		2 (Low spin)	3.0–3.3
Cr^{3+}	d^3	3	3.7–3.9
Mn^{2+}	d^5	5 (High spin)	5.7–6.0
		1 (Low spin)	2.0
Mn^{3+}	d^4	4 (High spin)	4.8–5.0
		2 (Low spin)	3.0–3.3
Mn^{4+}	d^3	3	3.7–3.9
Fe^{2+}	d^6	4 (High spin)	5.9–5.6
		0 (Low spin)	0
Fe^{3+}	d^5	5 (High spin)	5.7–6.0
		1 (Low spin)	2.2–2.5
Co^{2+}	d^7	3 (High spin)	4.3–5.2
		1 (Low spin)	2.0–2.7
Co^{3+}	d^6	0 (Low spin)	0
Cu^{+}	d^{10}	0	0
Cu^{2+}	d^9	1	1.8–2.1

5.A.4 Operation of the Evans–Johnson Matthey Balance for Solids

1. Turn the RANGE knob on the balance to $\times 1$ and allow the balance to warm up for 30 min.
2. Adjust the ZERO knob until the display reads 000. The zero should be readjusted if the range is changed.

> **NOTE:** ***The zero knob on the balance has a range of 10 turns. It is best to operate the balance in the middle of this range. This can be accomplished by turning the knob 5 turns from one end and then, ignoring the bubble level, adjusting the back legs of the balance until the digital display reads about zero. Once this is done at the beginning of the laboratory period, all further adjustments can be made with the knob on the front of the instrument.***

3. Place an empty tube of known weight into the tube guide and take the reading R_0.

> **NOTE:** ***The instrument can drift over short periods of time and should be rezeroed before each measurement.***

On the $\times 1$ setting the digital display should fluctuate by no more than ± 1. However, when you record R or R_0 take a "visual average" of this fluctuation and use this as your reading.

4. Carefully fill the opening at the top of the sample tube with the solid and tap the bottom of the tube gently on a hard surface to pack the sample. Obtain the mass, in grams. You must have at least 1.5 cm of solid to obtain a stable reading of R.
5. Rezero, place the packed sample tube into the tube guide on the top of the balance, and take the reading R. A negative reading indicates a diamagnetic sample.

NOTE: ***A critical part of the technique is correctly packing the well-powdered sample of the solid in the sample tube. To be sure you have the true value of R after the first reading, repeatedly tap the bottom of the tube firmly but gently on a hard surface (preferably not the table the balance is on) for about 30–60 s, then take another reading of R. Continue this until you have three values that agree within ±1. Also during the tapping process ensure that the solid forms an even surface in the tube and is not sloped to one side.***

6. If the reading is off-scale, change the RANGE knob to $\times 10$, rezero, and multiply the reading by 10.
7. Calculate the mass susceptibility using the equation

$$X_g = \frac{CL(R - R_0)}{1 \times 10^9 \text{ m}}$$

NOTE: ***Along with a recording of R, R₀, L, and m you should also determine the temperature to 0.1 °C with a thermometer placed or suspended near the balance.***

8. The sample may be removed by gently tapping the tube upside down on a piece of weighing paper. Do not tap too hard since the glass lip can be easily broken during this procedure. After the tube is empty it can then be rinsed with the appropriate solvent using a microliter syringe or disposable pipette with a fine tip. The solvent is shaken out and the process repeated until the tube is clean. Place the tube upside down in a small breaker to dry.

5.A.5 Operation for Liquids and Solutions

The general procedure for making measurements on liquids and solutions is somewhat similar to that used on solid samples, however, there are some important differences. With this in mind please note the following:

1. For liquids and solutions the full expression for mass susceptibility, including the correction for the volume susceptibility of displaced air, must be used (Eq. 5.5). In this expression the cross-sectional area must be determined as well as the mass of the solute, mass of the solvent, and volume of solution (to calculate the density). It is therefore best to prepare the solution by

weighing an empty vial or volumetric flask, weighing the container after the solute has been added, and weighing it again after the solution is prepared. The volume of solution can be determined directly by using a small volumetric flask (1–10.00 mL), or by preparing the solution in a vial and measuring the volume of the solvent added with a microliter syringe. As long as the amount of solute is small there should be no detectable change in the volume of liquid when the solution is prepared. You can therefore use the volume of added solvent as the volume of solution.

2. Place at least 2.5 cm of solution into the sample tube using a microliter syringe. Tap out any air bubbles.
3. The mass susceptibility of the solvent, X_0, can be measured separately as described in Ref. 5.
4. The measurement of sample length, L, on liquid or solution samples, should be taken from the bottom of the meniscus to the bottom of the sample (not the bottom of the tube). The diameter of the thin bore and standard sample tubes are 2.00 and 3.23 mm, respectively. This can be used to calculate the value of A.

5.A.6 Determining Magnetic Susceptibility by NMR Spectroscopy

There have been many studies detailing the data collection and analysis of solution NMR measurements of magnetic susceptibility. The use of a sealed melting point capillary as the center tube of a 5-mm coaxial cell unit is perhaps the best method for measuring magnetic susceptibility by NMR.[6] Both compartments must contain the solvent, plus 1–2% of an indicator species. The best combinations of solvent and indicator are water and 1–2% *t*-butyl alcohol, chloroform and tetramethylsilane (TMS), or acetone and TMS. It is best to place the solution with the paramagnetic solute in the melting point capillary. The appropriate amount of solvent can then be adjusted in the outer tube to cause the capillary to stand up in the center of the outer tube. The NMR tube should be spun as quickly as possible (~50 rps or greater) to minimize spinning sidebands.

In order to lock to a ^2H resonance on an FT NMR spectrometer, D_2O, $CDCl_3$, and CD_3COCD_3 must replace the water, chloroform, and acetone in both the tubes of the coaxial cell. However, *t*-butyl alcohol and TMS can still be used as the indicator species. The sweep width on an FT NMR should be expanded, so that there will be little or no "foldover" of the spinning sidebands that can make determination of $\Delta\nu$ difficult. Since the indicator species in the capillary tube will have a weak signal, it can often be missed, particularly when there are numerous spinning sidebands. It is important to check all the peaks in the spectrum by calculating their distance from the main resonance of the indicator. Any pairs of peaks on either side and at equal distances from the main resonance are probably sidebands. If there is any doubt, the spin rate can also be changed to see if the peaks change frequency. If the frequency changes, the peak was a spinning sideband.

The temperature of the probe should be determined at each session by finding the $\Delta\nu$ between the methyl and hydroxyl protons of methanol or ethylene glycol. The chemical shift difference, $\Delta\nu$, is then compared to a calibration chart supplied with the spectrometer and the temperature of the probe is recorded. It was noted that the calibration charts supplied with some instruments are incorrect,[7] and even under the best conditions, cannot give an accuracy higher than ± 1 °C. A more complete overview of NMR temperature measurements may be found in Ref. 8.

Prepare a solution of the paramagnetic compound using the solvent–indicator mixture described in Section 5.A.5. Place a small quantity, ~2 cm in height, in a melting point capillary tube. Seal the tube with a microburner ~1 cm above the liquid level. Place the capillary in an NMR tube containing the solvent–indicator mixture, and adjust the liquid level in the tube to make the capillary tube "stand up." Cap the NMR tube and obtain the spectrum.

REFERENCES

1. Drago, R. S., *Physical Methods in Chemistry*, Saunders: Philadelphia, 1977, p. 411–432 and 436–463.
2. Evans, D. F. *J. Chem. Soc.* **1959,** 2003.
3. Johnson, B. F. G.; Benfield, R. W., *Topics in Inorganic and Organometallic Stereochemistry*, G. L. Geoffroy, Ed., Wiley: New York, 1981. Also see Angelici, R. J., *Synthesis and Technique in Inorganic Chemistry*, 2nd ed., University Science: Mill Valley, CA, 1986.
4. Figgis, B. N., "Ligand Field Theory" in *Comprehensive Coordination Chemistry*, G. Wilkinson, Ed., Pergamon: Oxford, 1987, Vol. 1, Chapter 6, p. 213.
5. See a current *CRC Handbook of Chemistry and Physics*, CRC Press: Boca Raton, FL.
6. Loliger, J.; Scheffold, R. *J. Chem. Educ.* **1972,** 49, 646.
7. Van Geet, L. A. *Anal. Chem.* **1970,** 42, 679.
8. Sandström, J., *Dynamic NMR Spectroscopy*, Academic Press: New York, 1982.

General References

O'Connor, C. J., "Magnetochemistry—Advances in Theory and Experimentation" in *Progress in Inorganic Chemistry*, S. J. Lippard, Ed., Interscience: New York, 1982, Vol. 29, p. 204.

Figgis, B. N.; Lewis, T., "The Magnetic Properties of Transition Metal Complexes" in *Progress in Inorganic Chemistry*, F. A. Cotton, Ed., Interscience: New York, 1964, Vol. 6, p. 37.

5.B THERMAL ANALYSIS

5.B.1 Introduction

Thermal analysis[1] is a collective term for a group of techniques that measure some change in a physical property of a sample as a function of temperature. One of the most familiar examples of a thermal measurement is the determination of a melting point. Others that may be familiar include determination of boiling points, dehydration points, and isomer transition points. There are three common thermal techniques that will be covered here: differential scanning calorimetry (DSC, unit shown in Fig. 5.2), the closely related differential thermal analysis (DTA), and thermogravimetric analysis (TGA, unit shown in Fig. 5.3).

5.B.2 Differential Scanning Calorimetry and Differential Thermal Analysis

In DTA, a sample is placed in an inert heating pan, and is heated under a specific temperature program. The energy absorbed or given off by the sample is compared to that of a reference material (alumina, glass beads, etc.) undergoing the same heating program. The DTA thermogram is a plot of the difference in temperature between the two samples. The entire process takes place in a well-insulated cell. If the sample should melt or boil, for example, the phase change will occur at constant temperature. The reference will continue to heat up, however, and the temperature difference between them, $T_{sample} - T_{reference}$, will become negative, giving an endothermic peak. In DSC, a similar procedure occurs; however, heat is added to the sample or to the reference as necessary to keep both at identical temperatures. The DSC thermogram is a plot of the amount of heat necessary to accomplish these changes. Since when a compound

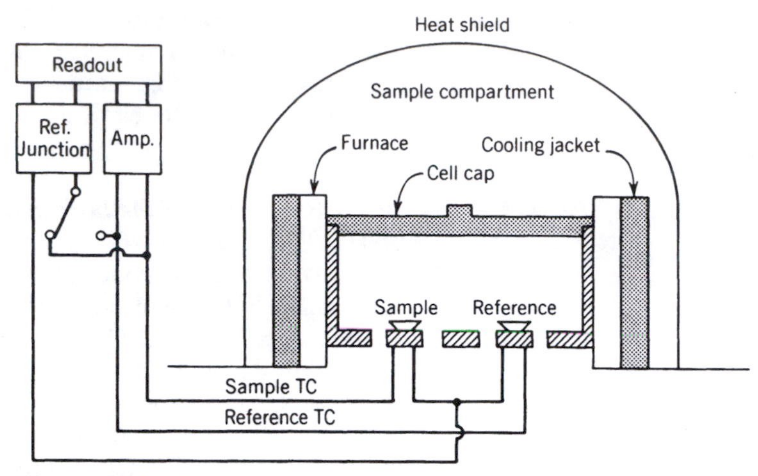

Figure 5.2. *Differential scanning calorimetry (DSC) cell.*[2]

melts or boils, an additional amount of heat must be added to the sample to maintain it at the same temperature as the reference, a heat input peak (also drawn downwards, as an endotherm) results.

5.B.3 Thermogravimetric Analysis

In TGA, the sample is placed in an inert boat (usually made of platinum), which is suspended from a quartz beam that serves as the fulcrum of a microbalance. The sample is heated under a specific temperature program, and its weight is continuously monitored as a function of temperature. A typical TGA instrument is shown in Figure 5.3.

5.B.4 Variables in Thermal Analysis

All thermal instruments operate following a temperature program, which on more advanced instruments, may be set by the operator. Variable parameters include the ramp rate (how many degrees per minute the instrument will heat up) and holding times at a particular temperature. Ramp rates can be set at different values for different temperature regions. If the sample's thermal behavior is already known, one can scan quickly through regions of no activity, and then slowly through regions where thermal processes occur. Normal temperature ranges for these instruments are from room temperature to about 1200 °C.

The atmosphere in the cells can also be varied, merely by connecting a cylinder to the cell and purging with the desired gas. The thermal response will obviously be different in different atmospheres. For example, a sample that might

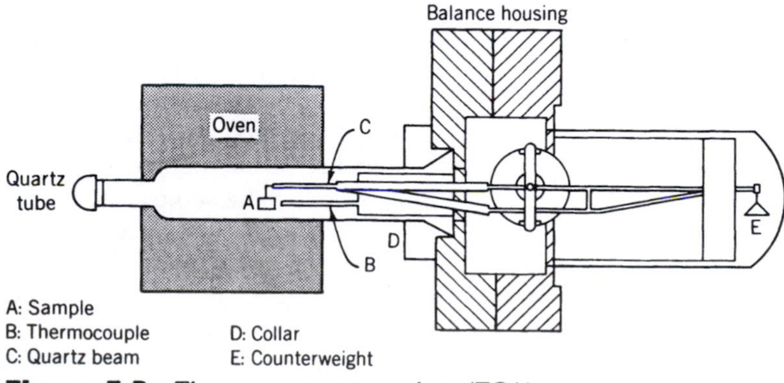

Figure 5.3. *Thermogravimetric analysis (TGA) apparatus.*

oxidize in an atmosphere of pure oxygen or air would undergo different behavior (which would be unobservable in an oxidizing atmosphere) in an atmosphere of nitrogen or argon.

5.B.5 Analysis of the Thermogram

The thermal properties of a sample are determined by measuring its temperature response. As a simple example, we consider the thermal analysis of copper sulfate pentahydrate, $CuSO_4{\cdot}5H_2O$. The TGA thermogram for the compound is shown in Figure 5.4 and the DSC thermogram is shown in Figure 5.5.

It is clear from both thermograms that between the temperature limits analyzed, three separate physical processes occur. In the DSC or DTA thermogram, an endothermic (downwards) peak appears centered at 70 °C, a second endotherm appears at 100 °C, and a third at 225 °C. One could, if desired, integrate the areas of the endotherms, and determine the amount of heat absorbed by the sample in all three instances. The three endotherms correspond to losses of water, a fact that will be more clearly seen with the TGA thermogram. In a DSC or DTA thermogram, exothermic peaks usually correspond to chemical reactions, polymerizations, or crystallizations that have occurred. Endothermic peaks are associated with phase transitions, dehydrations, reductions, and some decompositions. Occasionally, the baseline of a DSC or DTA thermogram will shift upwards or downwards at a particular temperature. Since the thermogram is a plot of heat input versus temperature, the base line value is a function of the heat capacity of the sample. A change in base line indicates that the sample has undergone a transition to a product that has a different heat capacity than the sample.

In the TGA thermogram of the copper sulfate pentahydrate, there are three physical processes that occur, each with loss of weight. The molecular weight of the parent compound is 249.5 amu. The first change corresponds to a ~14% weight loss, or a weight loss of approximately

$$(0.14 \times 249.5) = 35.9$$

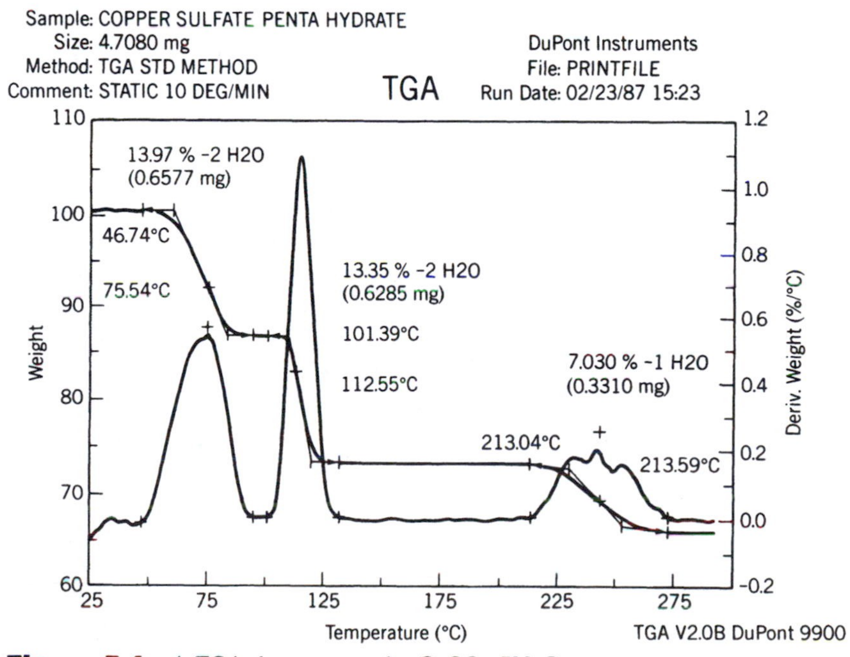

Figure 5.4. *A TGA thermogram for $CuSO_4{\cdot}5H_2O$.*

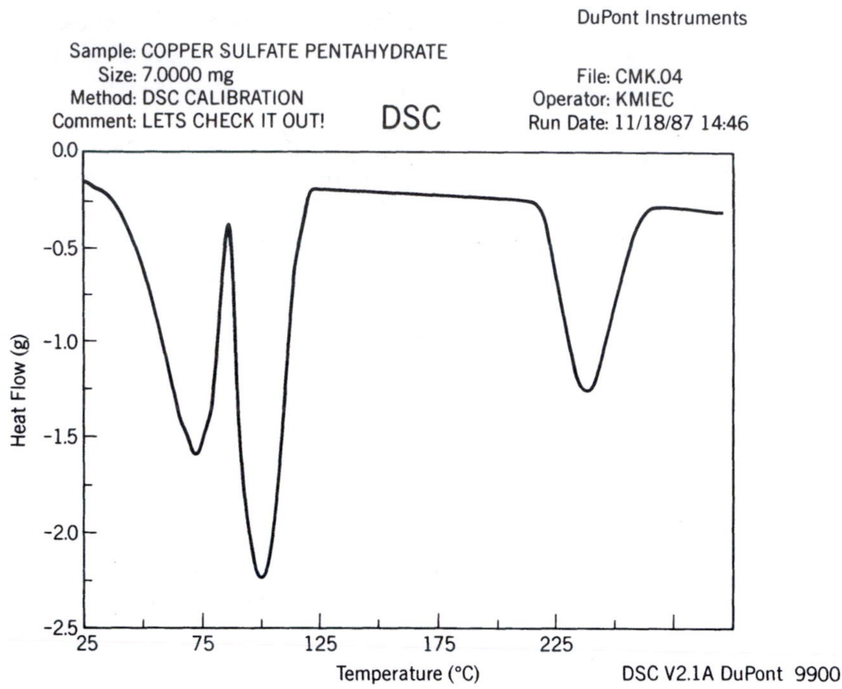

Figure 5.5. *A DSC thermogram for $CuSO_4 \cdot 5H_2O$.*

or ~36 amu. It is therefore apparent that two water molecules (2 × 18 amu) are being lost, and the compound is undergoing the reaction

$$CuSO_4 \cdot 5H_2O + \text{heat} = CuSO_4 \cdot 3H_2O + 2H_2O$$

At just slightly higher temperature, another ~13.4% of the weight is lost, corresponding to the reaction

$$CuSO_4 \cdot 3H_2O + \text{heat} = CuSO_4 \cdot H_2O + 2H_2O$$

Loss of the last water molecule (~7.03% weight loss) occurs at much higher temperature (~250 °C), and corresponds to the reaction

$$CuSO_4 \cdot H_2O + \text{heat} = CuSO_4 + H_2O$$

Because the first four molecules of water are lost at similar temperatures, it is likely that these molecules are similarly bound within the compound. Since copper(II) is usually four or six coordinate, we can reasonably conclude that these four water molecules are bound to the copper(II) ion, and the fifth is bound to the lattice.

Often, in TGA, we find the simple thermogram curve is plotted, as well as the derivative of this curve. This data can be quite useful in cases where more than one reaction occurs simultaneously. It is also much easier to see at what temperature the weight loss reaches a maximum when the derivative is plotted. Figure 5.6 shows the TGA thermogram for $CaC_2O_4 \cdot H_2O$, with the derivative plotted.

5.B.6 Theoretical Aspects

The temperature at which a molecule undergoes thermal change resulting from decomposition is a function of the lattice energy of its parent. The larger the lattice energy, the less likely the compound is to undergo decomposition. The

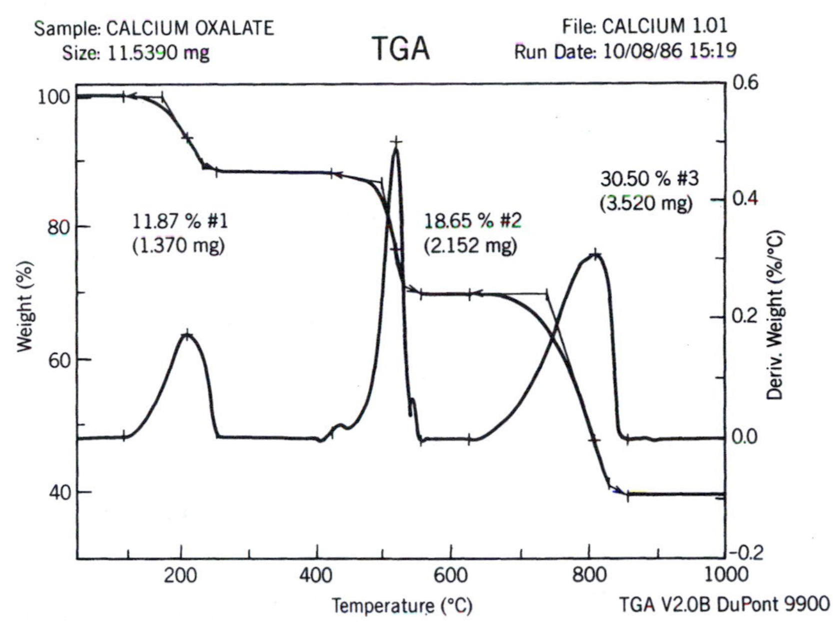

Figure 5.6. *A TGA thermogram for $CaC_2O_4 \cdot H_2O$, with derivative plotted.*

lattice energy, U, for an AB compound is calculated as

$$U = -\frac{Z^2e^2NA}{r_0}\left(1 - \frac{1}{n}\right)$$

where Z is the charge on the ions, e is the unit electrical charge, N is Avogadro's number, A is a constant known as the Madelung constant, which depends on the type of lattice, r_0 is the interionic distance, and the parenthetical term accounts for interionic repulsions. A number of trends are clear from the lattice energy equation.

1. As the charge, Z, on the ions increases, the lattice energy increases.
2. As the distance, r_0, between the ions decreases, the lattice energy increases. The distance would decrease when the ions get smaller in size.

Thermal analysis can be used to determine trends in lattice energy of a family in the periodic table; this is the basis of Experiment 2.

5.B.7 Applications

In addition to studying the previously mentioned group trends of lattice stabilities, thermal analysis has been used in a wide variety of areas in both inorganic and organic chemistry. Clays, ceramics, glasses, and other composite materials are extensively investigated by thermal means. Furthermore, thermal analysis is commonly used to obtain melting, boiling, and decomposition points, and to generate phase diagrams. Organic applications include purity studies for pharmaceuticals and determinations of physical characteristics of polymers.

REFERENCES

1. General References on Thermal Analysis include:
   (a) Dodd, J. W.; Tonge, K. H., *Thermal Methods*, Analytical Chemistry by Open Learning, Wiley: New York, 1987.
   (b) Wendlandt, W., *Thermal Analysis (Chemical Analysis Vol. 19)*, Wiley: New York, 1986.

(c) The American Chemical Society offers a short course on Thermal Analysis in Materials Characterization, directed by Dr. Edith Turi. Information may be obtained by writing to the American Chemical Society.
2. Diagram based on Figure 4.6*d* in Ref. 1a.

5.C VACUUM AND INERT ATMOSPHERE TECHNIQUES

5.C.1 Introduction

Many inorganic, organic, organometallic, and biochemical compounds are air-sensitive. A compound is said to be air-sensitive when it is unstable in the presence of moisture, carbon dioxide, or oxygen present in air. The preparation and manipulation of such compounds requires the use of vacuum or inert atmosphere techniques.[1,2]

In most situations, an inert atmosphere is created by replacing air by dry nitrogen. In some cases, where nitrogen interferes with the synthesis and manipulation of the desired product, argon is used. The inert atmosphere is created by various methods, the more common ones of which are described in the following sections. These techniques are based upon common laboratory methods, familiar to most chemists. They are meant for macro- or microscale work, involving the adaptations of common laboratory practices and common laboratory glassware. The choice of one particular method over the other depends on the conditions of the reactions, ease of handling the equipment, and other specific requirements for a particular experiment.

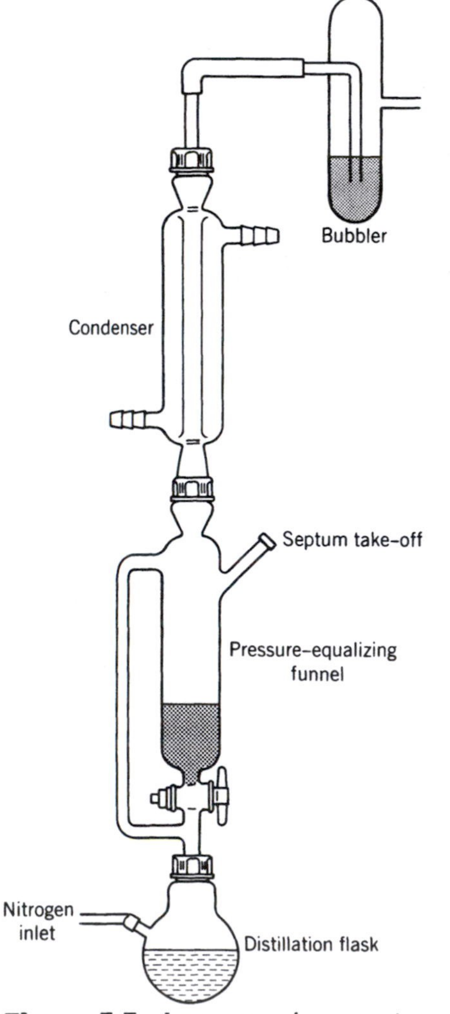

Figure 5.7. *Inert atmosphere reaction apparatus.*

5.C.2 Purging with an Inert Gas

One of the fastest and most convenient methods for handling air-sensitive compounds is to flush the air from the reaction vessel, using a steady stream of an inert gas (such as nitrogen or argon) obtained from a compressed gas cylinder or lecture bottle. See Section 1.B.4 for information on the safe handling of gas cylinders and lecture bottles. An example of an experimental apparatus employing a nitrogen purge is shown in Figure 5.7.

A cylinder of N_2 gas, fitted with a pressure regulator and needle valve (Fig. 5.8), is used as the source of the inert gas. The main valve on the cylinder is opened by rotating it counterclockwise. The gas pressure within the cylinder is

Figure 5.8. *Gas cylinder regulator. (Courtesy of Matheson Gas Products, Inc., Secaucus, NJ.)*

displayed on the main pressure gauge, located on the right side of the regulator. The pressure adjustment valve is then cautiously opened by turning the knob clockwise (so that it turns inwards). While turning the knob, carefully watch the pressure reading on the outlet pressure gauge, located on the left side of the regulator. For purging, it should read ~3–5 psig. This is the gauge pressure of the gas at the exit end of the regulator. To convert from gauge pressure to absolute pressure, the following relationship is used:

$$P_{\text{absolute}} = P_{\text{gauge}} + P_{\text{atmospheric}}$$

A Note of Caution: *Diaphragm regulators will sometimes stick when the pressure adjustment knob is turned to increase the pressure at the exit end. In such cases, there is a real danger of overpressurising the system, causing a hazardous situation. To avoid this,* **make sure that the outlet needle valve is closed**, *such that there is no flow of gas. Then, slowly open the regulator control valve until the desired pressure is attained. The regulator will now maintain this pressure during the period the gas is being used for the experiment. The flow rate of the gas is adjusted by manipulating the needle valve. This serves as the gas outlet from the cylinder and is connected using heavy walled Tygon or rubber tubing (having low gas permeability) to the system to be purged. If necessary, the purge gas can be prepurified and dried by passing it through oxygen scavenging and/or drying towers.*

The purge gas is now flushed through the reaction glassware, which has been cleaned and predried in an oven. To conserve the purge gas, several such assemblies can be arranged in series.

In the example shown in Figure 5.7, a modified reaction apparatus for generating a dry solvent is used. The apparatus is set up in such a manner so that there is one inlet for N_2 gas flow, and an outlet for the unobstructed flow of the gas through a bubbler containing mineral oil or mercury covered with a layer of mineral oil. The bubbler provides for the escape of the gas which is led (if necessary), using rubber or Tygon tubing, to a hood for safe venting. It also prevents the diffusion of air into the system by maintaining a positive pressure in the reaction vessel. For anhydrous reaction conditions, any ground-glass joints should be lightly greased (Teflon joints, if available, do not need greasing).

Many variations of this basic method are possible and depend on the reaction conditions, design of the experiment, and the glassware available in the laboratory. One only needs to be somewhat imaginative to develop a system adaptable for a particular reaction. The following points are worth noting:

1. Purging an apparatus with an inert gas may be carried out either in the empty vessel (before the addition of any chemicals) or after the vessel is charged with the solvent or reactants.
2. After flushing the apparatus with the purge gas, reduce the flow rate of the gas through the system. If this is not done, there is a strong possibility of entraining the solvent vapors from the reaction vessel, causing an accumulation of unwanted liquids inside the bubbler or tubing.
3. While the specific experiment will specify the procedure for purging the glassware, individual project work requires careful preplanning as to how to arrange the purging setup. In any event, do not hesitate to ask for help from your instructor.
4. During cooling periods, do not forget to slowly *increase* the flow rate of the purge gas in order to compensate for the volume reduction of the vapors on cooling. This will prevent oil from the bubbler or atmospheric gases from backing up into the system.

5. In situations where several N_2 flushing systems are to be set up and enough inert gas cylinders are not available, a balloon filled with N_2 may be used at the inlet end. However, diffusion of air through the membrane of the balloon may contaminate the inert gas.
6. For working under anhydrous reaction conditions, the use of dry glassware and incorporation of drying tubes into the system are strongly recommended. Addition of reactants (liquids–solids–solutions), filtrations under inert atmosphere, and other related procedures are described, as necessary, within the appropriate experiments.

5.C.3 Use of Manifold for Inert Gas or Vacuum

A simple manifold for inert gas and vacuum is very convenient for manipulating air- and moisture-sensitive compounds. The manifold (Fig. 5.9) can be used for such purposes as distillation under reduced pressure, sublimation, vacuum drying of reaction products, and isolation of desired compounds under inert atmosphere conditions. The inert atmosphere manipulations of compounds not only require the use of a manifold, but may also require specially designed Schlenk glassware. These include reaction tubes, filtration tubes with frits, tear flask with a side arm, and a score of other specialized glassware. The basic Schlenk techniques involving the use of microscale glassware is described below.

The system that is described here is available from Ace Glass (Vineland, NJ), or other commercial sources, and may consist of a microscale manifold fitted with a pressure release valve and a vacuum-gauge port, a mineral oil or mercury bubbler, a liquid N_2 cold "muck" trap, micro-Schlenk tubes, crystallization tubes, and the frame for holding the manifold. The manifold, illustrated in Figure 5.9, consists of two glass tubes, connected to each other through four-position, three-way high-vacuum stopcocks (Young's stopcocks are an excellent alternative). One tube acts as the N_2 line and the other serves as a vacuum line. Each major part of the vacuum system is now briefly described.

Rough Pump

A simple dual stage rotary pump capable of displacing 20 L of gas per minute or more is sufficient for routine laboratory work. This pump should be able to reduce the pressure to a level of 10^{-3}–10^{-4} torr inside the vacuum line, excluding the vapor pressure of Hg ($\sim 10^{-3}$ torr). Any direct drive pump having a larger capacity for displacing air will be useful in saving time.

Cold Trap

The pump is directly connected to a cold trap (also called a muck trap), which is in turn hooked to the vacuum line. Two small pieces of high-pressure rubber tubing are used to connect the rough pump to the trap. The use of rubber tubing adds flexibility to the system, and reduces the chances of breaking the line or trap when the latter is disconnected for emptying the condensed materials in it.

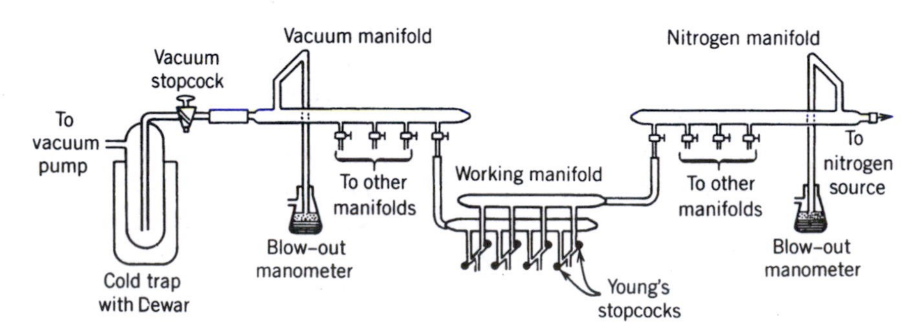

Figure 5.9. *Vacuum lines in series.*

A high vacuum three-way stopcock is used in between the vacuum line and the center tube of the trap. The three-way stopcock allows air to be bled into the cold trap while still being isolated from the vacuum line. This operation releases the outer tube of the trap, since it is no longer under reduced pressure. The trap is held in position using a clamp.

The Manifold

The manifold illustrated in Figure 5.9 consists of two narrow bore glass tubes 25 cm in length having a diameter of 10 mm. One of the tubes serves as a N_2 line, and the other acts as a vacuum line. They are connected to each other through three-way high vacuum stopcocks (or Young's stopcocks). One end of the N_2 tube is connected to a bubbler, which is in turn connected to a source of dry N_2 gas. The other end of the N_2 tube is attached by a long, narrow bore tube, through a stopcock, to a mineral oil or mercury open reservoir (sometimes called a blow-out manometer). The open end of the reservoir is led to an efficient hood for gas disposal. When the vacuum line is not in use, this stopcock should be closed. To ensure that the narrow bore tube does not move, and thereby break, secure the reservoir in a container filled with sand (for ballast).

CAUTION: *Before closing, make sure that the N_2 flow has been discontinued by shutting the main valve of the N_2 tank. Failure to do so may cause the manifold to explode* because of *the nitrogen pressure.*

The vacuum tube is attached to an efficient vacuum pump via a cold trap, as described earlier. The opposite end of the tube is totally sealed, or has a stopcock for attachment to a second manifold. A mercury manometer should also be attached to the vacuum line. This manometer is employed in the measurement of quantities of volatile materials, as well as in the detection of leaks in the system. Both arms of the manometer should be equipped with stopcocks.

The open ends of the stopcocks attached to the manifold are fitted with flexible vacuum tubing of convenient length (~3 ft is recommended). The choice and the selection of the sizes of tubing depends on the sizes of the glassware to be used in manifold manipulations. Heavy-wall butyl-rubber tubing is a good choice, because it is impermeable to oxygen and resists oxygen degradation.

The free end of the vacuum tubing may be sealed using a U-shaped, fire-polished glass rod. The U shape of the rod also allows the vacuum tubing to be hung out of the way. Whenever the vacuum line is used, make sure that the tubing has been flushed with N_2 gas. The habit of sealing the ends of the tubing helps prevent contamination by air.

One can test for leaks in the vacuum system by using a Tesla coil. A vacuum below 10^{-3} torr will not generate an electric discharge. A pinhole, however, will generate an electrical "spark" from the coil, which will jump to the location of the pinhole. The degree of vacuum in the line can also be measured electronically, or by using a McLeod gauge. If your laboratory is equipped with such devices, ask the instructor to explain their use.

Multiple Vacuum Lines in Series

In a laboratory where two or more vacuum manifolds are required, the use of distribution tubes (Fig. 5.9) provides an economical alternative to having several "stand-alone" vacuum systems. The distribution channel consists of two tubes fitted with stopcocks—one connected to a pump and the other to a N_2 tank. The working vacuum manifolds are then connected to each distribution channel as shown.

Use and Operation of Manifolds

1. Make sure that the manifold is set up correctly.
2. Cautiously open the N_2 line and allow a slow rate of flow of N_2 gas (60–120 bubbles·min^{-1}). (See Section 5.C.2 if gas cylinders are used.)
3. Make sure that all the stopcocks are closed.
4. Apply a liberal amount of vacuum grease (several types are commercially available) on the inner surface of the outer jacket of the cold trap, and attach it to the inner tube. Give a slight twist to the jacket to ensure that it is attached tightly to the inner tube. Enclose the cold trap with a liquid nitrogen Dewar. The Dewar should be supported on a Lab-Jack platform.

CAUTION: *Liquid N_2 may cause severe cold burns, if it is allowed to maintain contact with the same portion of skin for more than 1-2 s. It will boil and spurt vigorously when poured into an ambient temperature flask. It is therefore imperative that liquid N_2 be added slowly at first. When the boiling subsides (e.g., the trap has cooled), slowly fill the Dewar. Wrap the mouth of the flask with a rag or towel to reduce losses due to evaporation.*

5. Keeping the stopcock between the vacuum line and the cold trap closed, turn on the pump. Allow it to run for some time, so that the cold trap is totally evacuated.
6. Slowly open the stopcock between the cold trap and the vacuum line.

CAUTION: *If the manometer is open, the mercury may rapidly rise up the evacuated arm of the manometer, cracking the glass.*

7. When vacuum has been established, check for vacuum stability in the following manner: Simultaneously, open **both sides** of the Hg manometer to vacuum. After the manometer has equalized, close it on one side, and watch for changes in the level over time.

The manifold is now ready to be used for inert atmosphere manipulations. A more elaborate vacuum system is shown in Figure 5.10.

5.C.4 Cannula Techniques

Cannula techniques may be used instead of, or in addition to, the inert atmosphere manipulations described previously. A cannula is a long, flexible needle,

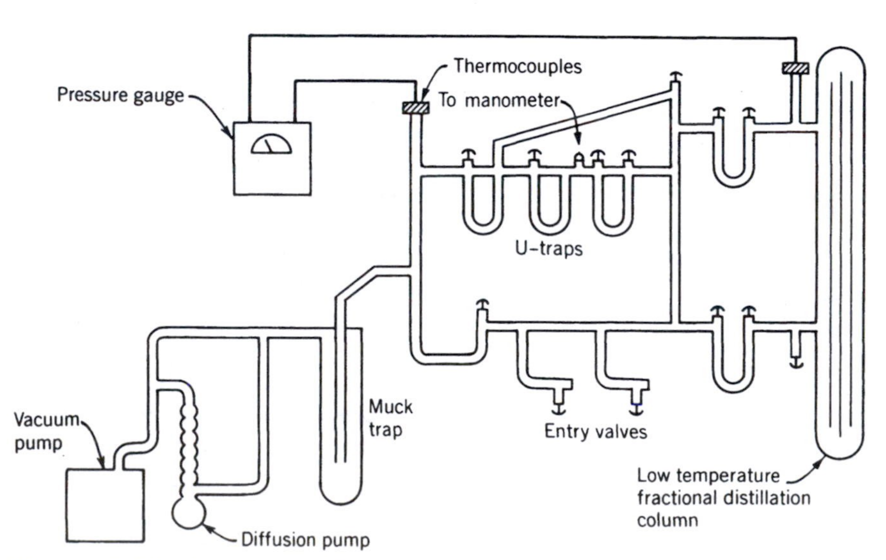

Figure 5.10. *Vacuum manifold.*

which can be connected to a nitrogen line, and inserted through a septum into a reaction flask. A second cannula can serve as a pressure outlet.

Cannulas can also be used to transfer and filter solutions from one container to another. The connecting head of the cannula is fitted with a small piece of cotton (which acts as a filter) and held in place with copper wire (or a similar material). The needle end of the cannula is inserted through the septum of the receiving vessel. By either pressurizing the head end, or evacuating the needle end, the medium can be transferred along the cannula. Additional information about the use of cannula techniques may be found in Ref. 3.

REFERENCES

1. Shriver, D. F.; Drezdzon, M. A., *The Manipulation of Air Sensitive Compounds*, 2nd ed., Wiley: New York, 1986.
2. Angelici, R. J., *Synthesis and Technique in Inorganic Chemistry*, 2nd ed., Saunders: Philadelphia, 1977, p. 43.
3. McNally, J. P.; Leong, V. S.; Cooper, N. J., "Cannula Techniques for the Manipulation of Air-Sensitive Materials" in *Experimental Organometallic Chemistry: A Practicum in Synthesis and Characterization*, ACS Symposium Series No. 357, A. L. Wayda and M. Y. Darensbourg, Eds., American Chemical Society: Washington, DC, 1987, p. 6. Also see Aldrich Bulletin AL-134, Aldrich Chemical Co., Inc., 1001 W. St. Paul Ave., Milwaukee, WI 53233.

5.D CRYSTALLIZATION TECHNIQUES

5.D.1 Introduction

The techniques of crystallization and recrystallization are used extensively for the isolation and purification of inorganic solids. Other techniques that may be used to purify solids include the various forms of chromatography (see Section 5.G), sublimation (see Section 5.H), and extraction (see Section 5.I).

The processes of crystallization and recrystallization occur in two stages:

1. The solid material is dissolved in a suitable solvent (or is already present in solution). This completely disrupts the crystal structure of the species involved.
2. The growth or regrowth of crystals is accomplished under conditions in which any unwanted impurities are left in solution.

Crystallization is an art! A proper crystallization often takes long periods of time and therefore patience and determination are virtues when using this technique. Impatience, leading to improper dissolution of solid or insufficient waiting times, is the most common reason for the failure of a recrystallization. In recrystallization, if difficulty is encountered in crystal formation, crystals of the compound are available for nucleation. Addition of a trace of pure crystals to the saturated solution will often cause crystallization to occur. This process is referred to as *seeding*. If crystallization still does not commence, the seed crystals should be crushed so as to expose fresh surfaces. Recrystallization of a substance should be repeated until physical properties, such as melting point, remain unchanged.

5.D.2 Crystallization from Solution

Crystallization from solution usually involves the following sequence of steps:

Step 1 The selection of the appropriate solvent for crystallization is critical for a successful purification operation. The compound should be soluble in the solvent at or near its boiling point, but relatively insoluble in the cold solvent.

Step 2 The solution containing the desired material is gently heated so that the solvent evaporates until crystals begin to appear.

Step 3 The solution is then allowed to cool. Since the solubility of most substances decreases on cooling, additional crystals form during this period. Slow cooling gives larger and usually purer crystals than does rapid cooling.

Step 4 The impurities should remain behind in the mother liquor. It is therefore important to leave enough of the mother liquor to hold the impurities in solution. Otherwise, the impurities crystallize with the desired product.

Step 5 If the impurities to be separated remain undissolved in the hot solution (Step 1), they must be filtered from the solution before it is cooled.

Other considerations to be aware of are:

1. Colored impurities can be removed from the hot solution by use of a decolorizing agent (usually charcoal in the form of a powder or the pelletized type) followed by filtration of the hot solution, if necessary.
2. The crystals must be isolated by filtration and subsequently washed to remove adhering solvent.
3. The crystals must be dried by some suitable means.

The efficiency of the separation of impurities from the major component by recrystallization depends directly on the choice of solvent. The recrystallization solvent is indicated in the majority of the experiments in this text.

Many inorganic compounds are water soluble and thus, crystallization from an aqueous solution may sometimes be used to isolate and purify these compounds. In a number of cases, for example, in the isolation of a salt such as tetraamine copper(II) sulfate monohydrate $[Cu(NH_3)_4SO_4{\cdot}H_2O]$, a second solvent that is miscible with water, but in which the compound itself is insoluble (ethanol), is added to the solution. Under these conditions, the copper salt is precipitated from solution. Solubility data for known compounds is available from a number of sources,[1] to assist you in considering what solvent to use in the isolation or recrystallization step. For an unknown compound, the approach recently described by Craig for determination of recrystallization solvents at the microscale level may be adopted.[2]

Crystallization of a given material from a solution is a selective process. It follows that very careful control of temperature and composition of the solution is critical if pure products are to be obtained. Furthermore, as indicated previously, exact knowledge of solubility relationships is required. Information on solubility relationships is obtained from solubility diagrams. These diagrams are like "travel guides" on the trip to accomplish selective crystallizations. For an in-depth treatment of this area, one should refer to a standard textbook on physical chemistry.[3]

Figure 5.11 shows the solubility of two salts, A and B, as a function of temperature. Line A is the solubility line for salt A, and line B is the solubility line for salt B. A 50:50 solution of mixed salt is prepared at a concentration of 14 $g{\cdot}L^{-1}$ of A and of B, at a temperature of 100 °C. This corresponds to point 1 on the diagram. At this point, we are below the solubility lines for both components, and both salts stay in solution.

Suppose the solution is now cooled. We would be moving horizontally left from point one, as shown in the figure. Eventually, the solubility line A is crossed, and salt A begins to precipitate out. Since we are still below line B, essentially none of salt B precipitates out with it. Thus, we have separated out essentially pure salt A.

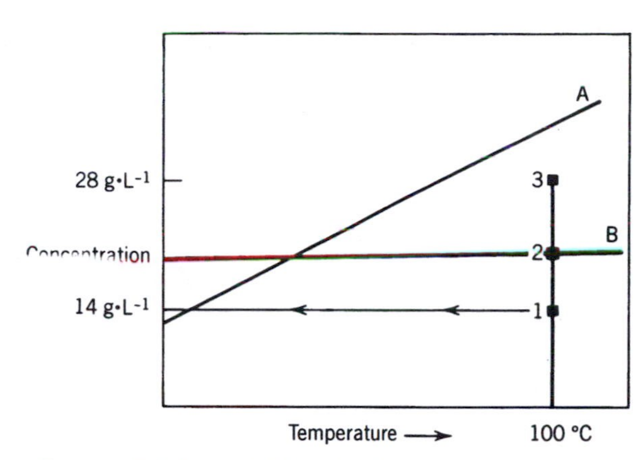

Figure 5.11. *Solubility chart.*

Alternatively, suppose that one half the water was boiled off (at 100 °C, of course), that is, the solution was concentrated. The concentration of each salt would now be doubled, to 28 g·L^{-1}. This is indicated by point 3. Notice that by going from point 1 to point 3 on the diagram, we would have crossed line B (at point 2). When this happens, salt B precipitates out. Since the solubility line A was not crossed, essentially none of salt A will precipitate. Thus, we will have separated out essentially pure salt B. This process illustrates the principle of crystallization. When one material is precipitated in the presence of another, the process is termed selective precipitation.

5.D.3 Isolation of Crystalline Products (Suction Filtration)

The process of collecting crystalline products at the microscale level usually involves the technique of filtration. The equipment used generally relates to the amount of material one is dealing with. The conventional procedures are discussed next.

When the amount of solid exceeds ~50 mg, a conventional porcelain Hirsch funnel with a filter paper disk, or a sintered glass or plastic Hirsch funnel fitted with a polyethylene frit is used to collect the material. This operation is done under vacuum, generally by use of a water aspirator, and called suction or vacuum filtration. A typical arrangement is shown in Figure 5.12. When using this arrangement, always securely clamp the filter flask to a support to prevent tipping. Many times a valuable product is lost by not observing this simple rule.

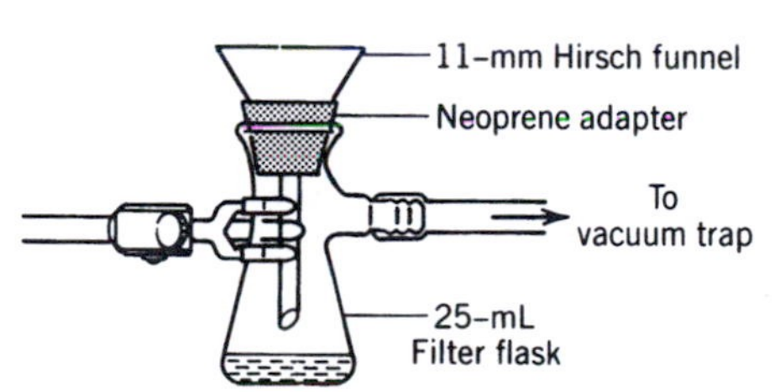

Figure 5.12. *Vacuum filtration apparatus.*

A water trap (Fig. 5.13) must *always* be placed between the filter flask and the aspirator. This is to prevent water from backing up into the filter flask, and perhaps destroying the product. Be aware that when several persons are using the same water line, the water pressure can change at any time. It is recommended that the system be opened to the atmosphere by loosening the screw clamp on the trap or by removing the tubing from the filter flask before the water is turned off.

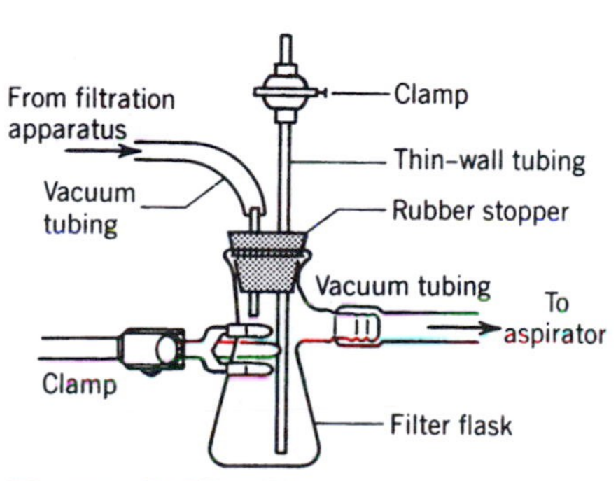

Figure 5.13. *Water trap.*

5.D.4 The Craig Tube Method

When the amount of solid material to be purified is in the range of 10–100 mg, the Craig tube is a convenient apparatus to use. A typical Craig tube is shown in Figure 5.14. The following steps are involved in a Craig tube recrystallization:

1. Place the sample to be recrystallized in the bottom section of the Craig tube. As a safety precaution, place this in a 10-mL beaker to prevent tipping.
2. Add the solvent (0.3–1 mL) of choice and dissolve the sample by heating

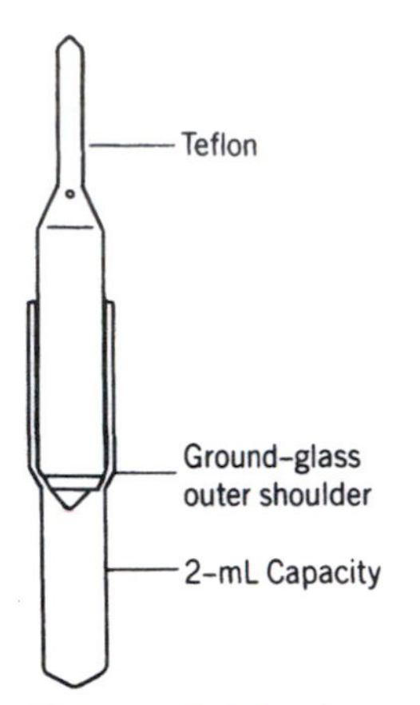

Figure 5.14. *Craig tube.*

the tube in a sand bath or aluminum block. Stir the mixture rapidly with a microspatula (roll the spatula rod between your fingers). This operation greatly aids the dissolution process and protects against boilover. If the solid material does not all dissolve, add additional solvent dropwise, in stages, until complete dissolution occurs.

NOTE: *If the material to be recrystallized needs to be decolorized, carry out this initial dissolution step in a small test tube (10 × 75 mm), add the decolorizing agent, and then transfer the hot solution to the Craig tube using a Pasteur filter pipet that was preheated with hot solvent.*

3. Concentrate the hot solution to the point of saturation by gentle boiling in the sand bath. Constant agitation of the solution with a microspatula during this short period will avoid the use of a boiling stone and prevent boilover. The appearance of crystals on the microspatula just above the solvent surface often indicates that the saturation point is near.
4. Place the upper section of the Craig apparatus in the tube, stand the assembly in a small beaker or Erlenmeyer flask, and allow the system to cool in a safe place. Upon cooling, crystallization usually commences. If it does not, add a seed crystal just at the surface of the solution, or scratch the inside of the tube with a glass rod. Slow cooling is desired since it usually produces larger crystals.
5. After the system reaches room temperature, place the assembly in an ice bath to promote further crystallization.
6. To isolate the crystals, place a round-bottom test tube (16 × 125 mm) down over the Craig tube assembly. Hold the Craig tube apparatus firmly up into the test tube and invert the total system 180° (see Fig. 5.15).
7. Place the assembly into a centrifuge (do not forget to balance the centrifuge) and spin the mother liquor away from the crystals. The mother liquor collects in the bottom of the test tube; the crystals remain in the upper section of the Craig apparatus above the insert.
8. Remove the assembly from the centrifuge and disassemble the Craig tube. The crystals may be recovered and dried by several methods (see Section 5.D.9).

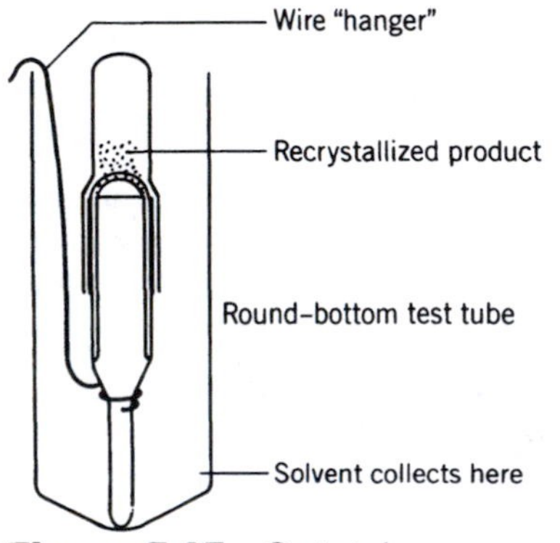

Figure 5.15. *Craig tube arrangement for centrifugation.*

The cardinal rule in carrying out the purification of small quantities of solids is **keep the transfer to an absolute minimum!** The Craig tube is very helpful in this regard.

5.D.5 Recrystallization Pipet

Landgrebe[4] recently described an alternative to the Craig tube method. This approach utilizes a modified Pasteur pipet as a recrystallization tube. The method works well for quantities of 10–100 mg, as long as the volume of solvent used in the recrystallization does not exceed 1.5 mL, the capacity of a Pasteur pipet. The sequence is described as follows:

1. Prepare a recrystallization tube (Fig. 5.16) by pushing a plug of cotton (copper wire is used) into the Pasteur pipet so that the cotton resides 1–2 cm below the wider bore of the pipet.

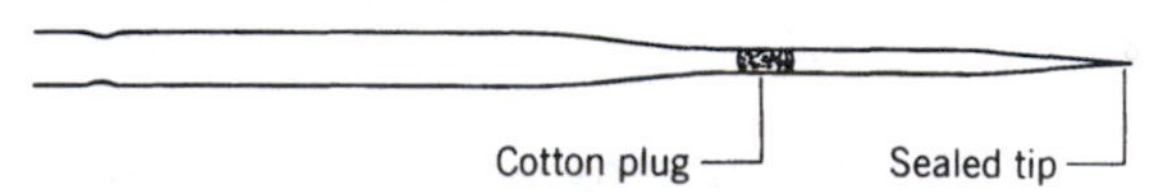

Figure 5.16. *Recrystallization tube. (Reprinted by permission from Landgrebe, J. A., J. Chem. Educ.* **1988,** 65, *460.)*

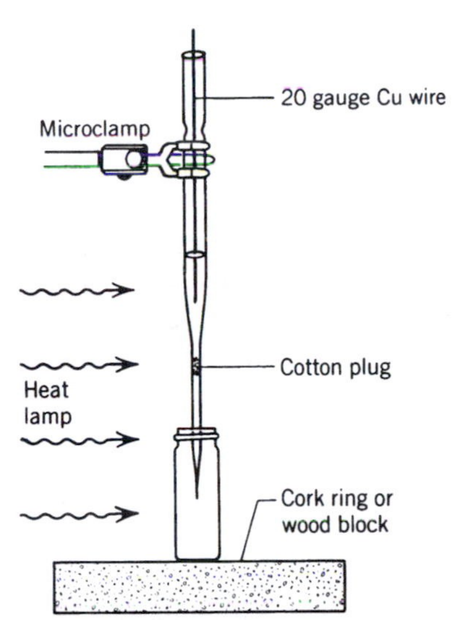

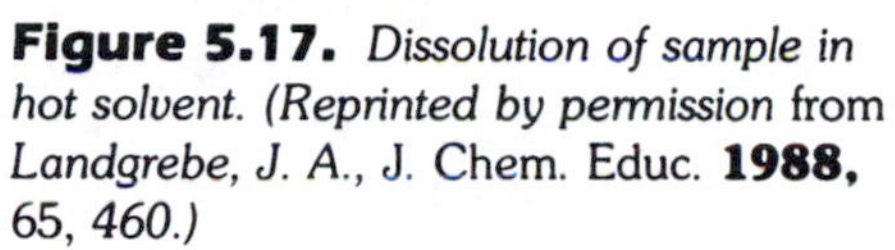
Figure 5.17. *Dissolution of sample in hot solvent. (Reprinted by permission* from *Landgrebe, J. A., J.* Chem. Educ. **1988,** 65, *460.)*

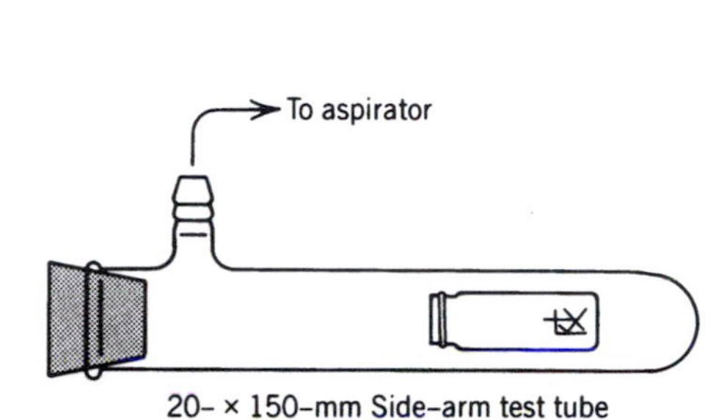

Figure 5.18. *Drying a sample. (Reprinted by permission* from *Landgrebe, J. A.,* J. Chem. Educ. **1988,** 65, *460.)*

2. Seal the lower part of the tube below the cotton plug with a microburner. Pull the glass so that a very narrow tip is formed. This allows the tip to be broken easily at a later stage of the operation.
3. Place the solid into the tared tube, reweigh to determine the weight of solid, and then clamp the tube near the top in a vertical position. Arrange a tared vial so that the bottom tip of the recrystallization tube protrudes ~1 cm into it (Fig. 5.17).
4. Add an appropriate amount of solvent to the tube using a Pasteur pipet. Stir the suspension with a copper wire and arrange a heating lamp ~6–8 cm from the tube.
5. When the solid has dissolved, remove the vial, snap the tip off the tube, and quickly replace the vial under it. Continue to warm the solution being filtered (do not boil). If the filtration is too slow, gently apply pressure using a pipet bulb.
6. After crystallization is complete, the mother liquor may be removed using a Pasteur filter pipet. Cold, fresh solvent may be added to wash the crystals, and the wash solvent then removed using the Pasteur filter pipet as before. Dry the crystals as discussed in Section 5.D.9 (Fig. 5.18).

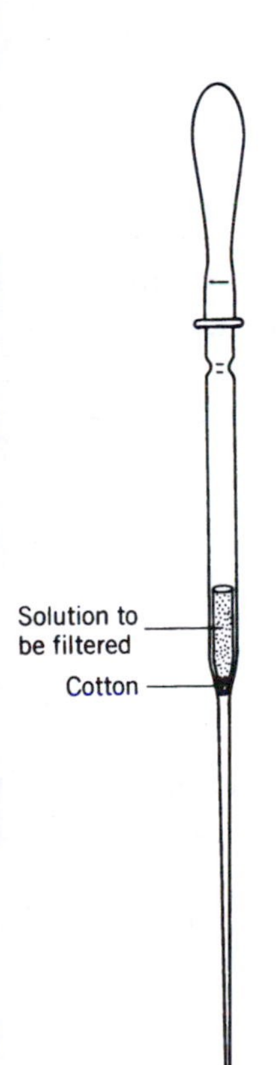

Figure 5.19. *Filtering Pasteur pipet.* From *Williamson, K. W.,* Microscale Organic Experiments, *Heath: Lexington, MA. Reprinted by permission of D. C. Heath Co., Lexington, MA.)*

5.D.6 Removal of Suspended Particles from Solution

At the microscale level, a modified Pasteur pipet can often be used to remove solid materials from small quantities of solution. Figure 5.19 illustrates the arrangement.

Transfer the solution to be filtered to the filter pipet using another Pasteur pipet. If the solution does not flow freely through the cotton plug, pressure may be applied by squeezing the bulb and thus forcing the solution through the filter. The collected solid and plug are generally rinsed with additional solvent. Remember to always clamp the filter pipet securely in a vertical position and collect the filtrate in a suitable container.

5.D.7 Washing of the Collected Crystals

Washing of collected crystals on a Hirsch or sintered glass funnel is an important procedure. First, never let the filter cake dry out. This is to prevent the cake from cracking, which leads to channel formation. This results in the wash liquid passing through the cracks and not permeating the crystalline mass. An additional precaution is to use several small portions of wash liquid instead of one large amount. It is best to allow the wash liquid to be sucked through the filter cake very slowly, and when the level of the liquid is ~1 mm above the cake, a second portion of wash liquid may be added, and so on.

Water is the wash liquid of choice in many inorganic preparations. Often the solid is washed with water and then with another solvent to remove the water. Acetone is a good solvent for this purpose, since it is completely miscible with water and has a relatively low boiling point. Another procedure is to wash the solid with alcohol followed by ether. In this case, the alcohol removes the water and the ether removes the alcohol. The ether, in turn, is very volatile and evaporates rapidly when the cake is spread out to dry. Ground-glass jointed fritted filters are available to filter air-sensitive materials under vacuum.

5.D.8 Decolorization

Decolorization is a technique that uses the phenomenon of adsorption to remove colored impurities during the process of recrystallization. In this process, the impure solid is dissolved in the recrystallizing solvent, a small amount of decolorizing agent added, and the suspension heated briefly to boiling. The decolorizing agent is then removed by filtration. When heating the solution, remember to stir briskly to prevent bumping or boilover. Never add the powdered carbon to a boiling solution; boilover will likely occur.

Activated carbon is the adsorbent used in the majority of cases. The carbon has no polar groups and does not participate in hydrogen-bonding interactions. The degree of adsorption depends on the degree of the polarizability of the compound(s) being removed as an impurity from the solution. Compounds that cause coloration in a system are usually highly unsaturated or aromatic in nature and have a high molecular weight. This is important in inorganic chemistry in the fields of organometallics and coordination compounds. Organic compounds and solvents are used in many of these preparations. Thus, impurities from this source caused by decomposition, and so on, occasionally introduce impurities into the reaction system that can cause coloration.

At the microscale level, two forms of decolorizing charcoal are normally used. The older form (Norit) is an unwashed, steam activated form (pH 9–11) of carbon available as a powder. The smallest amount of carbon to accomplish the task should be used. A good rule of thumb is *<10 mg of carbon/500 mg of compound.* When this form of carbon is used, since the particles are so fine, it is removed from the hot solution by gravity filtration through fluted filter paper arranged in a glass funnel.

An alternate form of decolorizing carbon is pelletized Norit. This has the advantage of being easily removed by gravity filtration, suction filtration, or the solution may be drawn into a Pasteur filter pipet (see above) and transferred from the pellets. This is a convenient approach when small amounts of material are involved. It is effectively used in combination with the Craig tube or recrystallization tube technique of filtration and crystallization. Although this form of Norit is not as effective in removing impurities from a colored solution as the powered form (due to less surface area), it has the advantage of being a faster and cleaner method.

An additional option for decolorization is to use a Pasteur pipet as a "chromatographic" column (see Section 5.G) packed with Norit, silica gel, alumina, or a combination of these adsorbents. The column packing adsorbs the impurities and the purified solution is eluted from the column. It is important to realize that additional solvent must be used, not only to keep the desired material in solution, but also to wash the column after the original solution has passed through. An additional concentration step is often necessary to concentrate the filtrate to a volume where crystallization can commence.

5.D.9 Drying Techniques

The temperature and other conditions under which a compound should be dried vary widely. The conditions generally depend on the type of water to be removed from the crystals. There are four types of bound water that should be removed if a sample is to be considered totally dry. These are:

1. Solution water, as in a moist salt.
2. Adsorbed water, as held by capillary or surface forces.
3. Water of crystallization, as in a hydrate (see Experiment 2).
4. Occluded water, as that held mechanically in cracks or crevices in the crystals.

A solid that is recrystallized from, or washed with, a volatile organic solvent can usually be air-dried by spreading the crystals on a clay tile or on filter paper. This is sufficient for most cases. The tile or paper absorbs traces of water or residual solvent.

It is possible to further dry some materials by placing them in an oven set at a temperature below their melting point. Care must be taken not to heat the crystals to the point that decomposition can occur. Note that one advantage of using a sintered glass filter is that the filter plus the crystals can be placed directly in the oven. In this manner, the number of transfers is reduced (see the vacuum technique, which follows). This is also true for the Craig tube method of collection. If the lower section was previously tared, the crystals can be left to air dry to constant weight (wrap a piece of filter paper over the open end secured by a rubber band to avoid dust collecting on the product while drying) or the tube can be placed in a 10-mL beaker and then in the oven. The yield then can be calculated directly.

In order to obtain an absolutely dry product, a vacuum desiccator (room temperature conditions), vacuum line, or a vacuum oven (elevated temperature conditions) is used. Air-sensitive solids may be dried in this manner. Such materials as calcium chloride, t.h.e. SiO_2, sulfuric acid, or phosphorous pentoxide (P_4O_{10}) are commonly used as desiccants.

An alternative to the vacuum desiccator is to place a small vial containing the material to be dried into a side armed test tube (Fig. 5.18).[4] The tube is stoppered, the side arm is connected to the water aspirator (remember the trap), and the pressure is reduced to about 16 mm. This operation can be done at room temperature, or the tube can be heated (carefully) with an infrared lamp, or immersed in a heated water bath, sand bath, or aluminum block. Materials may also be dried using a vacuum manifold (Section 5.C).

REFERENCES

1. **(a)** *CRC Handbook of Chemistry and Physics*, 1st student ed., Weast, R. C., Ed., CRC Press: Boca Raton, FL, 1989.
   (b) Dean, J. A., Ed., *Lange's Handbook of Chemistry* 13th Ed., McGraw–Hill: New York, 1985.
   (c) *International Critical Tables of Numerical Data, Physics, Chemistry and Technology*, Washburn, E. W., Ed., Vol. I–VII, McGraw–Hill: New York, 1926–1937.
2. Craig, R. E. *J. Chem. Educ.* **1989,** *89*, 88.
3. For example, Moore, W. J., *Basic Physical Chemistry*, Prentice–Hall Inc: Englewood Cliffs, NJ, 1983.
4. Landgrebe, J. A. *J. Chem. Educ.* **1988,** *65*, 460.

5.E DETERMINATION OF MELTING POINTS

5.E.1 Introduction

Since many inorganic compounds are solids, one of the key physical properties used in their characterization is the melting point. The melting point is defined as the temperature at which the solid and liquid phases of a substance are in equilibrium at a pressure of one atmosphere. Not only can this physical property be used to characterize a substance, but it may also be used to assess its purity. This occurs because the melting point of a substance is generally reproducible. A pure solid compound should melt over a narrow temperature range of about 0.5–2.0 °C. If a wider range for the melting point is observed, it is a good indication that the material is impure. We note here that thin layer chromatography (see Section 5.G.2) is also a very effective technique for determining the purity of a solid material.

5.E.2 Theory

In a crystalline structure, molecules are arranged in a regular, fixed pattern. Addition of energy to a crystalline substance, usually in the form of heat, causes this fixed array to rearrange to the more random liquid state. The melting point of a substance then, is the temperature at which this process is completed. The randomness (disorder) of a system increases as a system passes from the solid to the liquid state. The molecules in the liquid are much freer to move than when they are held in the crystal lattice of the solid state. The amount of randomness in a system can be determined quantitatively and is referred to as the entropy, S, of the system. Entropy is a very important thermodynamic concept. An increase in entropy is the major driving force in many physical (such as melting) and chemical processes.

A solid at a given temperature has a finite vapor pressure. As the temperature is increased by the input of energy (heating), the vapor pressure of the solid increases. At this point, both the solid and liquid are in equilibrium with the vapor, and, at the melting point, are in equilibrium with each other (Fig. 5.20). In this process, at some temperature, the kinetic energy of some of the molecules becomes great enough to overcome the intermolecular forces holding them in the crystal lattice and the solid begins to melt.

When the melting point is reached, the temperature of the system remains constant as additional heat is applied, until all of the solid has melted. The quantity of heat required to change a given amount of solid to the liquid state is known as the heat of fusion, ΔH_{fusion}, of the material. It should be noted that the quantity of heat liberated during crystallization (freezing) is equal to that absorbed during fusion (melting).

A vapor pressure–temperature diagram, generally called a phase diagram, for both the solid, liquid, and gas phases of a typical inorganic compound, is given in Figure 5.21. As seen from this figure, the normal melting point of the substance is that temperature at which the vapor pressure of the pure solid and the pure liquid are equal, and the pressure is one atmosphere, represented by point F.

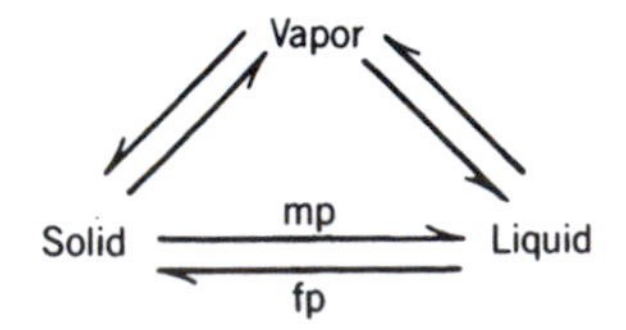

Figure 5.20. *Solid and liquid in equilibrium.*

Melting points are affected by several factors. In general, the melting point of a compound increases as the molecular weight increases. Superimposed upon this trend is the effect of intermolecular attraction. Water (MW = 18), a compound of low mass, has an unusually high melting point, 273 K. By comparison, methane, a compound of similar mass (MW = 16) has a melting point of 90 K. The difference is explained by intermolecular forces of attraction. Water molecules interact through relatively strong intermolecular forces called hydrogen bonds. Methane molecules, on the other hand, interact via relatively weak van der Waals' forces. The stronger the intermolecular forces, the higher the melting point. Conversely, if the intermolecular forces are weak, even high molecular weight compounds can have very low melting points. A rather extreme example

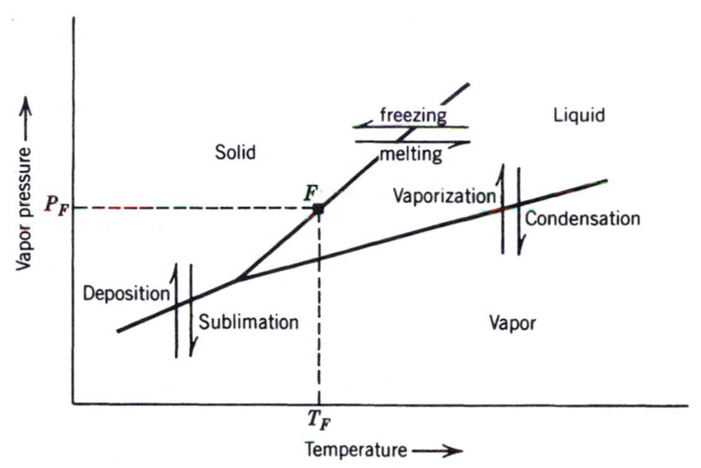

Figure 5.21. *Phase diagram for a solid–liquid system.*

is given by UF_6 which, because of extremely weak intermolecular forces, is a gas at room temperature and is used to separate uranium isotopes for manufacture of enriched nuclear fuels.

5.E.3 Mixture Melting Point Determination

The presence of a small quantity of an impurity in a solid sample that is soluble in the sample's liquid phase generally has the effect of depressing the melting point of the sample. In cases where two compounds have nearly the same melting point, this fact may be used as an identification tool in the form of a mixture melting point determination. This is a useful technique when samples of the pure compounds are available to obtain reference data. In this technique, an experimental product is mixed with a pure sample of the product to determine if the appropriate melting point is obtained. If the two materials are different, a lower melting point will be obtained.

A melting point composition diagram for a binary mixture (sometimes called a eutectic diagram) is obtained experimentally by preparing mixtures of the two pure compounds, A and B, containing various percentages of the components. The melting points of the mixtures are then determined. The result is a diagram similar to that shown in Figure 5.22.

The phase diagram shows that the pure lower melting solid, A (at a composition 100% A or 0% B), has a melting point of Y. The pure higher melting

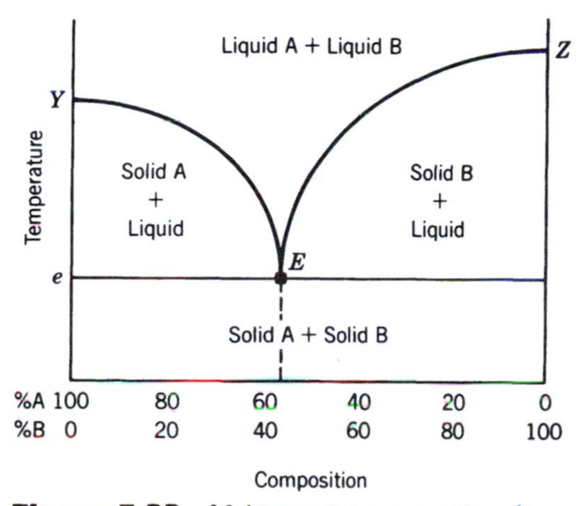

Figure 5.22. *Melting point composition diagram.*

Figure 5.23. *Thomas–Hoover melting point apparatus.* From *Mayo, D. W.; Pike, R. M.; Butcher, S. S.,* Microscale Organic Laboratory, 2nd. *ed., Wiley: NY., 1989. (Reprinted by permission of John Wiley & Sons, New York.)*

solid, B (at a composition 100% B or 0% A), has a melting point of *Z*. Mixtures of A and B melt at lower temperatures, reaching a minimum at point *E*. This is known as the eutectic composition, and the temperature is the eutectic temperature. Note that the eutectic point is not necessarily at the 50:50 ratio of the components—in Figure 5.22, the eutectic point is at 57% A or 43% B.

In practice, it is advisable to prepare a number of mixtures that contain various proportions of the substance to be identified and the substance suspected of being identical to it. Pure samples of both substances should also be available.

Many melting point apparatus are capable of taking the melting point of several samples simultaneously. The Thomas–Hoover apparatus shown in Figure 5.23 is ideal for this purpose. Not only is this faster, but better results are obtained, as the rate of heating will be the same for all the samples. It is desirable to determine the melting points of the mixtures and of the pure substances simultaneously.

If the unknown species and the known sample are identical, the melting point of a mixture of the two in any proportion will be the same as for each individual species (i.e., no depression of the melting point). If they are different from one another, the melting point of the mixture will usually, but not necessarily always, be lower than that of either of the pure compounds.

The mixture melting point approach to identification of chemical compounds is not absolute. It is highly recommended that other forms of analysis also be utilized before identification is firmly established.

5.E.4 Corrected Melting Points

Since the melting point of an inorganic solid is one of the criteria used for establishing its identity, it is important that the thermometer in the melting point device be calibrated so that accurate comparisons with literature values may be made. This is easily done by observing the melting points of several pure compounds (some typical inorganic materials are listed in Table 5.3. Organic compounds may also be used) over the temperature range of operation.

A calibration curve, shown in Figure 5.24, is prepared by plotting the observed melting point of the standard against the actual value. Using the plot consists of projecting the observed melting point (*A*) horizontally to the curve (*B*) and then dropping a perpendicular to the *x* axis to obtain the corrected value (*C*) as shown. This technique corrects for variations in the thermometer and also provides a stem correction.

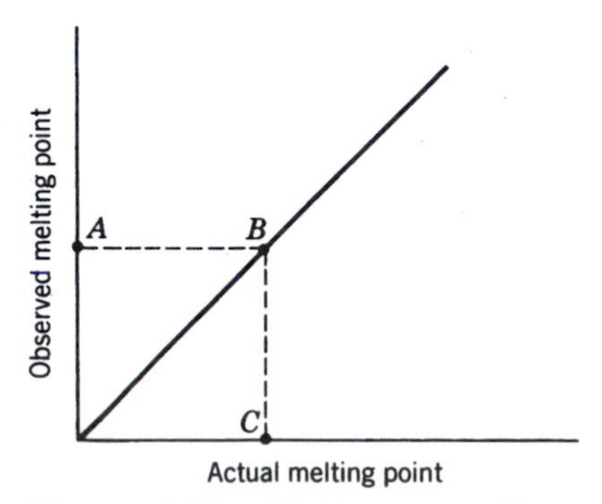

Figure 5.24. *Melting point calibration curve.*

Table 5.3 Inorganic Temperature Calibration Standards

Compound	Melting Point (°C)
Ice	0
Ferric chloride hexahydrate	37
Ferric nitrate nonahydrate	47
Sodium hydrogen sulfate monohydrate	58.5
Urea sulfate	75–77
Magnesium nitrate hexahydrate	89
Zinc acetylacetonate hydrate	136–138 (decompose)
Ferrocene	174–176
Copper sulfate (anhydrous)	200 (decompose)
Potassium hydrogen sulfate	214
Bis(2,4-Pentanedionato)nickel(II)	230
Sodium hydrogen tartrate	253 (decompose)
Sodium nitrate	271
Potassium hydrogen phthalate	295–300 (decompose)
Potassium nitrate	334

5.E.5 Determination of the Melting Point Range

The melting point range of an inorganic solid can be determined by introducing a small amount of the substance into a thin-walled capillary melting point tube, which is sealed at one end.[1] These tubes are commercially available and are ~1 mm in width and 6 cm in length. In loading the capillary tube, a small amount of the compound (powdered if the crystals are not fine enough to fit into the tube) is placed on a clean surface, such as a clay tile or filter paper. A small amount of the compound is then introduced into the tube by tapping the open end of the tube onto the solid material (Fig. 5.25*a*). The material is then packed into the bottom of the tube by rubbing the vertically held tube with a file (Fig. 5-25*b*), by dropping the tube (closed end down) down a glass tube (Fig. 5.25*c*), or by tapping the tube rapidly (carefully so as not to snap the tube) on the bench top.

The determination of the melting point requires enough sample so that equilibrium can be established between the solid and liquid phases and the temperature measured. Usually, ~1–2 mm of sample packed in the tube bottom is sufficient.

When taking a melting point, the oil bath or heating block can be heated rapidly to about 15 °C below the actual melting point, if that value is known. The temperature is then increased at the rate of about 2 °C/min. This is to allow the material to melt before the temperature can rise above the true melting

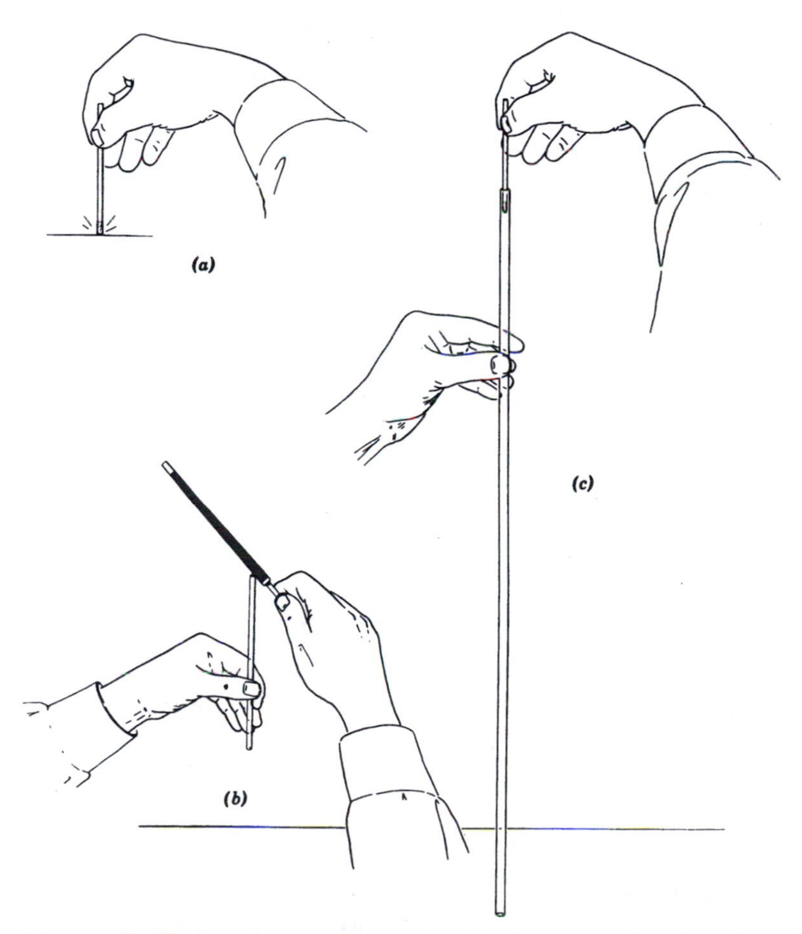

Figure 5.25. *Loading a capillary tube:* (a) *charging and packing* (b and c). *From Shriner, R. L.; Fuson, R. C.; Curtin, D. Y.; Morrill, T. C.,* The Systematic Identification of Organic Compounds, *6th ed., Wiley: New York, 1980. (Reprinted by permission of John Wiley & Sons, New York.)*

point. If this rule is not observed, high readings are generally the result. If you do not know the true melting point of the solid you are working with, time may be saved in the long run by taking a preliminary reading with a fairly rapid heating rate. A second run can then be made at the slower rate to obtain a more accurate value.

Always report melting points as a temperature range. This range is the thermometer reading taken as the first drop of liquid appears to that taken where the solid is totally melted and only liquid is present.

There are several conditions that require special handling in the determination of the melting point of specific substances: when the compound readily undergoes sublimation, when decomposition occurs at the melting point, when the substance is hygroscopic, and when the substance decomposes upon exposure to air.

Sublimation occurs when the substance passes directly into the gas phase without going through the liquid state upon being heated (see Fig. 5.21). The most direct approach is to seal the open end of the capillary tube directly in the flame of a microburner about one third of the way along the tube from the open end. An alternate approach to combat the problem of sublimation (and also decomposition) is to use the evacuated melting point tube method.[2] In this technique, the tube is prepared from a Pasteur pipet and evacuated before being sealed.

Problems with melting points for hygroscopic compounds are overcome by loading the melting point tube in a glovebag previously flushed with dry N_2 gas (Fig. 5.26). Before removing the melting point tube from the glovebag, the open end of the tube should be closed with a little stopcock grease.[3] The tube is then removed from the glovebag and sealed as described previously.

(*a*) Glovebag connected to nitrogen gas cylinder and purged.

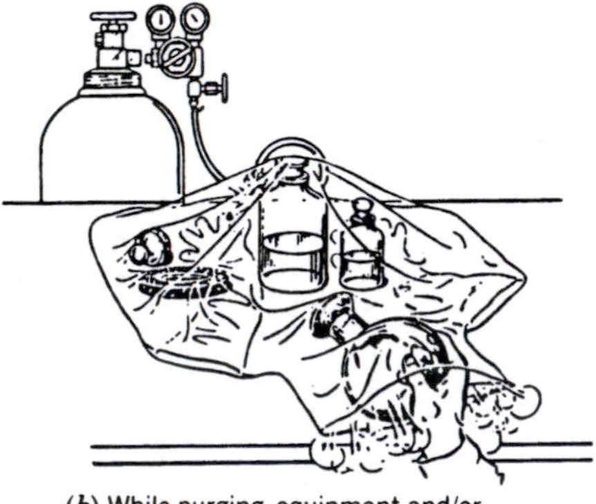

(*b*) While purging, equipment and/or chemicals placed in the glovebag.

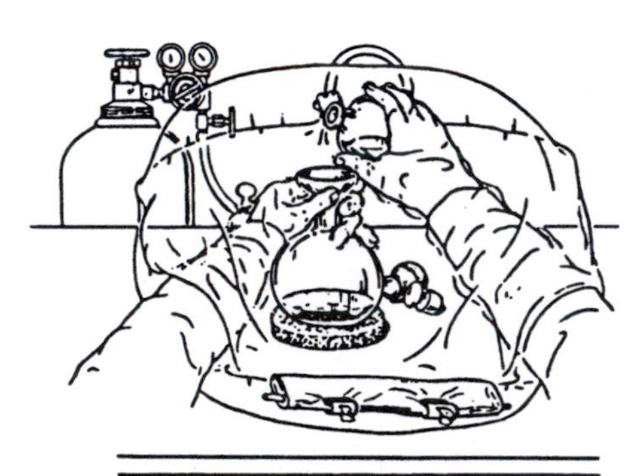

(*c*) After purging, the equipment sleeve is folded and closed with clips. Gas inflates the glovebag. Hands inserted and task performed.

Figure 5.26. *Use of a glovebag. (Courtesy of Instruments for Research and Industry, I²R, Inc., 108 Franklin Ave., Cheltenham, PA 19012.)*

5.E.6 Melting Point Apparatus

There are many different methods that can be used to determine the melting point of a solid inorganic compound. Only the most common techniques (and only those using the micromelting point method) will be presented here. Most approaches are convenient and easy to use, and with care, very accurate results can be obtained.

The Thiele Tube

The Thiele tube is commonly used in general chemistry laboratories to obtain melting points. It is an old technique, inexpensive, and fairly accurate if used carefully. The tube contains an oil bath, which is heated with a Bunsen burner trained on the side arm. The convection currents of the hot oil distribute the heat throughout the system, so stirring is not necessary. The melting point capillary is attached to the thermometer as shown in Figure 5.27.[1a]

The Mel-Temp Apparatus

This device employs a metal block on which the rate of heating is controlled using a voltage regulator (Fig. 5.28). A thermometer is inserted into a hole in the metal block, and gives the temperature of the block and of the sample capillary tube.

> **CAUTION:** ***The metal block can cause burns if touched while still hot! Multiple capillary sample tubes may be observed in one heating cycle. Instructions for the operation of this device will be supplied by your instructor or they can be found in Ref. 1.***

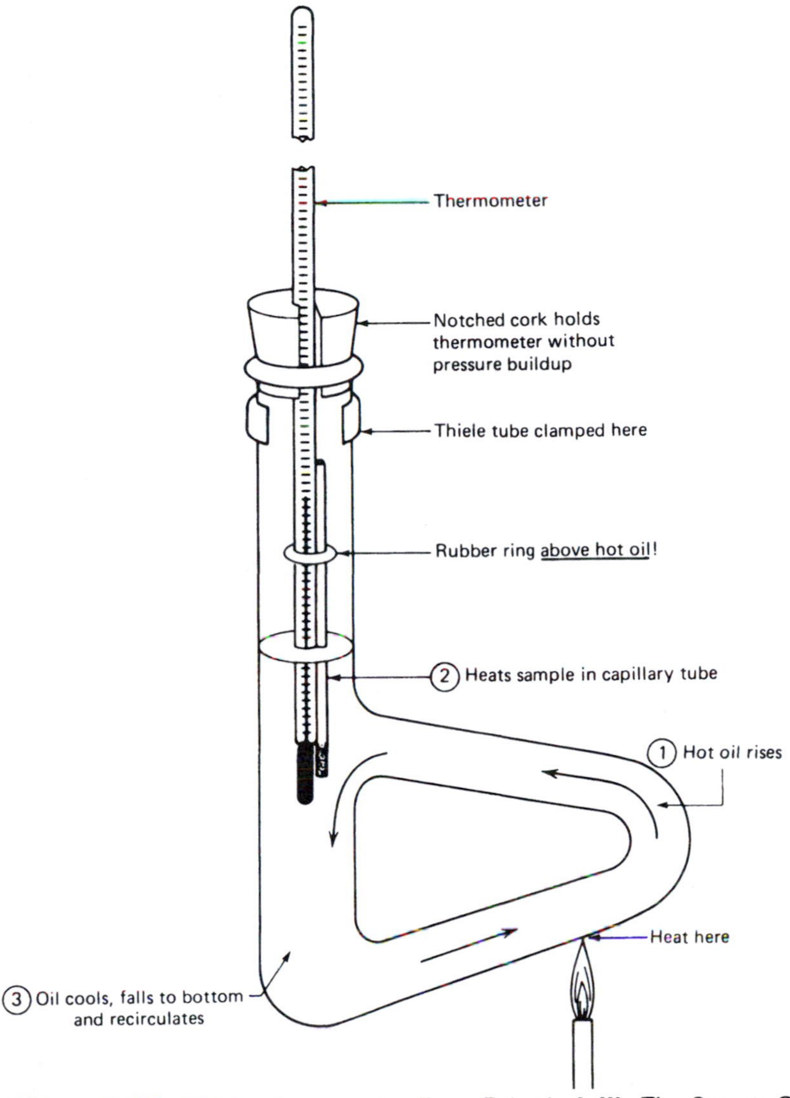

Figure 5.27. *Thiele tube apparatus.* From *Zubrick, J. W.,* The Organic Chem Lab Survival Manual, *2nd ed., Wiley: New York, 1988. (Reprinted by permission of John Wiley & Sons, New York.)*

The Fisher–Johns Apparatus

This apparatus is another of the metal block type, but a capillary tube is not used to hold the sample. The sample is placed between two thin circular glass cover slides, placed in a depression on the metal block (Fig. 5.29). The advantage of this technique is that melting point readings on a very small sample may be observed. The rate of heating of the block is controlled by a voltage regulator. The lighted sample area is observed through a small magnifying glass.

CAUTION: *As with the Mel-temp apparatus, be aware that the hot metal block can cause burns!*

Forceps should be used to handle the cover slides, as moisture from your fingers could ruin a melting point determination. The main disadvantage of the Fischer–Johns technique is that it is slow. Only one sample can be done at a time and further time is lost in waiting for the hot block to cool. Detailed instructions for the operation of the Fisher–Johns melting point apparatus will be supplied by your instructor or may be found in Ref. 1.

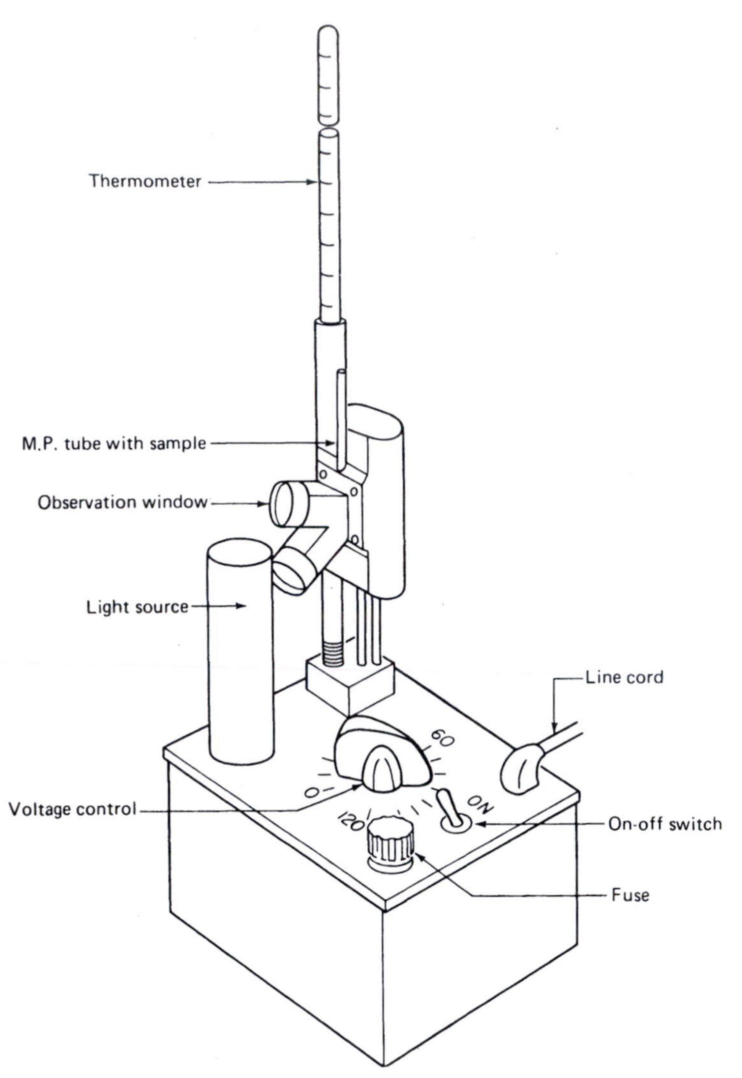

Figure 5.28. *Mel-Temp apparatus.* From *Zubrick, J. W.,* The Organic Chem Lab Survival Manual, *2nd ed., Wiley: New York, 1988. (Reprinted by permission of John Wiley & Sons, New York.)*

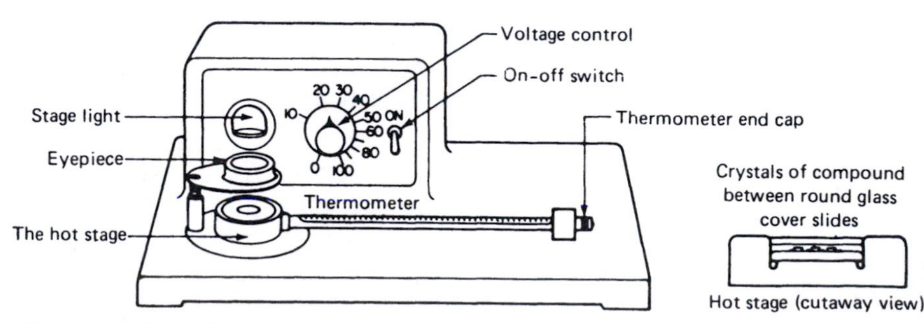

Figure 5.29. *Fisher–Johns apparatus.* From *Zubrick, J. W.,* The Organic Chem Lab Survival Manual, *2nd ed., Wiley: New York, 1988. (Reprinted by permission of John Wiley & Sons, New York.)*

The Thomas–Hoover Apparatus

This device is an electrically controlled instrument employing a stirred and heated oil bath (Fig. 5.23). The voltage control is very sensitive and thus the rate of heating of the oil bath may be controlled to very precise limits. The melting point capillary tube is placed in a holder so that the melting of the sample can be viewed through a small magnifying glass. Multiple melting point determinations may be made using this device, the Uni-Melt stage accepting up to seven tubes in one heating cycle. The Thomas–Hoover apparatus is quite expensive and should be treated with great care. Detailed instructions for the operation of the unit will be supplied by your instructor or may be found in Ref. 1.

REFERENCES

1. General references relating to melting point procedures include:
  (a) Zubrick, J. W., *The Organic Chem Lab Survival Manual*, 2nd ed., Wiley: New York, 1988, Chapter 9.
  (b) Skau, E. L.; Arthur, J. C., Jr., *Technique of Chemistry*, Weissberger, A. and Rossiter, B. W., Eds., Wiley–Interscience: New York, 1971, Vol. 1, Part 5, Chapter 3.

2. Mayo, D. W.; Pike, R. M.; Butcher, S. S., *Microscale Organic Laboratory*, 2nd ed., Wiley: New York, 1989, p. 41.

3. Angelici, R. J., *Synthesis and Technique in Inorganic Chemistry*, 2nd ed., Saunders: Philadelphia, 1977, p. 105.

5.F CONCENTRATION OF SOLUTIONS

5.F.1 Introduction

The removal of solvent from chromatographic fractions or other solutions collected during workup of reaction products can be carried out by several different techniques, which are discussed in the following sections.

5.F.2 Evaporation Techniques

If it is not necessary to recover the solvent you wish to remove from a solution, several evaporation methods are available in the laboratory. The simplest approach is to place the vial or flask, containing a boiling stone, in a warm sand bath in the hood. The rate of evaporation can be controlled by the temperature of the bath. If a boiling stone is not placed in the container, the solution should be agitated by twirling a microspatula in the solution between the thumb and forefinger. This operation prevents bumping and speeds up the evaporation process.

An alternate method is to conduct a gentle stream of air (if the compound to be isolated is not sensitive to oxidation) or an inert gas such as nitrogen over the surface of the solution while warming the sample in a sand bath. This is usually done at a hood station where several blunt-ended syringe needles or Pasteur pipets can be attached to a manifold leading to a tank of compressed gas. Gas flow to the individual syringe needles or pipets is easily controlled by Hoffman clamps (Fig. 2.4) on the tubing used to arrange the manifold. Always test the gas flow with a blank vial of solvent. Do not leave the heated vial in the gas flow after the solvent is removed. This is particularly important in the isolation of liquids since low boiling products may be lost! Tare the container before placing the solution to be concentrated into the container. Evaporation of solvent to a constant weight of the container is the best indication of total solvent removal.

5.F.3 Removal of Solvent Under Reduced Pressure

The concentration of solvent under reduced pressure is very efficient. It reduces the time of solvent removal in microscale experiments dramatically, compared

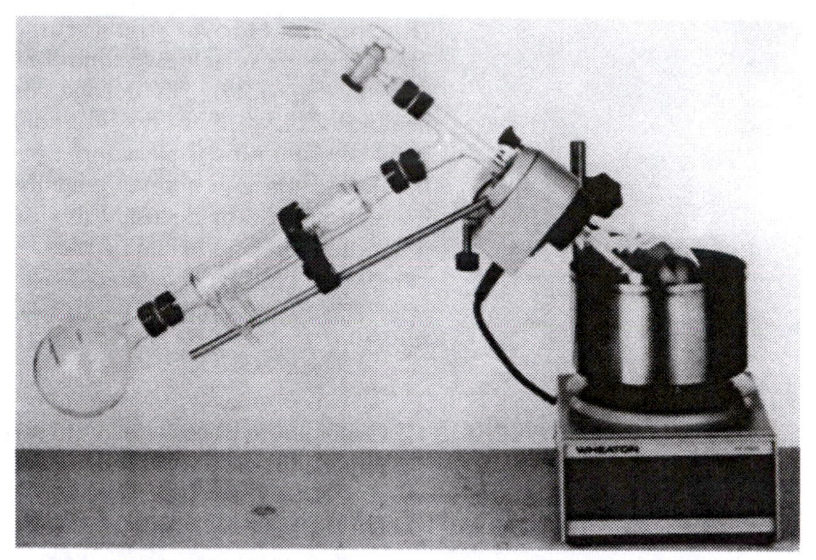

Figure 5.30. *Rotary evaporator. (Courtesy of Wheaton, Millville, NJ.)*

to the distillation or evaporation techniques. A rotary evaporator (Fig. 5.30) may be employed in carrying out this method of concentration. The use of this commercially available device makes it possible to recover the solvent removed during the operation. Rotary evaporators are available for microscale and larger scale work.[1]

The rotary evaporator is a motor-driven device that rotates the flask containing the solution to be concentrated under reduced pressure. The rotation motion aids the evaporation process, as well as minimizes bumping of the solution. The rotary flask may be warmed in a water or oil bath, thus controlling the rate of evaporation.

In microscale work it is important to remember never to pour the recovered product, if it is a liquid, from the rotary flask. Always use a Pasteur pipet to accomplish this transfer.[2]

REFERENCES

1. Buchi Model Rotary Evaporators are available from Brinkman Instruments, Cantiaque Road, Westbury, NY 11590; Wheaton, 1000 N. Tenth Street, Millville, NJ 08332 offers micro- and semimicro Spin-Vap Rotary Evaporators having 0.1–250-mL capacity. Ace Glass, P.O. Box 688, Vineland, NJ 08360 also offers a micro Rotary Evaporator.
2. Mayo, D. W.; Pike, R. M.; Butcher, S. S., *Microscale Organic Laboratory*, 2nd ed., Wiley: New York, 1989.

5.G CHROMATOGRAPHY

5.G.1 Introduction

Many methods of analysis measure properties that are not unique to a particular substance. In the case of infrared (IR) spectroscopy, for example, the presence of a peak at a particular frequency is not indicative of any particular compound (although it may be indicative of a particular functional group). An unknown sample cannot be analyzed by IR spectroscopy unless it has first been separated into its individual components. It is possible, however, to perform quantitative analysis on a mixture using IR spectroscopy if the individual components of the mixture are known. Nuclear magnetic resonances spectroscopy, thermal analysis, and magnetic susceptibility measurement are other examples of nonspecific techniques. The most widely used method for accomplishing chemical separation of mixtures is called chromatography.[1,2]

The Russian botanist, M. Tswett, is credited with the first experimental work in this field in 1906. The ether extract of green leaves was passed through a column packed with $CaCO_3$, and the separated species appeared as colored bands on the column; thus the name chromatography (*chroma*—Greek for color). Since this initial work, chromatography has become one of the most important tools in analytical procedures. The number of systems and techniques used has become quite varied. In this section we will touch upon only three types: gas chromatography (GC), liquid chromatography (LC), and thin-layer chromatography (TLC).

Two factors are common to all types of chromatography: a stationary phase and a mobile phase. The mixture being chromatographed is separated as it is carried through the stationary phase by the flow of the mobile phase. The components that are being separated must be soluble in the mobile phase and these components must also interact with the stationary phase based upon some type of property. Such interactions occur when the materials dissolve in the stationary phase, are absorbed by it, or chemically react with it.

Some generalities can be made about the two phases: the stationary phase may be a liquid or a solid, and may be held in some type of a container. In column chromatography (see Experiment 28) the stationary phase is a ground solid held in a narrow tube, such as a buret or, for microscale work, a modified Pasteur filter pipet. The use of a finely divided solid stationary phase, spread on a glass or plastic plate, is referred to as planar chromatography or thin-layer chromatography.

Other arrangements commonly used in column chromatography include the solid being treated with a liquid and held in the column; the liquid now becomes the stationary phase. The liquid being used can also form a thin layer on the inner walls of a capillary tube. The mobile phase may be a liquid or a gas, which progresses through the stationary phase by gravity, capillary action, or by applied pressure.

The basic concept of chromatography involves a sample of material being dispersed in the mobile phase, and being placed or introduced at the beginning of the chromatographic column. The solute is immediately distributed between the mobile and stationary phases. More mobile phase is added and pushes the fraction of sample in the mobile phase further down the column. As it reaches a new portion of the stationary phase, further partitioning occurs. At the same time, the new mobile phase and stationary phase at the head of the column undergo partitioning (some sample on the stationary phase will enter the mobile phase). This process continues throughout the length of the column. The sample only moves along (down) the column while it is in the mobile phase. The rate at which a particular component migrates is proportional to the fraction of time it spends in the mobile phase. Since different components spend different amounts of time in the mobile phase, they will travel at different rates, and therefore separate.

The mobile phase is referred to as the eluent, and the process of the sample moving down the column is called elution. As the components are eluted individually from the end of the column, they pass through a detector, and if the detector is not destructive, they can be collected. The detector responds in some manner to the presence of the eluted component. The output signal is then plotted as a function of time, producing a symmetrical peak. Such a plot is called a chromatogram. The chromatogram can be used for both qualitative and quantitative analysis.

The chromatogram in Figure 5.31 shows that component A preferentially stays in the stationary phase and lags behind in the elution process. Component B preferentially stays in the mobile phase. Therefore, it has the greater rate of migration and is the first to elute off the column. At short distances along the

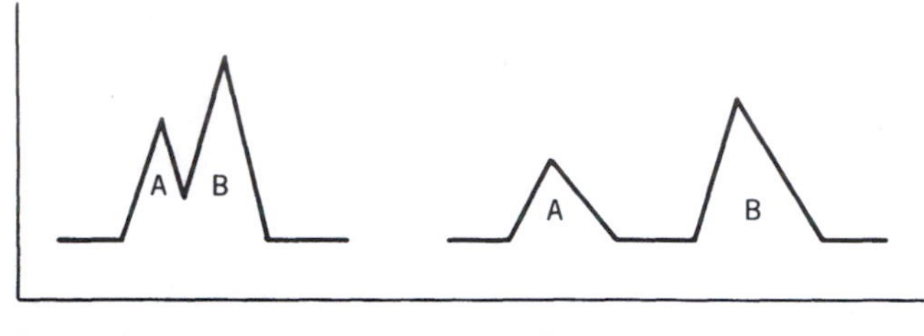

Figure 5.31. *Schematic of a chromatogram.*

column, the two components in the mixture have not yet completely separated. This results in peak overlap. As distance along the column increases, the peaks separate more and more, until they are distinct. It should be noted that some band or zone broadening has occurred.

Qualitative analysis is based upon one piece of data: The length of time the component requires to elute from a particular column, called the retention time. Since many compounds will have essentially the same retention times, chromatography is not a good technique for the identification of total unknowns. Quantitative results can be obtained by an analysis of the peak areas in the chromatogram. More elaborate chromatographs electronically integrate the peak areas, but simpler instruments do not. In such cases, the peak area can be measured by a variety of methods, such as (a) cutting out each peak with scissors and taking the weight ratio, (b) using a planimeter, or (c) calculating the area by measuring the peak height and the peak width at the half-height. The peak area is proportional to the amount of the particular component. Qualitative analysis of mixtures is the major use of chromatography in industry.

5.G.2 Thin-Layer Chromatography

Thin-layer chromatography (TLC)[3–5] is a technique that offers an inexpensive, fast, easy, and powerful method for determining the purity of a given inorganic substance. Although TLC does not have the universal utility in inorganic chemistry that it enjoys in organic chemistry, it can be a valuable separation technique for inorganic and organometallic derivatives (see Experiment 40). It also offers a rapid method for determining elution solvents for use in column chromatography. The method involves the same basic principles as column chromatography, being a form of solid–liquid separation. A differential partitioning of the components in a mixture occurs by their being distributed between a stationary phase (the thin-layer plate) and a mobile phase (the developing solvent). In TLC, the stationary phase is generally a thin layer (~250 μ thick) of silica or alumina spread on a solid support of glass or plastic. Both types are now commercially available, although they may be prepared in the laboratory if desired. It is also possible to do ion-exchange TLC. One potential source of trouble in inorganic analyses is that silica gel usually contains some metallic impurities (Na^+, Mg^{2+}, Ca^{2+}, and Fe^{3+}), which may interfere. Specially washed forms of silica gel HR are available and in general are preferable for inorganic work.

The real advantage of TLC is the very small amount of material required for analysis. In some cases, detection of as little as 10^{-6} mg has been accomplished. On the other hand, one disadvantage is that it is not readily amenable to large scale separations. One distinct difference between column and TLC chromatography is that in the column technique, the eluting solvent migrates down the column while in TLC, the solvent migrates up the plate. Since the absorbent layer of silica or alumina is very thin, evaporation of the eluting solvent can readily occur. Therefore, it is immediately clear that TLC analysis will be quite useful for the analysis of nonvolatile solids. This fact makes TLC particularly attractive in the inorganic area. The sequence of operations is described next.

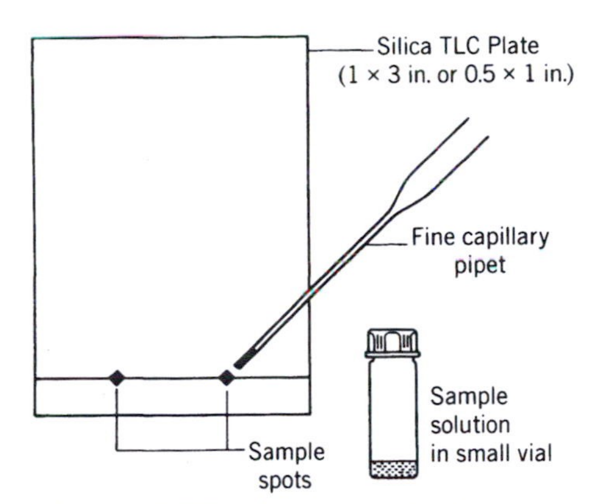

Figure 5.32. *Sample application to a TLC plate.*

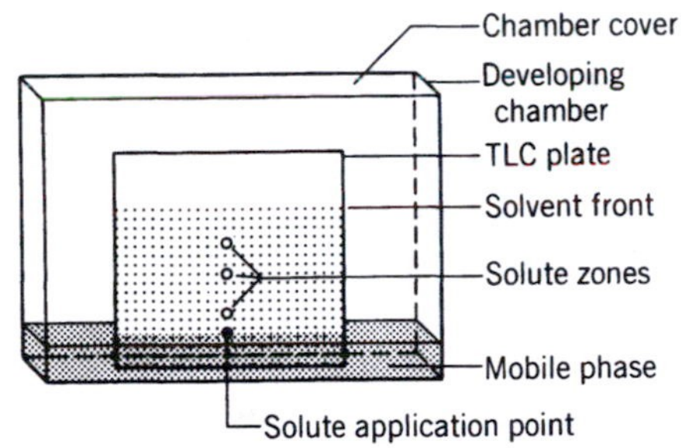

Figure 5.33. *Developing chamber and TLC plate.*

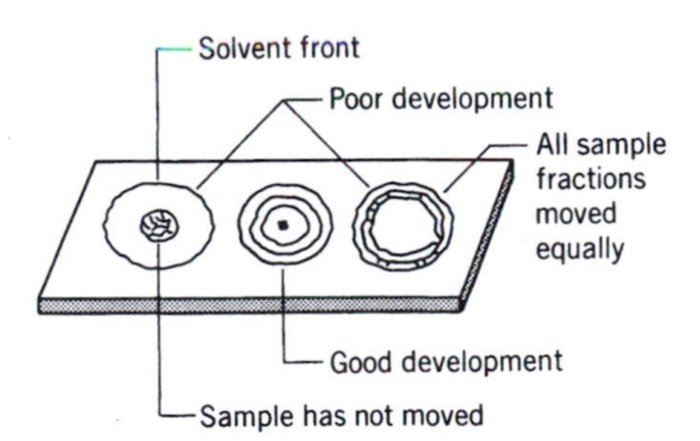

Figure 5.34. *Solvent testing for TLC.*

Thin-layer chromatography plates are obtained as plastic-backed sheets, which allow one to cut the original sheet into very economical 1- × 3-in. strips, or for very small plates, into 1- × 0.5-in. strips. Also commercially available are glass-backed plates, which can be scored with a glass cutter and carefully snapped to obtain the desired size. If you prepare your own plates, it is easier to start with the glass size desired.

A pencil line is drawn parallel to the short side of the plate, so that when the plate is placed in the developing chamber, the solvent does not come above this line. One or two or more points, evenly spaced, are then marked on the line. The sample (~1 mg) to be analyzed is placed in a small vial and dissolved in several drops of solvent. A fine microcapillary pipet is prepared by heating the center of a melting point tube and pulling it quickly to form, after being broken, two fine capillary pipets. The pipet is dipped into the sample solution, and the fine end is pressed lightly to the plate on the line (pencil dot) to deliver a small fraction of the solution from the pipet to the plate (see Fig. 5.32).

The chromatogram is developed by placing the spotted thin-layer plate in a developing chamber (a screw-capped wide-mouth jar or a beaker with aluminum foil as the cover may be used for this purpose), which contains a small amount of developing solvent (see Fig. 5.33). The sample spot(s) on the TLC plate initially must be positioned above the solvent line. The plate should only be handled with forceps, never with your fingers. The container should quickly be covered in order to maintain an atmosphere saturated with the developing solvent. The elution solvent will rapidly ascend the plate by capillary action, and spotted material will elute vertically up the plate. Resolution of mixtures into individual spots along the vertical axis occurs by precisely the same mechanism as in column chromatography. As the solvent line nears the top of the plate, the plate should be removed from the container, and the solvent line should be marked quickly with a pencil before the solvent evaporates.

A rapid method for determining the best TLC developing solvent to use is to make several spots on a plate of the solution to be chromatographed using a micropipet. A clean micropipet containing a test solvent is then used to touch one spot. The solvent will diffuse outward in a circle and the compound(s) will also migrate. In this manner, a series of solvents can quickly be evaluated (see Fig. 5.34). Lists of mobile phases for inorganic TLC work are available.[3]

If the spots themselves are colored, they can be seen immediately on the plate. They should be lightly circled with a pencil and the R_f value(s) of the components should be determined (see below). If the compounds are colorless and therefore do not show visually on the plate, an indirect method must be used. There are now commercially available TLC plates containing a fluorescent indicator. The spots on the developed plate, when irradiated with an ultraviolet (UV) lamp, appear dark against the fluorescing silica gel, and can thus be seen. Each spot should be outlined lightly with a pencil so that a permanent record is made of the chromatogram.

An alternate method used to detect colorless substances is to place the developed plate in an iodine vapor chamber for a few seconds. Iodine forms a reversible complex with most organic substances. Thus, dark spots will develop in those areas containing sample material. On removal from the iodine chamber, the spots should immediately be marked by pencil, because they will fade rather rapidly. Certain colorless compounds can be detected by spraying them with special solutions containing a reagent(s) that reacts with them to form a colored compound. This technique is very selective. Specific information is found in Refs. 3–5.

The elution characteristics of a particular species is reported as an R_f value. This value is a measure of the distance traveled by a substance up the plate during the development of the chromatogram relative to the solvent movement.

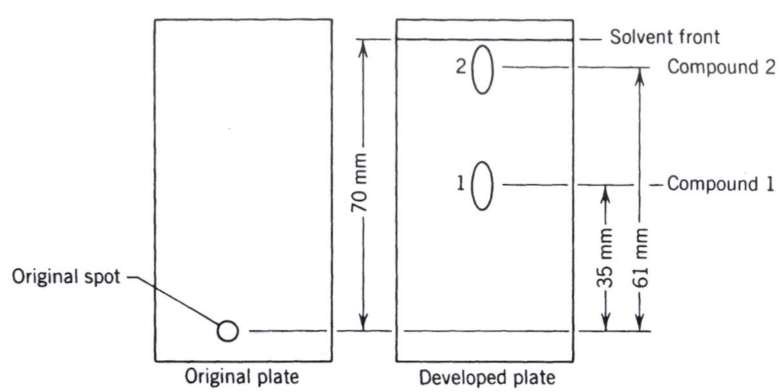

Figure 5.35. *Developed TLC plate.*

A sample calculation is shown in Figure 5.35. In this example, the R_f value for Compound 1 would be $\frac{35}{70}$ = 0.50, and for Compound 2 it would be $\frac{61}{70}$ = 0.87.

5.G.3 Gas Chromatography

Gas chromatography (GC)[6,7] makes use of a gaseous mobile phase in contact with a solid or liquid stationary phase. The basic components of a GC are shown in Figure 5.36.

Gas chromatography sees tremendous use in the area of organic chemistry. In inorganic chemistry, its use was somewhat more problematic, for several reasons:

1. The sample must pass through the system in the gas phase and many inorganic materials only volatilize at high temperatures. Gas chromatographs cannot operate at temperatures much above 500 °C (and then only with special columns—see below), as the high temperature will be enough to volatilize the solid or liquid stationary phase, which will then flood the detector preventing detection of the sample components.
2. The sample must not react with the solid or liquid stationary phase or with the carrier gas. Since many of the more volatile inorganic materials are also quite reactive, this is a major (but not insurmountable) difficulty. In extreme cases, it is necessary to operate the GC at low temperatures, and to choose inert solid or liquid stationary phases, such as Teflon and fluorocarbon oils.
3. The sample must be thermally stable when heated above its boiling point. Many inorganic compounds do not boil, decomposing instead.

The main use of GC in inorganic chemistry is in the investigation of organometallic compounds, which generally have lower boiling points, although it is

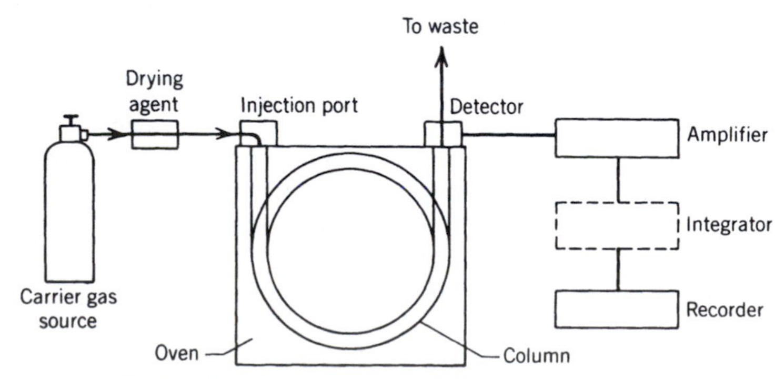

Figure 5.36. *Block diagram of a GC.*

possible to investigate elemental gases and vapors, binary inorganic compounds, metal chelates, and various derivatized inorganic species.[6]

Of the binary inorganic compounds, the most obviously amenable to GC analysis are the very volatile hydrides of the boron, silicon, and nitrogen families. Different difficulties appear here, namely, that inorganic hydrides tend to be very reactive, explosively so in many cases. It is necessary to have extremely inert substrates, such as Teflon, and to operate at low temperatures to avoid the reaction of the analyte with the chromatographic system. The more volatile halides (metal fluorides, some chlorides, and bromides) may also be studied in this manner.

Organometallic compounds form the largest class of inorganic materials studied by GC. Of the σ-bonded organometallics, those of Groups 13–17 (Groups IIIA–VIIA) are particularly amenable. In fact, carborane–silicone polymers (trade name: Dexsil®) are used as stationary phases because of their high temperature stability. Other similar compounds, such as inorganic alkoxides and silicones have also been extensively investigated. The main organometallic studies have centered on metal carbonyls and metal aryl compounds. Experiment 40 is an example of this, using GC to separate ferrocene derivatives. Some industrial work has been done investigating organometallic catalysts in gasolines.

Metal chelates (see Experiments 22–23 for examples of their preparations) are easily separated, as long as they are coordinatively saturated with the chelating ligand. The fluorine derivates of metal acetylacetonates are even more volatile and thermally stable.

Components of the Gas Chromatograph

The carrier gas used as the mobile phase must be chemically inert with respect to the system being separated. The most common gases used are helium, argon, nitrogen, and hydrogen. The inlet pressure of the gas is generally set between 10 and 50 psi and the flow rates at 20–50 $mL \cdot min^{-1}$. It is common practice to place a drying agent or molecular sieve between the gas source and injection port in order to remove any water and particulate impurities from the carrier gas.

The injection port allows for the introduction of the sample onto the column. For liquid samples, the injection port is generally heated to 20–50 °C above the boiling point of the sample, to ensure that the analyte does not condense. On most GC units, a $\frac{1}{4}$-in. o.d. column is used. This normally limits sample sizes to 1–10 μL. Much smaller sample sizes must be injected onto the column when a capillary column is used (in the range of 10^{-3} μL). This is accomplished by use of a sample splitter in the system, with most of the sample being discarded. For gas samples, the sample is introduced onto the column by a rotary sampling valve. A solid can be introduced into the GC by first preparing a solution of the material, and then injecting the solution onto the column as a normal liquid sample. In a different procedure, the solid can be sealed in a small, thin-walled glass vessel that is placed at the head of the column and then broken.

There are two common types of columns used in GC: packed columns and capillary columns. Capillary columns (0.3–0.5 mm i.d.) have their inside wall coated with a liquid stationary phase. The length of these columns can range up to 100 m. The ratio of the stationary phase/mobile phase volumes of 100:300 leads to high column efficiency. Because of the small bore of the column, small sample sizes are necessary. The injection port therefore needs to be fitted with a sample splitter, as mentioned previously.

Packed columns are generally 1–8 mm i.d. and in the range of 2–20 ft long. The volume ratio is 15:20 and results in a lower column efficiency than the capillary columns provide. Nonetheless, with thinner and longer columns, excellent efficiencies can be obtained.

The solid supports used in packed columns are commercially available and are most often made of diatomaceous earth. Firebrick is a durable substance with a large surface area (~4 $m^2 \cdot g^{-1}$), highly active but not used for polar compounds. Firebrick trade names include C22®, Sterchamol®, and Chromasorb P®. Kieselguhr has a smaller surface area, but is not as reactive and therefore used for polar compounds. Trade names include Celite®, Chromasorb W®, Embacel®, and Celatom®.

The stationary liquid phase must have certain properties.

1. Low volatility. The phase must generally have a boiling point at least 200 °C above the temperature at which the column oven will operate.
2. The phase must be thermally stable over the operating temperature range.
3. Chemical inertness with respect to the samples to be analyzed.
4. Good selectivity as a solvent for samples that are to be separated.

Almost any type and size of column needed for a particular application is available commercially, and all are relatively inexpensive. Making a specialty column is also not difficult. A slurry of a solution of the liquid stationary phase is prepared in a volatile solvent with the solid support material. The coated support slurry is then placed into the column, and the column coiled to the proper dimension in order to fit the oven chamber. During the addition of the slurry, the column is agitated in order to prevent channeling of the packing.

There are a wide variety of liquids available for use in GC. The choice of which liquid to use is frequently obtained by trial and error, although the temperature needed to volatilize the analyte and the reactivity of the analyte narrows the list of choices considerably. Bonded phases, wherein the liquid stationary phase is chemically bonded onto a silica support, can theoretically overcome temperature difficulties, but such packings are expensive.

The oven temperature is generally set close to the boiling points of the liquids being separated. With average eluent flow rates of 50 $mL \cdot min^{-1}$ and 8-ft columns, the elution time is generally between 2 and 20 min. Samples containing components with a wide range in boiling points are usually best separated using a programmed oven temperature ramping system, rather than using an isothermal oven temperature.

The detection system must react rapidly to the presence of the sample and give reproducible results even to low concentrations of eluted solute. It must also produce uniform response to the variety of solute materials being analyzed. The three most common types of detectors used in GC are the thermal conductivity detector (TCD), flame-ionization detector (FID), and the electron capture detector (ECD), although more expensive detectors such as FT IR spectrometers and mass spectral detectors are also available.

The TCD (also known as a katharometer) measures changes in thermal conductivity of the gas stream as it exits the chromatographic column. The gas stream consists of the eluent gas or a mixture of eluent gas and solute. Most systems use two detectors—one sample and one reference—so that the thermal conductivity of the carrier gas (eluent) is canceled and changes in the thermal conductivity depend only on the solute material exiting from the column. The two most commonly used carrier gases are helium and hydrogen, due to their high thermal conductivity. Small amounts of sample in the carrier gas therefore lead to a significant decrease in the thermal conductivity of the exiting gas. The detector temperature therefore rises in comparison to the detector temperature in the reference column. This ΔT is the recorded signal in the chromatogram. Advantages of the TCD are that it is simple and rugged in design, inexpensive, quite accurate, nonselective, and it does not destroy the sample.

The FID measures current produced from the ionic intermediates formed upon combustion of the eluted compounds in a H_2–air flame. The use of the FID gives a highly sensitive method of analysis and can be used for low concentrations of sample (such as in the use of a capillary column). This process destroys the sample, preventing collection and use of the sample in further studies.

Electron capture detectors function by having the exiting eluent N_2 gas pass over a β emitter. As the β particles hit the N_2 carrier gas, a stream of electrons is produced. This results in a constant standing current. As a compound passes through the detector, the β particles are absorbed, causing a reduction in the constant standing current. This change in current is used as the signal to the recorder. Use of the ECD is a highly sensitive method for compounds with electronegative functional groups.

5.G.4 Liquid Chromatography

Liquid chromatography (LC) makes use of a liquid mobile phase in contact with a solid or liquid stationary phase. Liquid chromatographic techniques originally involved columns ranging from 10 to 50 mm in diameter and from 50 to 500 cm long, with a solid support particulate size of 150–200 μm in diameter. Solvent flowed through the column by gravity, with average rates of 0.1 $mL{\cdot}min^{-1}$, and no additional pressure was needed. Smaller particle size slowed the flow rate too much for efficiency or practical use. Such techniques are still commonly used and called column chromatography.

Ion and Ion Exchange Chromatography

Ion exchange chromatography[8] is a special case of column chromatography. An interchange of ions occurs between a liquid mobile phase and an insoluble solid stationary phase that is in contact with that solution. Many substances, both naturally occurring and synthetic, act as ion exchangers. Among the naturally occurring types are clays and zeolites. Ion exchangers see extensive use in water softening, water deionization, solution purification, and ion separation. Synthetic ion-exchange resins are high molecular weight polymers, containing a large number of ionic functional groups per polymer unit. There are two general forms of resin: cationic and anionic. In a cationic exchange resin, the ionic functional group is usually a sulfonic acid ($R{-}SO_3H$) or carboxylic acid ($R{-}CO_2H$). The exchange process is as follows (for a sulfonic acid resin):

$$n(RSO_3^-)(H^+) + M^{n+} = (RSO_3^-)_nM^{n+} + nH^+$$

The metal cation therefore replaces the hydrogen ion on the resin. Anionic resins usually contain either quarternary ammonium or amine groups. The exchange process is

$$n(RNH_4^+)(OH)^- + A^{n+} = (RNH_4^+)_n(A^{n-}) + nOH^-$$

The anionic species replaces hydroxide ion on the resin.

The exchange reactions are equilibria and depend on the affinity of the resin for the cation (or anion) in the solution. This in turn depends on the size and charge of the solvated ion. The greater the charge/size ratio of the ion passing through the column, the greater the affinity for the resin. The order of affinity among cations for the cationic resin in dilute solutions was demonstrated to be

$$Th^{4+} > Fe^{3+} > Al^{3+} > Ba^{2+} > Pb^{2+} > Sr^{2+} > Ca^{2+} > Fe^{2+} > Co^{2+} >$$

$$Mg^{2+} > Ag^{+} > Cs^{+} > Rb^{+} > NH_4^{+} > K^{+} > Na^{+} > H^{+} > Li^{+}$$

Care must be exercised when comparing the charge/size ratios in determining placement on this list. For example, in water, both sodium and potassium ions would be hydrated. Since sodium has the larger charge/size ratio, the hydrated radius of sodium is larger than that of potassium. It is the hydrated sodium and hydrated potassium ions that pass through the column. Both hydrated ions have the same charge. Thus, the smaller hydrated potassium ion would be attracted to the resin more than the larger hydrated sodium ion. The potassium ion would spend more time attached to the solid, and the sodium ion would spend more time in solution. The sodium ion, being less strongly held by the column, would move more rapidly, and would therefore elute first. Thus, in dilute solution, potassium has a higher resin affinity than sodium. In concentrated solution, the reverse is true as the ions are not effectively hydrated. Experiment 28 uses ion exchange chromatography to separate the various oxidation states of vanadium.

As was the case in GC it is generally desirable to have some sort of detector at the end of the column for routine investigation. It would obviously be best to have a detector that would be sensitive to any ion passing through the resin, thus a conductivity meter would be an obvious choice. Unfortunately, the eluting solvent is also conductive and present in much greater amounts than the ion of interest. The ion signal therefore would be lost. In 1970, it was first suggested that a second ion-exchange column could be used to remove the conductivity of the eluent solvent, leaving only the ion signal of interest. This column is called a suppressor column. For cation analysis (where HCl is a typical eluent), the suppressor column is basic in order to neutralize the HCl.

$$\text{resin } OH^- + H^+Cl^- = \text{resin } Cl^- + H_2O$$

Any cation that enters this column will be converted to the hydroxide.

$$\text{resin } OH^- + M^+A^- = \text{resin } A^- + M^+OH^-$$

A similar process is carried out with an acidic suppressor column for the analysis of anions. Needless to say, the suppressor column must be regenerated periodically. This two column arrangement is called ion chromatography,[8] and is patented by Dow Chemical under the Dionex Corporate subsidiary. To allow smaller column particle sizes and faster column throughput, commercial ion chromatographs pump the eluent through the column in the same manner as is done in high-performance liquid chromatography, which is discussed in the following section.

High-Performance Liquid Chromatography

Improvements in liquid chromatographic methods could only occur as particle sizes decreased, but this required methods other than gravity flow to have the eluent pass through the column. Particle sizes of 5–10 μm are now available and the liquid must be forced through the column by some type of pump. The original nomenclature used for this process was high-pressure liquid chromatography (HPLC).[9] The letters have remained the same, but the name was changed. The abbreviation HPLC now stands for high-performance liquid chromatography. The basic components of an HPLC unit are shown in Figure 5.37.

Since the liquid mobile phase is being pumped through a column of small diameter solid particles, certain precautions must be taken. Solvents must be absolutely free of particulate matter in order to prevent clogging of the column. Filtering the eluent prior to use is essential. Most solvents commonly used in HPLC are commercially available and are indicated as HPLC grade. Solvent should also be free of dissolved gas, as the dissolved gas bubbles can lead to zone or band broadening and also can interfere with the performance of the detector.

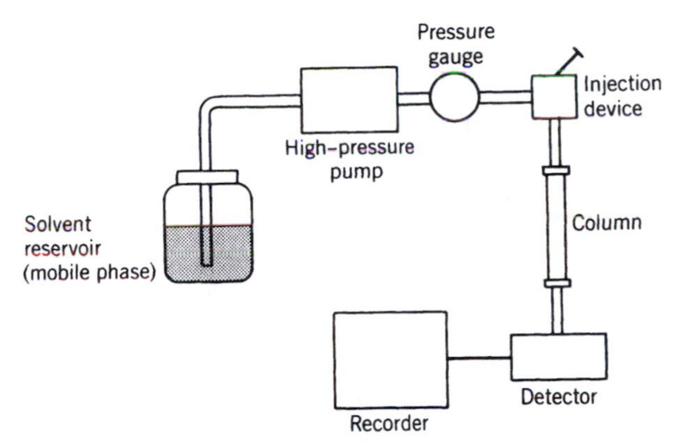

Figure 5.37. *Schematic for HPLC.*

The solvent can be either a pure liquid or a mixture of liquids. If a pure liquid or a constant composition mixture is used, it is referred to as an isocratic system. If the solvent composition is changed during the separation procedure, it is referred to as a gradient system. In order to be able to run a gradient system, the HPLC unit requires at least two pumps. Obviously, a gradient system in general will allow for improvement in the separation of complex mixtures.

Most columns available in HPLC units require pumps capable of pressures of 1000–6000 psi and capacity for flow rates of 1–3 $mL \cdot min^{-1}$. It is imperative that the pump provide a stable flow, without pulsation, in order not to disturb the detector. Two types of pumps are commonly used.

1. **Screw-Driven Syringe:** This type of pump is pulse free, but has a low capacity. It is difficult to change solvents or use as a gradient system.
2. **Reciprocating Pump:** The most commonly used pump for HPLC, but it gives a pulsed flow. This problem is easily solved by the use of baffles to dampen the flow.

Since the pumps have small volumes, it is easy to set up a gradient system using some type of mixing chamber.

High-performance liquid chromatography columns are small in diameter and have lengths of approximately 15–150 cm. Columns are usually made of stainless steel, in order to withstand the high pressure put on the eluent. The materials used for packing the column vary, and include finely divided silica gel with a particle size of 5–10 μm, small glass beads ~40 μm in diameter coated with a layer of silica gel, alumina, and ion-exchange resins of 1–3 μm. Currently, the most widely used stationary phases are the bonded phase packings. These are silica gel particles that were reacted with organic compounds producing a nonpolar surface.

$$\text{silica gel—Si—OH} + \text{ClSiR}_3 = \text{silica gel—Si—O—SiR}_3$$

A precolumn is used as a protective device to filter out impurities and thereby prolong the lifetime of the column. The precolumn is packed with the same material as the column, but with a larger particle size. This helps prevent a pressure drop.

Sample injection generally is accomplished by either a rotary or a slider valve, although syringes can be used. Sample values are also automated, for unattended operation.

There are several detectors available for HPLC, the most common of which are the UV–Visible detector and the differential refractive index (RI) detector.

The UV–Visible detector is a low-pressure mercury lamp that emits essentially monochromatic radiation at 254 nm. This UV radiation is focused on the sample and reference cells (the reference cell contains air), and the difference in signal provides the chromatogram. Since many compounds do not absorb at 254 nm, UV–Visible detectors are available for other wavelengths.

The RI detector responds to all samples (as all materials have a refractive index), but is much more expensive than a simple UV detector. It cannot be used for gradient elution and therefore is not as heavily used.

REFERENCES

General Background

1. Macdonald, John C., Ed., *Inorganic Chromatographic Analysis* (*Chemical Analysis*, Vol. 78), New York: Wiley, 1985.
2. Zweg, G.; Sherma, J., Eds., *Handbook of Chromatography*, Chemical Rubber: Cleveland, OH, 1972.

Thin-Layer Chromatography

3. Sherma, Joseph, "Thin-Layer Chromatography of Inorganic Ions and Compounds," in *Inorganic Chromatographic Analysis* (*Chemical Analysis*, Vol. 78), J. C. Macdonald, Ed., New York: Wiley, 1985, Chapter 7.
4. Hamilton, R.; Hamilton, S., *Thin Layer Chromatography*, Wiley: New York, 1987. (Part of the "Analytical Chemistry by Open Learning" series.)
5. Touchstone, J. C.; Dobbins, M. F., *Practice of Thin Layer Chromatography*, Wiley: New York, 1978.

Gas Chromatography

6. Uden, P. C., "Gas Chromatography of Inorganic Compounds, Organometallics, and Metal Complexes," in *Inorganic Chromatographic Analysis* (*Chemical Analysis*, Vol. 78), J. C. Macdonald, Ed., New York: Wiley, 1985, Chapter 5.
7. Willet, J. E., *Gas Chromatography*, London: Analytical Chemistry by Open Learning, 1987.

Liquid Chromatography

8. MacDonald, J. C., "Ion Chromatography," in *The Best of American Laboratory*, International Scientific Communications Inc.: Fairfield, CT, 1983, p. 343.
9. Lindsay, S., *High Performance Liquid Chromatography*, London: Analytical Chemistry by Open Learning, 1987.

5.H SUBLIMATION

5.H.1 Introduction

A material is said to sublime when it enters the gas phase directly from the solid phase, without melting. Some materials sublime at atmospheric pressure (CO_2, or dry ice, is the best known example of this), but most sublime when heated under reduced pressure. Substances that can be purified by sublimation are those that do not have strong intermolecular attractive forces. Molecular covalent solids are usually amenable to sublimation, whereas ionic solids are not. Ferrocene, prepared in Experiment 40, is a typical molecular compound. Camphor, iodine, and water are additional examples.

Sublimation is a technique that is especially suitable for the purification of solid substances at the microscale level. It is particularly advantageous when the impurities present in the sample are nonvolatile under the conditions employed. Since many inorganic preparations begin with ionic salts, this condition is often met. Sublimation is a relatively straightforward method in that the impure solid

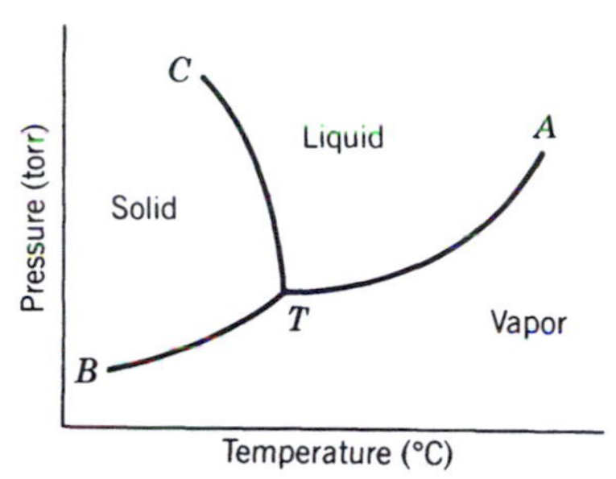

Figure 5.38. *Single-component phase diagram.*

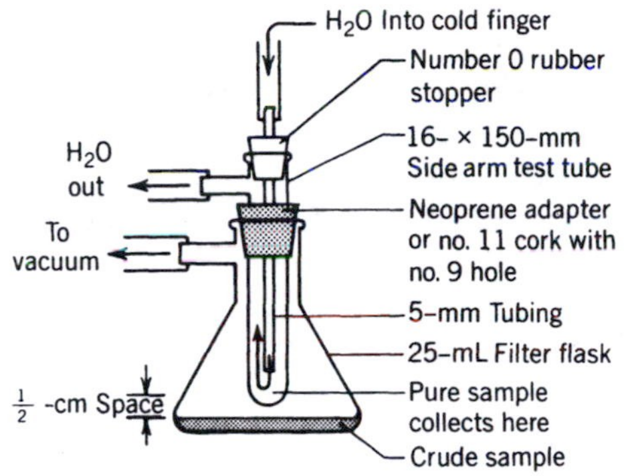

Figure 5.39. *Sublimation apparatus.*

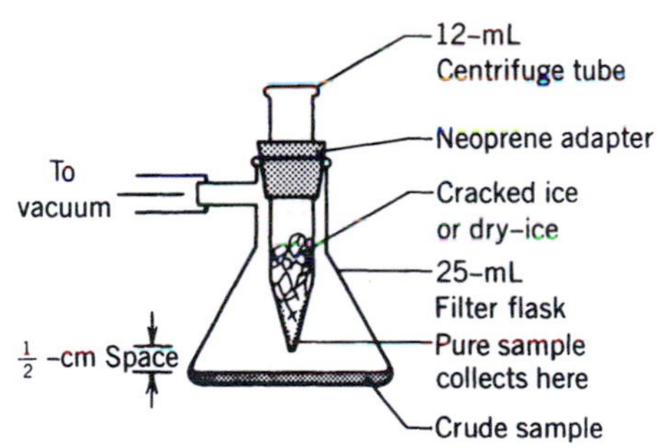

Figure 5.40. *Vacuum sublimator.*

need only be heated at reduced pressure until it sublimes, and then condensed onto a cold surface. Mechanical losses are therefore quite minimal.

5.H.2 Theory

The processes of sublimation and distillation are closely related. In a solid, all the particles do not possess the same energy. There is a distribution of energy and we refer to the particles as having an average energy. Consequently, some particles will have a lower energy, and some will have a higher energy than the average value. If a particle with higher than average energy is near the surface of the solid crystal, it may overcome the attractive forces of its neighbors and escape into the vapor phase, creating a vapor pressure. Eventually, in a closed system, a dynamic equilibrium results, the rate of sublimation being equal to that of condensation.

Sublimation is an endothermic process, whose enthalpy is called the heat or enthalpy of sublimation, $\Delta H_{\text{sublimation}}$. This enthalpy is the sum of the heats of fusion and vaporization.

$$\Delta H_{\text{sublimation}} = \Delta H_{\text{fusion}} + \Delta H_{\text{vaporization}}$$

Since the magnitude of the attractive forces varies for different substances, the equilibrium vapor pressure differs from one substance to another. If the attractive forces are small (ferrocene, I_2, and CO_2) the escaping tendency is large and the vapor pressure tends to be high. If the binding forces are stronger, such as in the large number of inorganic salts having ionic attraction in their crystalline structure (NaCl, KF, etc.) the escaping tendency is low and therefore these substances tend to have low vapor pressures and cannot be easily sublimed.

The vapor pressure of solids varies directly with temperature. An increase in temperature results in an increase in energy. The escaping tendency is thus increased, which results in a higher vapor pressure for the system.

A typical single-component phase diagram is shown in Figure 5.38, relating the solid, liquid, and vapor states of a substance with temperature and pressure. Where two of the areas (solid, liquid, or vapor) touch, there is a line along which the two phases exist in equilibrium. The line *BT* is the sublimation vapor pressure curve of the substance in question, and only along line *BT* can solid and vapor exist together in equilibrium. At temperatures and pressures along the *BT* curve, the liquid state is thermodynamically unstable. Where the three lines representing pairs of phases intersect (point *T*), all three phases exist together in equilibrium. This point is called the triple point. Every substance has a triple point, fixed by the nature of the attractive forces present in the crystalline system. Reference 1 contains an extensive list of vapor pressures of selected inorganic and organometallic solids.

5.H.3 Sublimation Technique

Generally, when using sublimation as a purification process, the procedure is carried out under reduced pressure so that the substance can be sublimed at a lower temperature.[2] Decomposition is less likely under these conditions. The impure material is placed in a sublimation chamber, the pressure reduced, and heat is applied if necessary. The sublimed material, now in the vapor phase, is then condensed on a cooled surface. The collected solid is removed from the apparatus and characterized in an appropriate manner. Typical sublimation apparatus are shown in Figures 5.39 and 5.40.

REFERENCES

1. Shriver, D. F.; Drezdzon, M. A., *The Manipulation of Air Sensitive Compounds*, 2nd ed., Wiley: New York, 1986.

2. Zubrick, J. W., *The Organic Chem Lab Survival Manual*, 2nd ed., Wiley: New York, 1988.

5.1 SOLVENT EXTRACTION

5.1.1 Introduction

Solvent extraction is a technique frequently used in the inorganic and organic laboratory to separate or isolate a desired species from a mixture of compounds or from impurities.[1] It is very useful since the technique is rapid and gives clean separations. Organic chelating agents such as 8-hydroxyquinoline (oxine) and diphenylthiocarbazone (dithizone) (shown in Fig. 5.41) find widespread application for extraction separation of various metal ions.

Nitrate salts can be selectively separated by extraction with ether. For example, uranium(VI), UO_2^{2+}, is easily separated from lead and thorium by ether extraction of a 1.5*M* HNO_3 solution saturated with NH_4NO_3. As a further example, it was known for many years that the Fe^{3+} ion can be extracted into ether from highly concentrated hydrochloric acid solutions. This extraction process finds use in the separation of large quantities of iron prior to the determination of other elements present in ferrous alloys.

Solvent extraction methods are readily adapted to microscale work and are an important technique whose use can often lead to separation and isolation of pure products. Since small quantities are easily manipulated in solution, the technique is applicable over a wide concentration range.

5.1.2 Theory

A given solute, placed in a mixture of two immiscible solvents, will distribute (partition) itself in a manner that is a function of its solubility between the two solvents. For example, a solute X will be distributed between two immiscible solvents (1 and 2) according to the following equilibrium distribution as established by Nernst in 1891.

$$[X]_2 \rightleftharpoons [X]_1$$

and

$$K_{eq} = K_{dist} = \frac{[X]_1}{[X]_2}$$

where [X] is the concentration of the solute in moles per liter ($mol \cdot L^{-1}$).

The ratio of the concentration of the solute in each solvent is a constant, K_{dist}, for a given system at a given temperature. This equilibrium constant expression is designated the distribution (dist) coefficient (also referred to as the partition coefficient). In thermodynamic terms, it is actually the ratio of the activities of the solute in the two phases that is constant, not the concentrations. However, molar concentration units are often used in calculations to obtain approximate results without serious error.

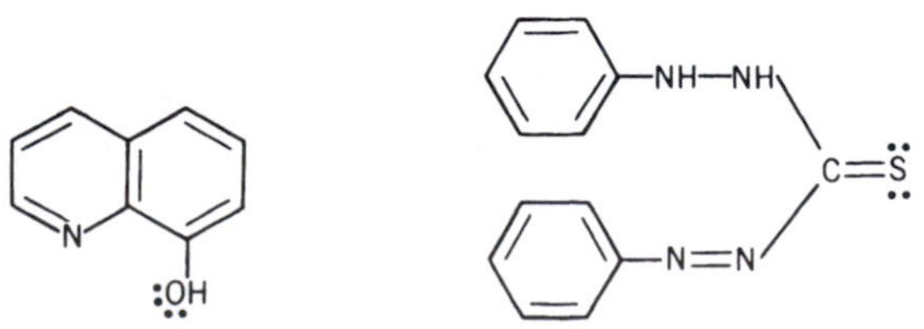

N
:OH
NH—NH
C=S
N=N

Figure 5.41. *Structures of 8-hydroxyquinoline and diphenylthiocarbazone.*

The basic equation used to express the coefficient K_{dist} in simple systems where the solute exists as a single species in both solvent phases[2] is

$$K_{dist} = \left[\frac{g}{100\ mL}\right]_1 \bigg/ \left[\frac{g}{100\ mL}\right]_2$$

This expression uses grams per 100 mL or grams per deciliter ($g{\cdot}dL^{-1}$), but grams per liter ($g{\cdot}L^{-1}$), parts per million (ppm), or molarity are also valid. The distribution coefficient is dimensionless, so that any concentration units may be used, provided the units are the same for both phases. If equal volumes of both solvents are employed in the extraction, the equation reduces to the ratio of the masses of the given solute in the two solvent phases.

$$K_{dist} = \frac{[g]_1}{[g]_2}$$

Determination of the distribution coefficient for a particular species in various immiscible solvent combinations can often given valuable information to aid in the isolation and purification of the species using extraction techniques.

Let us now look at a typical calculation for the extraction of an inorganic solute Q from an aqueous solution using diethyl ether as the organic solvent. We will assume that the K_{dist} ether–water value (distribution coefficient of solute Q between ether and water) is 3.5 at 20 °C.

If an aqueous solution containing a total of 100 mg of Q in 300 μL of water is extracted at 20 °C with 300 μL of ether, the following expression holds:

$$K_{dist} = \frac{[Q]_{ether}}{[Q]_{water}} = \frac{\text{weight } Q_{ether}/300\ \mu L}{\text{weight } Q_{water}/300\ \mu L}$$

Since the weight of Q in the water plus the weight of Q in the ether add up to 100 mg, the above relationship can be expressed as follows:

$$K_{dist} = \frac{\text{weight } Q_{ether}/300\ \mu L}{(100 - \text{weight } Q_{ether})/300\ \mu L}$$

Solving for the weight Q_{ether}, one obtains 77.8 mg; the value for the weight Q_{water} is 22.2 mg. Thus, after one extraction with 300 μL of ether, 77.8 mg of Q (77.8% of the total) is removed by the ether and 22.2 mg (22.2% of the total) remains in the water layer.

The question often arises whether it is preferable to make a single extraction with the total quantity of solvent available, or to make multiple extractions with portions of the solvent. The second method is usually preferable in terms of efficiency of extraction. To illustrate this, consider the following:

In relation to the foregoing example, let us now extract the 100 mg of Q in 300 μL of water with two 150-μL portions of ether instead of one 300-μL portion as previously done.

For the first 150-μL extraction

$$K_{dist} = \frac{\text{weight } Q_{ether}/150\ \mu L}{(100 - \text{weight } Q_{ether})/300\ \mu L}$$

Solving for the weight of Q_{ether}, one obtains 63.6 mg. The amount of Q remaining in the water layer is then 36.4 mg.

For the second extraction (from the water layer, now containing a total of 36.4 mg of Q),

$$K_{dist} = \frac{\text{weight } Q_{ether}/150\ \mu L}{(36.4 - \text{weight } Q_{ether})/300\ \mu L}$$

Solving for the weight of Q_{ether}, one obtains the value of 23.2 mg, and the weight of Q_{water} is 13.2 mg.

The two extractions, each of 150 μL of ether, removed a total of

$$63.6 \text{ mg} + 23.2 \text{ mg} = 86.8 \text{ mg}$$

of compound Q (86.8% of the total). The remainder of Q left in the water layer is then 100 mg − 86.8 mg or 13.2 mg (13.2% of the total).

Based on the above calculation, it is easily seen that the multiple extraction technique is the more efficient one. Whereas the single extraction removed 77.8% of Q, the double extraction increased this to 86.8%. To extend this relationship, multiple extractions with one third of the total quantity of the ether solvent in three portions would be even more efficient. You might wish to calculate this extension to prove the point. Of course, there is a practical limit to the number of extractions one would perform, based on time and degree of efficiency realized.

With many inorganic extractions, as in the use of an organic chelating reagent to remove a metal ion from an aqueous solution, a more definitive term to express the relationship is the distribution ratio, *D*. This ratio defines the concentration of all the species of the solute in the two phases. For example, let us examine the system using 8-hydroxyquinoline, OxH (Fig. 5.41), as a chelating agent to extract the Cu^{2+} ion from an aqueous solution.

We must realize that the reagent is a weak acid that ionizes in water; the metal ion displaces the proton on the hydroxyl group and the resulting chelate is a neutral species. The overall reaction may be depicted as follows:

$$2OxH + Cu^{2+} = (Ox)_2Cu + 2H^+$$

There are actually four equilibria involved in this extraction process.

Step 1 The chelating reactant, OxH, distributes between the aqueous (aq) and the organic (org) phases.

$$K_{dist} = \frac{[OxH]_{org}}{[OxH]_{aq}}$$

Step 2 The OxH, a weak acid, ionizes in the aqueous phase.

$$OxH = Ox^- + H^+$$

$$K_a = \frac{[H^+][Ox^-]}{[OxH]}$$

Step 3 The metal ion, Cu^{2+}, reacts with the anion, Ox^-, to form the neutral chelate.

$$Cu^{2+} + 2Ox^- = (Ox)_2Cu$$

$$K_{chelation} = \frac{[Ox_2\,Cu]}{[Cu^{2+}][Ox^-]^2}$$

Step 4 The neutral chelate distributes between the organic and aqueous phases.

$$[(Ox)_2\,Cu]_{aq} = [(Ox)_2\,Cu]_{org}$$

$$K_{dist} = \frac{[Ox_2\,Cu]_{org}}{[Ox_2\,Cu]_{aq}}$$

We can derive the relationship between the distribution ratio, D, and the distribution coefficient, K_{dist}, in the following manner: We will assume that the neutral chelate, $(Ox)_2Cu$, is more soluble in the organic phase and that it is largely undissociated. If so, the expression for D is

$$D = \frac{[Ox_2\,Cu]_{org}}{[Cu^{2+}]_{aq}}$$

Substitution of the correct expressions from Steps 1–4 to include all the K values leads to

$$D = \frac{K_{dist}(Ox_2\,Cu)K_{chelate}K_a}{K_{dist}(OxH)}\left(\frac{[OxH]_{org}}{[H^+]_{aq}}\right)$$

Several important aspects of this equation are apparent.

1. The distribution ratio, D, is independent of the Cu^{2+} ion concentration if the chelate, $(Ox)_2Cu$, is completely soluble in the organic phase.
2. The extraction is pH dependent. An increase in the hydrogen ion concentration lowers the extraction efficiency.
3. The extraction efficiency is directly dependant on the chelate reagent OxH concentration. An increase in this concentration increases the efficiency of the extraction.
4. In general, a stable chelate formation (large $K_{chelate}$) and a higher K_a value for the chelate reagent increases the extraction efficiency. The drawback is that chelate stability usually decreases as the chelating reagent becomes more acidic. Therefore, these relationships must be considered in concert when evaluating a chelating reagent.
5. As a general rule, the nature of the organic solvent does not seriously affect the outcome of the extraction process.

5.1.3 Extraction Procedures: Simple Extraction

Use of the Separatory Funnel

The separatory funnel (Fig. 5.42) is an effective device for extractions carried out at the semimicro- and the macroscale level. Many of you are familiar with this device from your work in the General or Organic Chemistry laboratory. The mixing and separation are done in the funnel itself, in one operation. It is important to note that the separatory funnel is not an effective device for extractions at the microscale level, because of the small quantities involved.

The solution to be extracted is added to the funnel, making sure the stopcock is in the closed position. The funnel is generally supported in an iron ring attached to a ring stand. The proper amount of extracting solvent is now added (~$\frac{1}{3}$ of the volume of the solution to be extracted is a good rule to follow) and the stopper placed on the funnel.

NOTE: *The size of the funnel should be such that the total volume of solution is less than three fourths of* the total volume *of the funnel. The stopcock and*

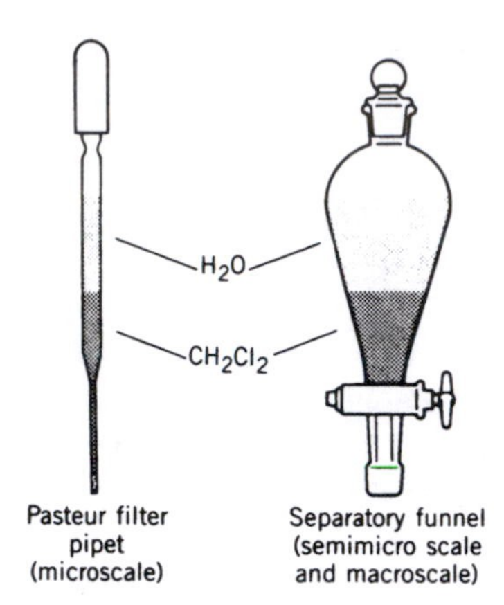

Figure 5.42. *Devices used in extraction procedures.* From *Mayo, D. W.; Pike, R. M.; Butcher, S. S.,* Microscale Organic Laboratory, *2nd ed., Wiley: New York, 1989. (Reprinted by permission of John Wiley & Sons, New York.)*

stopper must be lightly greased to prevent sticking, leaking, or freezing. If Teflon stoppers and stopcocks are used, greasing is not necessary since they are self-lubricating.

The funnel is removed from the ring stand, the stopper rested against the index finger of one hand, and the funnel held in the other with the fingers positioned so as to operate the stopcock (Fig. 5.43*a*). The funnel is shaken (Fig. 5.43*b*), carefully inverted, and the liquid is then allowed to drain away from the stopcock. The stopcock should be cautiously opened to release any pressure buildup (Fig. 5.43*c*).

NOTE: *Make sure the stem of the funnel is up and that it does not point at anyone.*

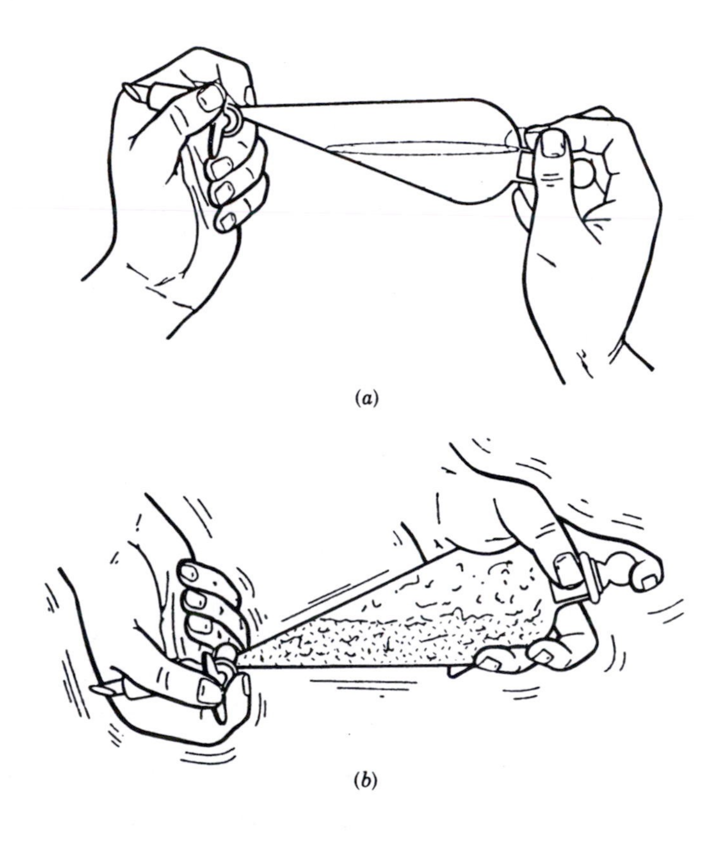

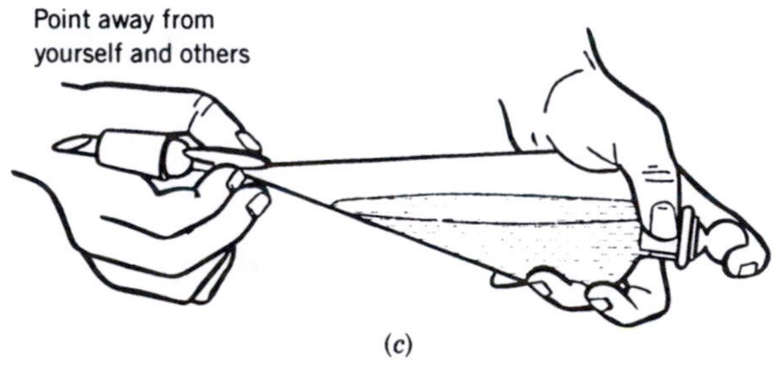

Figure 5.43. *Use of a separatory funnel.* (a) *Efficient method of holding a separatory funnel.* (b) *Shaking separatory funnel.* (c) *Venting separatory funnel.* (From Modern Experimental Organic Chemistry, *4th ed., by Roberts, R. M.; Gilbert, J. C.; Rodewald, L. B.; Wingrove, A. S., copyright © 1985 by Saunders College Publishing, a division of Holt, Rinehart and Winston, Inc., reprinted by permission of the publisher.)*

The stopcock is closed, the funnel shaken for several seconds, the funnel positioned for venting, and the stopcock opened to release pressure buildup. This process is repeated several times. After the final sequence, the stopcock is closed and the funnel returned upright to the iron ring.

The layers are allowed to separate, the stopper removed, the stopcock opened gradually, and the bottom layer is then drained into a suitable container (the upper layer can be removed by pouring from the top of the funnel).

When aqueous solutions are extracted with a less dense solvent, such as ether, the bottom aqueous layer is drained into the original container from which it was poured into the funnel. Once the top ether layer is removed from the funnel, the aqueous layer can then be returned for further extraction. In order to minimize losses, it is wise to rinse this original container with a small portion of the extracting solvent, which is then added to the funnel. When the extracting solvent is denser than the aqueous phase, for example, methylene chloride, the aqueous phase is the top layer and is therefore retained in the funnel for subsequent extractions.

Pasteur Filter Pipet Extractions

At the microscale level the Pasteur filter pipet (Fig. 5.42) replaces the separatory funnel. In this situation, the extraction process consists of two operations: (1) mixing of the organic and aqueous layers and (2) separation of the two layers after the mixing process.

1. **Mixing.** The organic solvent or solution is added to a conical vial or centrifuge tube containing the aqueous phase. The vial or tube is capped and shaken to thoroughly mix the two phases. (An alternative is to use a Vortex mixer.) The mixing operation should be interrupted at periodic intervals so that the cap may be loosened to release any pressure that may develop.

   After the mixing operation, the container is placed in a beaker and the two phases allowed to separate. A sharp boundary should be evident.
2. **Separation.** At the microscale level the two phases are separated with a Pasteur filter pipet (in some cases a Pasteur pipet can be used), which acts as a miniature separatory funnel. The separation of the phases is shown in Figure 5.44.

   Multiple extractions are performed to ensure complete separation. The following steps outline the general method (refer to Fig. 5.44).

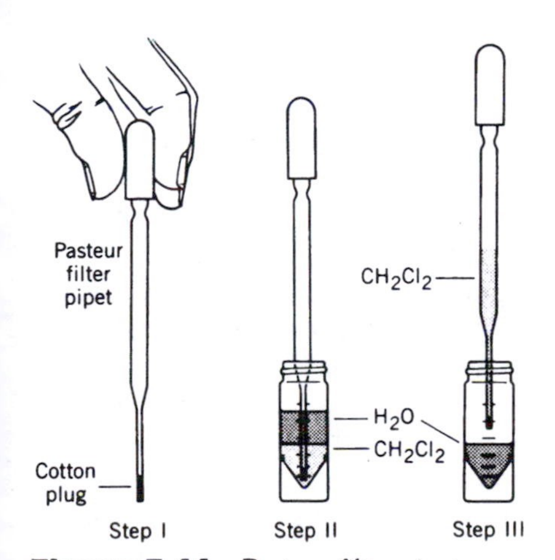

Figure 5.44. *Pasteur filter pipet separation of two immiscible liquid phases. The lower methylene chloride layer contains the product. From Mayo, D. W.; Pike, R. M.; Butcher, S. S.,* Microscale Organic Laboratory, *2nd ed., Wiley: New York, 1989. (Reprinted by permission of John Wiley & Sons, New York.)*

1. The pipet bulb is squeezed to force air from the pipet and then inserted into the mixture to a position close to the bottom of the container. Remember to hold the pipet in a vertical position.
2. Carefully allow the bulb to expand, drawing only the lower methylene chloride layer into the pipet. This is done in a smooth, steady manner, so as not to disturb the boundary between the layers. With practice, one can judge the amount that the bulb must be squeezed so as to just separate the layers. An effective device to use in this operation is a "pipet pump" of the commercially available type.[3]
3. Hold the pipet in a vertical position, place it over an empty vial or other container, and gently squeeze the bulb to transfer the methylene chloride solution into the vial. A second extraction can now be performed after addition of a second portion of methylene chloride to the original vial. The identical procedure is repeated. In this manner multiple extractions can be performed, with each methylene chloride extract being transferred to the same vial; that is, the extracts are combined.

In the case when a diethyl ether–water extraction is performed, the ether layer is less dense and thus is the top phase. The procedure followed to separate

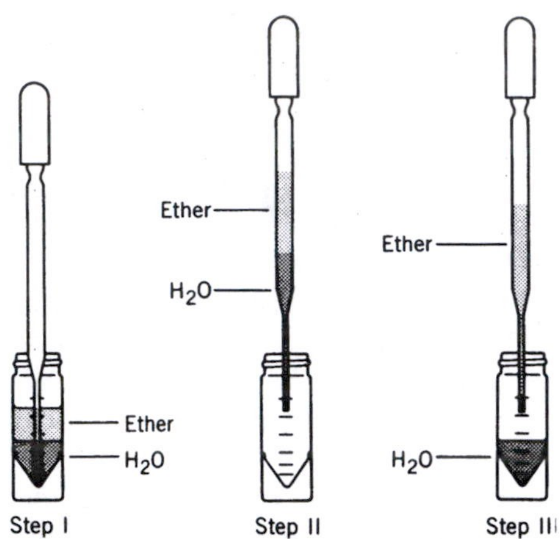

Figure 5.45. *Pasteur filter pipet separation of two immiscible liquid phases. The top ether layer contains the product.* From *Mayo, D. W.; Pike, R. M.; Butcher, S. S.,* Microscale Organic Laboratory, *2nd ed., Wiley: New York, 1989. (Reprinted by permission of John Wiley & Sons, New York.)*

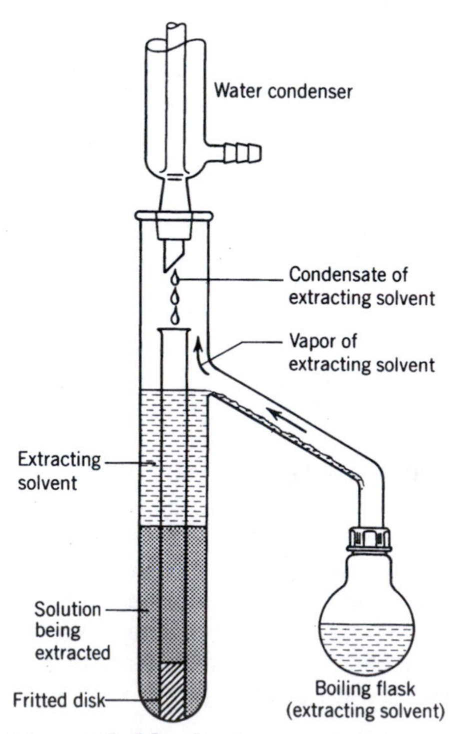

Figure 5.46. *Continuous extraction device using lighter than water extracting solvent.*

the water–ether phases is identical to that outlined above with the exception that it is the top layer that is transferred to the new container. The steps are shown in Figure 5.45.

1. Both phases are drawn into the pipet as outlined in the previous extraction (Steps 1 and 2). Try not to allow air to be sucked into the pipet since this will tend to mix the phases. If mixing does occur, allow time for the boundary to reform.
2. Return the bottom aqueous layer to the original container by gently squeezing the pipet bulb.
3. The separated ether layer is then transferred to a new vial.

5.1.4 Continuous Extraction

Continuous extraction of liquid–liquid systems is also possible. Figure 5.46 illustrates a type of apparatus often used to extract various species from aqueous solutions using less dense, immiscible organic solvents such as diethyl ether.

The important aspect of this approach is that the extraction can be carried out with a limited amount of solvent. Furthermore, the number of individual extractions that would have to be performed to accomplish the same task would be prohibitive. The actual extraction is carried out by allowing the condensate of the extracting solvent, as it forms in the condenser, to percolate through the solution containing the material to be extracted. The extracting solvent, containing a portion of the dissolved material to be extracted, is then returned to its original flask. In this manner the desired material eventually is collected in the boiling flask and is isolated by concentration of the collected solution.

Continuous extraction apparatus is also available that uses an arrangement that makes it possible to have the extraction solvent denser than the solution being extracted.

5.1.5 Drying of the Wet Organic Layer

It is important to realize that the organic extracts separated from aqueous phases are wet. Before evaporation of the solvent to isolate the desired species, or before further purification steps can be taken, the extracts must be dried to remove any residual water. This is conveniently achieved with an inorganic anhydrous salt such as magnesium, sodium, or calcium sulfate. These materials readily form insoluble hydrates, thus removing the water from the wet organic phase. The hydrated solid can then be removed from the dried solution by filtration or decantation. Table 5.4 summarizes the properties of some of the more common drying agents used in the laboratory.

There are two basic requirements for an effective solid drying agent: (a) it should not react with the material in the system and (b) it must be easily and completely separated from the dried liquid phase. The amount of drying agent used depends on the amount of water present and on the capacity of the solid desiccant to absorb water. If the solution is wet, the first amount of drying agent will clump (molecular sieves and t.h.e. SiO_2 are an exception). In this case, additional drying agent is added until the agent appears mobile on swirling the container. Swirling the contents of the container (by hand or by using a magnetic stirring apparatus) increases the rate of drying, since it aids in establishment of the equilibrium for hydration.

$$\text{drying agent} + n\text{H}_2\text{O} = \text{drying agent}\cdot\text{H}_2\text{O}$$

The drying agent may be added directly to the container containing the organic extract or the extract may be passed through a Pasteur filter pipet packed with the drying agent. A funnel fitted with a cotton, glass wool, or polyester plug to hold the drying agent may also be used.

Table 5.4 Properties of Common Drying Agents

Drying Agent	Formula of Hydrate	Comments
Sodium sulfate	$Na_2SO_4 \cdot 10H_2O$	Slow in absorbing water and is inefficient. Loses water above 32.4°C
Magnesium sulfate	$MgSO_4 \cdot 7H_2O$	One of the best. Can be used with nearly all organic solvents
Calcium chloride	$CaCl_2 \cdot 6H_2O$	Relatively fast drying agent. However, reacts with many oxygen and nitrogen containing compounds
Calcium sulfate	$CaSO_4 \cdot \frac{1}{2}H_2O$	Very fast and efficient. However, notice that it has a low dehydration capacity
Silica gel	$(SiO_2)_n$	High capacity and efficient. Commercially available t.h.e. SiO_2 drying agent is excellent.[a]
Molecular sieves	$[Na_{12}(Al_{12}Si_{12}O_{48})] \cdot 27H_2O$	High capacity and efficient. Use the 4-Å size[b]

[a] Available from EM Science, Cherry Hill, NJ 08034–0395.
[b] Available from Aldrich Chemical Co., Inc, 940 West Saint Paul Ave., Milwaukee, WI 53233.

5.1.6 Drying of Organic Solvents

While it is never desirable to use wet solvents in a chemical reaction, some reactions are more likely to fail because of the presence of water than others. The most familiar case is with the Grignard reaction, where the glassware and solvent [tetrahydrofuran (THF) or ether] must be bone-dry in order to ensure success. It is therefore imperative, in many instances, to dry organic solvents prior to their use.

There are two common procedures for drying organic solvents. The first involves the distillation of the wet solvent. In the case of toluene, for example, distillation results in an azeotrope being initially formed, with the first portion of the distillate consisting of a mixture of toluene and water. This azeotrope will distill at a temperature below that of toluene itself. When all of the water has distilled in this manner, pure toluene will remain in the still pot. Experiment 4 employs a modification of this technique to remove water produced in the course of a reaction, which could decompose the desired product. Other solvents that can be dried in this manner include benzene, xylene, and carbon tetrachloride.[4]

The second method involves the distillation of the wet organic solvent from a drying agent. Ethers are notorious for absorbing and containing large quantities of water. Unless a fresh bottle of solvent is used, one runs a large risk of experimental failure with these solvents. Various drying agents may be used for this purpose, of which two will be mentioned here. If a quick drying is desired, the ether solvent may be distilled from sodium borohydride. For a more efficient drying, ethers may be distilled from finely divided sodium metal (sodium ribbon from a sodium press is best). Benzophenone is commonly added to such mixtures as a dryness indicator. When bone-dry, the solution is blue to black in color. Increasing wetness changes the color of the solution to green and (for very wet solutions) brown.

The dry ethers prepared in this manner must be kept under an atmosphere of dry N_2 (or dry air), as they will quickly and readily reabsorb water.

REFERENCES

1. For a general overview of the process see:
   (a) Day, R. A., Jr.; Underwood, A. L., *Quantitative Analysis*, 5th ed., Prentice–Hall: Englewood Cliffs, NJ, 1986.
   (b) Skoog, D. A.; West, D. M., *Analytical Chemistry: An Introduction*, 4th ed., Saunders: Philadelphia, 1986.
2. For further examples of calculations on the extraction of metal ions using chelating reagents see:

(a) Kennedy, J. H., *Analytical Chemistry: Principles*, Harcourt Brace Jovanovich: New York, 1984.
(b) Christian, G. D., *Analytical Chemistry*, 4th ed., Wiley: New York, 1986.
3. These pipet pumps may be obtained from Bel-Art Products, Pequannock, NJ 07440-1992.
4. Adams, R.; Johnson, J. R.; Wilcox, C. F., *Laboratory Experiments in Organic Chemistry*, 6th ed., Macmillan: New York, 1970.

5.J CONDUCTIVITY MEASUREMENTS

In ionic compounds of the main group elements, it is usually a trivial matter to deduce the number of ions per mole present in infinitely dilute solution. The ionic compounds are viewed as dissociating completely in the dilute solution (although as the concentration of solute rises, the degree of ionization changes drastically), and thus $Ca(NO_3)_2$ would be expected to consist of three ions: one Ca^{2+} ion and two nitrate (NO_3^-) ions.

In transition metal complexes, the situation is not nearly as simple. A given anion may be a part of the complex (in which case it generally does not dissociate) or it may be present as a counterion (in which case it does). Werner, in 1912, investigated the octahedral complex $Co(en)_2Cl_3$ (en = ethylenediamine) which could have three different potential ligand arrangements in aqueous solution:

$[Co(en)_2(H_2O)_2]Cl_3$	4 ions
$[Co(en)_2(H_2O)Cl]Cl_2$	3 ions
$[Co(en)_2Cl_2]Cl$	2 ions

There is no way of knowing in advance which of the above formulas is correct. (The situation is complicated further in that optical activity is also possible in the above case.) The number of ions constituting the complex is best determined by measuring the conductivity of the solution of that compound. This conductivity measurement allows one to tell how many ions (cations and anions) are present in solution when an ionic product is dissolved in water.

Those ionic compounds that are soluble in water and conduct electric current in aqueous solution are called **electrolytes**. The dissolution process consists of complete dissociation of ionic compounds into mobile cations and anions. There are many compounds which, though soluble in water, do not exhibit any conductivity. These are termed **nonelectrolytes**. There is still another group of compounds that exhibit conductance in solutions only when that solution is quite dilute. Such compounds are known as **weak electrolytes**. Solutions that contain large numbers of mobile ions (cations and anions from the soluble ionic compounds) conduct current well, and solutions that contain only a few ions (acetic acid) or relatively immobile ions show poor conductivity.

The conductivity of a solution varies with the number, size, and charge of the ions constituting the solution.[1] The viscosity of a solution also affects the conductivity, by affecting the mobility of the ions. Ions of different species in solution will therefore show different conductivities. If, by means of a chemical reaction, we replace one ionic species by another having a different size and/or charge, we would observe a corresponding change in conductivity of the resulting solution.

The conductivity, L, of a solution is represented by the equation

$$L = Bc_i\alpha_iZ_i$$

where B is a constant that depends on the size and the geometry of the conductance cell, c_i is the concentration of individual ions in solution, α_i is the equivalent ionic conductance of individual ions, and Z_i is the charge of the ions.

In practice, although the conductance of a solution is more useful in dealing with electrolyte solutions, it is the resistance of a solution that is experimentally measured. The conductance is calculated from the resistance. The resistance of a solution is determined by inserting two electrodes into a solution. The resistance, R, is proportional to the distance, d, between the two electrodes and inversely proportional to the cross-sectional area, A, of the solution enclosed between the electrodes.

$$R = \rho d/A$$

The term ρ is called the **specific resistance** or more simply, the **resistivity**. The ratio d/A is usually referred to as the **cell constant**, K. Thus, the above relation becomes

$$R = K\rho$$

The **conductance**, L, of a solution is defined as the reciprocal of the resistance, and the **specific conductivity**, k, is defined as the reciprocal of the specific resistance.

$$k = 1/\rho = (d/A)(1/R) = KL$$

In practice, the cell constant, K, is determined for any cell by measuring the conductivity of a 0.0200M KCl solution at 25 °C, for which the specific conductivity, k, is 0.002768 Ω^{-1}.

The total conductivity of a solution arises from several sources, the largest of which is the ions. The self-ionization of a solvent contributes as well, but in practice is small enough to be neglected in all but the most careful measurements.

A very useful quantity is the **equivalent conductivity**, Λ. It is defined as the value of the specific conductivity, k, contributed by one equivalent of ions of either charge. More specifically, it is defined as the conductance of a solution containing one gram-equivalent of an electrolyte placed between electrodes separated by a distance of 1 cm. If c is the concentration of the solution in gram-equivalents per liter, the volume of the solution in cubic centimeters per equivalent (cm^3/equiv) is equal to $1000/c$. The equivalent conductance, Λ, is then given by

$$\Lambda = \frac{1000\ k}{c}$$

Substituting for k,

$$\Lambda = \frac{1000\ LK}{c}$$

Another frequently used quantity in conductance measurements is the **molar conductance**, Λ_m, defined as the conductance of a one cubic centimeter volume of solution that contains one mole (or formula weight) of the electrolyte. If M is the concentration of the solution in moles per liter, then the volume in cubic centimeters per mole is $1000/M$. The molar conductance is then given by

$$\Lambda_m = \frac{1000\ k}{M}$$

By comparing the molar conductance measured for a particular compound with that of a known ionic compound, we can estimate the number of ions produced in a solution. A range of values of molar conductances for 2–5 ions at 25 °C in water is given below.[2,3]

Number of Ions	Molar Conductances (cm^{-1} mol^{-1} Ω^{-1})
2	118–131
3	235–273
4	408–435
5	~560

As an example, some experimental values of molar conductance obtained for a series of platinum(IV) complexes are given below.

Complexes	Molar Conductance (cm^{-1} mol^{-1} Ω^{-1})	Number of Ions
$[Pt(NH_3)_6]Cl_4$	523	5
$[Pt(NH_3)_5Cl]Cl_3$	404	4
$[Pt(NH_3)_4Cl_2]Cl_2$	229	3
$[Pt(NH_3)_3Cl_3]Cl$	97	2
$[Pt(NH_3)_2Cl_4]$	0	0
$K[Pt(NH_3)Cl_5]$	109	2
$K_2[PtCl_6]$	256	3

The equivalent conductivity increases with increasing dilution due to the lessened interionic forces between ions (less ion pairing is the classical way of stating this). The conductivity for strong electrolytes at low concentration is linearly related to dilution according to the equation

$$\Lambda = \Lambda_0 - B\sqrt{c}$$

where Λ_0 is the **limiting equivalent conductance** (the equivalent conductance at infinite dilution) and B is a constant. The term Λ_0 may be obtained by extrapolating a plot of Λ (y axis) versus $\sqrt{c}$ (x axis) to zero concentration. For weak electrolytes, Λ_0 may be obtained from the equation

$$\Lambda_0 = \alpha_0^+ + \alpha_0^-$$

where α_0 is the equivalent conductance for cations and anions at infinite dilution. Tables of values of α_0 may be found in any physical chemistry textbook.

5.J.2 Experimental Procedure

Conductivity measurements require the use of two instruments: the conductivity cell and the conductivity meter or bridge. The cell constant, K, is first determined by measuring the conductivity, L, of an accurately prepared 0.0200*M* KCl solution for which the value of specific conductivity, k, is known to be 0.002768 Ω^{-1}

$$K = k/L$$

The molar conductivity of any compound is determined in the following way.

Table 5.5 Molar Conductances for Nonaqueous Solutions[a]

Solvent	Dielectric Constant	Two Ions	Three Ions	Four Ions	Five Ions
Nitromethane	35.9	75–95	150–180	220–260	290–330
Nitrobenzene	34.8	20–30	50–60	70–80	90–100
Acetone	20.7	100–140	160–200		
Acetonitrile	36.2	120–160	220–300	340–420	
N,N-Dimethylformamide	36.7	65–90	130–170	200–240	
Methanol	32.6	80–115	160–220		
Ethanol	24.3	35–45	70–90		

[a] Conductance units: $cm^{-1}mol^{-1}\Omega^{-1}$.

1. A 1×10^{-3} *M* solution of the compound of interest is prepared.
2. The conductivity is measured in the same cell for which the cell constant has been previously determined, as described previously.

NOTE: *Since many complexes undergo significant decomposition in solution over time, it is advisable to measure the conductance immediately after preparing the solution.*

3. The cell should be rinsed before each measurement.

5.J.3 Nonaqueous Solutions

Molar conductances of compounds can also be determined in solvents other than water. Obviously, the degree of dissociation of any electrolytes in a solvent other than water will be different from that in water. Molar conductances at 25 °C for electrolytes undergoing dissociations producing 2–5 total ions in other solvents are given in Table 5.5.

REFERENCES

1. Adamson, A. W., *Physical Chemistry*, 2nd ed., Academic Press: New York, 1979.
2. Geary, W. J. *Coord. Chem. Rev.*, **1971**, 7, 81.
3. Angelici, R. J., *Synthesis and Technique in Inorganic Chemistry*, 2nd ed., Saunders: Philadelphia, 1977.

Chapter 6
Spectroscopy

6.A INTRODUCTION

In an atom, molecule, or ion, there are various energy levels corresponding to various physical changes that may be undergone by the particle. Generally, the lower energy state (or states) is occupied and the higher energy state (or states) is unoccupied. Suppose that the difference in energy between the highest occupied state and the lowest unoccupied state is ΔE. By irradiating the particle with energy corresponding to ΔE, a transition between these states may be induced.

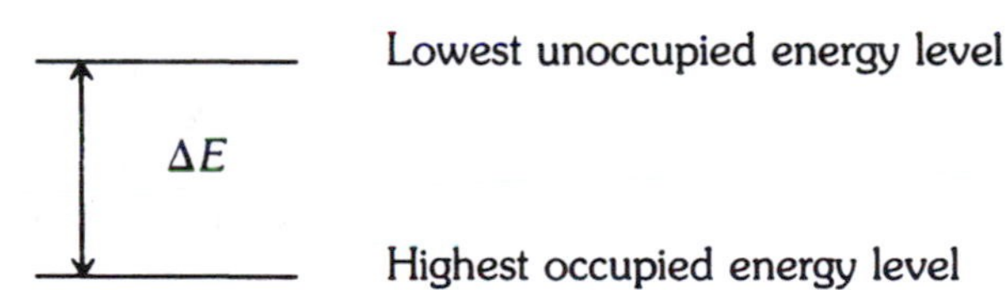

The energy of the radiation is related to the frequency of the radiation via the equation

$$\Delta E = h\nu$$

where h is Planck's constant, 6.626×10^{-34} J·s, and ν is the frequency in hertz.

Frequency is related to wavelength in the following way:

$$\nu = c/\lambda$$

where c is the speed of light, 2.998×10^{8} m·s^{-1}. Thus,

$$\Delta E = h\nu = hc/\lambda = hc\omega$$

where ω is the wavenumber, defined as the reciprocal of the wavelength.

The frequency range for the transition will depend on the energy difference. Some of these ranges are listed in Table 6.1. Both IR and Visible spectroscopy

Table 6.1 Spectroscopy Frequency Ranges

Region	Wavelength (m)	Energy	Change Excited
Gamma (γ) ray	$<10^{-10}$	$>10^6$ kJ·mol^{-1}	Nuclear transformation
X-ray	10^{-8}–10^{-10}	10^4–10^6 kJ·mol^{-1}	Inner-shell electron transitions
Ultraviolet (UV)	4×10^{-7}–10^{-8}	10^3–10^4 kJ·mol^{-1}	Valence shell electrons
Visible	8×10^{-7}–4×10^{-7}	10^2–10^3 kJ·mol^{-1}	Electronic transitions
Infrared (IR)	10^{-4}–2.5×10^{-6}	1–50 kJ·mol^{-1}	Bond vibrations
Microwave	10^{-2}–10^{-4}	10–1000 J·mol^{-1}	Molecular rotations
ESR	10^{-2}	10 J·mol^{-1}	Electron spin reversals
NMR	10	0.01 J·mol^{-1}	Nuclear spin reversals

are types of absorption spectroscopy. The major difference between the two types is the frequencies at which they operate, and the consequent energies involved.

The energy of an atom or molecule depends on which orbitals in the system are occupied by electrons. The separation of these electronic energy levels is large, on the order of tens of kilojoules per mole. Visible spectroscopy operates at frequencies corresponding to the energy of electronic transitions from the molecular ground state to excited states, usually from one *d* orbital to another in the case of transition metals.

Molecules have other kinds of energy as well. The bond length of a molecule is not constant. It undergoes vibration similar to the situation of two weights connected by a spring. Only certain vibrational energies are allowed. The separation of these vibrational levels is on the order of a few kilojoules per mole. This gives rise to IR spectroscopy.

6.A.1 Spectrometer Components

There are several components that are common to most types of instruments used to obtain Visible or IR spectra. First, a source of radiant energy is necessary. In the case of IR spectroscopy, the source is usually a silicon carbide rod heated to approximately 1200 °C (called a Globar) or a zirconium oxide–yttrium oxide rod (called a Nernst glower) heated to 1500 °C or a nichrome wire. In the case of Visible spectroscopy, the source is a tungsten filament lamp. In these instruments, the source emits continuous radiation over a fairly wide range of wavelengths. In general, however, a narrower bandwidth is desired to obtain greater resolution and sensitivity.

The narrow bandwidth is achieved through the use of a monochromator. The polychromatic radiation emitted by the source is resolved into its individual wavelengths. In simplest terms, the monochromator consists of an entrance slit to allow the source radiation in, a collimating lens or mirror, a prism or grating, a focusing lens, and an exit slit. A simple prism monochromator is shown in Figure 6.1. All parts of the monochromator must be transparent in the spectral region of interest, so as not to interfere with the desired signal.

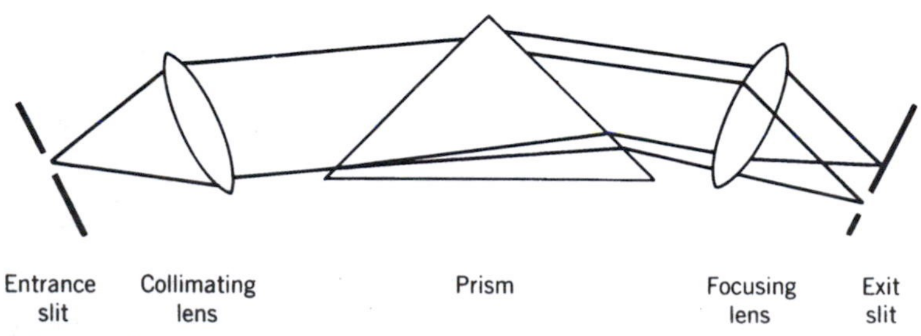

Figure 6.1. *A prism monochromator.*

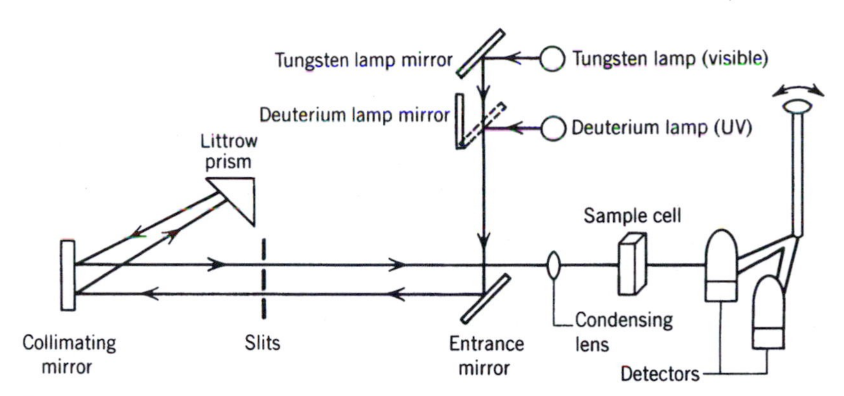

Figure 6.2. *UV-Visible single-beam spectrometer.*

The radiation, which is now monochromatic, then strikes the absorption cell, and the material within it undergoes a transition to a different energy state. When the material relaxes back to the ground state, it releases energy, which is measured by a detecting device, and the resulting signal is then displayed by an oscilloscope or recorder. A schematic diagram of a single-beam Visible spectrophotometer is shown in Figure 6.2.

There are two basic types of instruments: single or double beam. In a double-beam instrument, a beam splitter splits the source radiation prior to the absorption cells. One beam goes to the sample and another goes to a blank or reference cell. The two beams are continuously compared and deviations in the source due to whatever cause can be compensated for automatically. Given the high stability of most sources, a double-beam instrument (which is more expensive) is not necessary for many purposes.

6.B VISIBLE SPECTROSCOPY

6.B.1 Introduction and Theory

The Visible spectrophotometer is the most commonly used instrument in academic laboratories and the third most commonly used in industry.[1] One of the more appealing aspects of inorganic chemistry (transition metal chemistry in particular) is that many compounds have absorptions in the Visible region, and therefore appear colored. The Visible range of the electromagnetic spectrum is depicted in Figure 6.3.

Visible absorptions usually involve electronic transitions between *d* or *f* electron energy levels, although charge-transfer bands (see Experiment 28, VO_2^+) also frequently occur in the Visible region. The simplest example of the *d–d* type of transition is for a species with a single *d* electron, such as $[Ti(H_2O)_6]^{3+}$. The *d* orbitals in octahedral complexes are divided into two energy groups, shown in Figure 6.4. Other geometries split the *d* orbitals in a different manner.

In a vacuum, the five *d* orbitals are all of equal energy—that is, they are said to be degenerate. This is only true, however, in the free ion. Thus, in Ti^{3+}, for

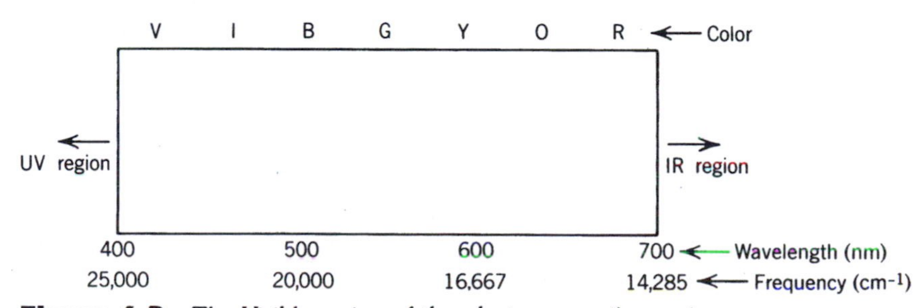

Figure 6.3. *The Visible region of the electromagnetic spectrum.*

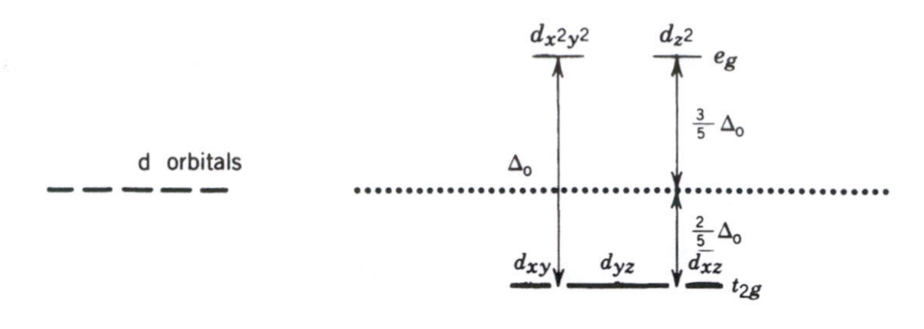

Figure 6.4. *Octahedral crystal field splitting.*

example, the d orbitals are all degenerate, but in $[Ti(H_2O)_6]^{3+}$, they are not. The d_{xy}, d_{yz}, and d_{xz} orbitals have their lobes directed between the Cartesian axes, and the $d_{x^2-y^2}$ orbitals have their lobes directed on the Cartesian axes.

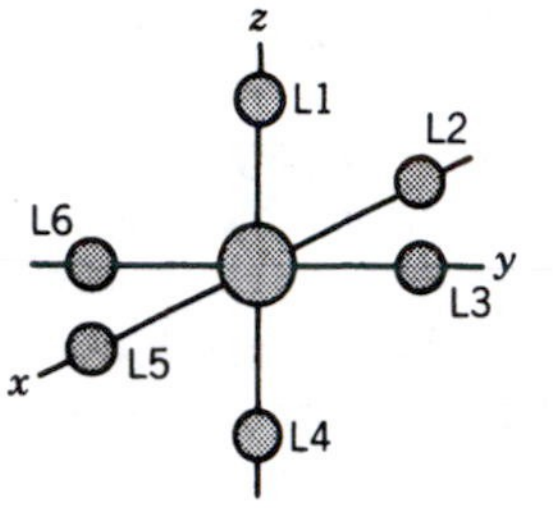

In an octahedral molecule or ion of formula ML_6, the metal can be viewed as being positively charged, and the ligands as being negatively charged. The structure of this arrangement is shown in Figure 6.5. The d orbitals contain electrons, and are thereby negatively charged. If any orbital lobe is directed at a negatively charged ligand, there will be a repulsion between them. This repulsion destabilizes these orbitals, which consequently rise in energy. The $d_{x^2-y^2}$ and d_{z^2} orbitals fall into this category. The d_{xy}, d_{yz}, and d_{xz} orbitals are not directed at the ligands, and there is little repulsion. These orbitals are therefore more stable than the $d_{x^2-y^2}$ and d_{z^2} orbitals. In fact, they are stabilized relative to the d orbitals of the free ion.

Figure 6.5. *Orientation of ligands in an octahedral complex.*

The d_{z^2} orbital lobes are pointed directly at ligands 1 and 4. The $d_{x^2-y^2}$ orbital lobes are pointed directly at ligands 2,3,5, and 6. It can be shown that the two orbitals are equally destablized, and that the d_{xy}, d_{yz}, and d_{xz} orbitals are equally stabilized. The orbital energies are thereby split in the manner shown in Figure 6.4. Since the total energy of the five orbitals must be conserved, the relative stabilizations can be calculated. The t_{2g} orbitals must be stabilized (call this x) by the same amount that the e_g orbitals are destabilized (call this y). Thus,

$$3x = 2y$$

and since

$$x + y = \Delta_o$$

the two equations may be solved to obtain $x = \frac{2}{5}\Delta_o$ and $y = \frac{3}{5}\Delta_o$. The t_{2g} orbitals are therefore stabilized by $\frac{2}{5}$ the value of Δ_o and the e_g orbitals are destabilized by $\frac{3}{5}$ the value of Δ_o.

In the octahedral case, the lower energy level is known as the t_{2g} level, and the upper level is the e_g level. The energy difference between these levels is given the symbol Δ_o, the o standing for octahedral. Since the titanium complex has only one electron, it will be found in the lower energy orbital, the t_{2g} level. Common terms used in designations of energy states are listed in Table 6.2.

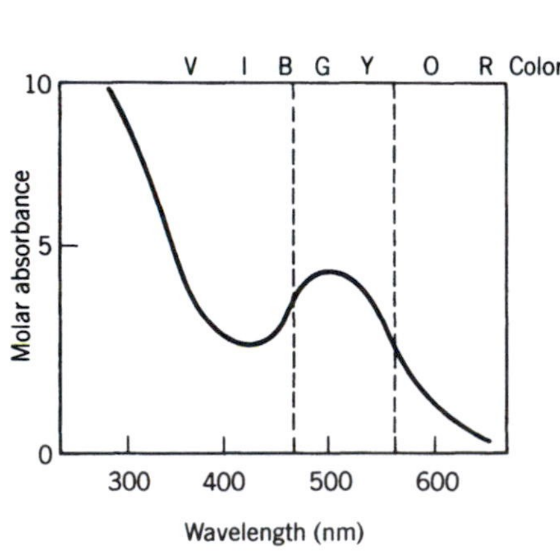

Figure 6.6. *Visible spectrum of $[Ti(H_2O)_6]^{3+}$.*

6.B.2 The Visible Spectrum

If energy at the appropriate frequency (corresponding to energy Δ_o) is applied from the source of a Visible spectrophotometer, the single electron of $[Ti(H_2O)_6]^{3+}$ will undergo a transition from the lower (t_{2g}) energy level to the higher (e_g) energy level. The Visible spectrum of $[Ti(H_2O)_6]^{3+}$ is shown in Figure 6.6.

Table 6.2 Energy State Terms

a and *b*	Singly degenerate orbital
e	Doubly degenerate orbital
t	Triply degenerate orbital
g	*gerade*, phase of the orbital symmetric with respect to inversion
u	*ungerade*, phase of the orbital not symmetric with respect to inversion

The maximum absorbance occurs at about 510 nm. This is in the green-yellow region of the Visible spectrum. Keep in mind that this is the color absorbed. The color of a compound depends on the color transmitted, which is the complement of the absorbed color. Thus, the complex appears to be red-violet in color. Since the absorption wavelength depends on ΔE between the t_{2g} and e_g levels, Δ_o for this complex can be calculated as follows:

$$\Delta_o = h\nu = hc/\lambda$$

$$\Delta_o = (6.626 \times 10^{-34}\ \text{J·s})(2.998 \times 10^{8}\ \text{m·s}^{-1})/(510 \times 10^{-9}\ \text{m})$$

$$\Delta_o = 3.9 \times 10^{-19}\ \text{J·molecule}^{-1} = 230\ \text{kJ·mol}^{-1}$$

This falls well within the range 100–1000 kJ found in Table 6.1 for Visible absorptions.

6.B.3 Molar Absorbance and Color

The strength of the molar absorbance in a given material depends on whether the quantum mechanical selection rules for electron transitions are obeyed. There are two selection rules that are of interest here.

1. If a molecule's geometry has a center of symmetry, transitions from one centrosymmetric orbital to another are forbidden (symmetry forbidden rule).
2. If the number of unpaired electrons in a molecule changes upon electronic transition, the transition is forbidden (spin forbidden rule).

If neither selection rule is violated, the electronic transition is allowed. If one rule is violated, the transition is forbidden and if both rules are violated, the transition is doubly forbidden.

The octahedral geometry has a center of symmetry. Transitions between two *d* orbitals are therefore forbidden by the symmetry rule. The only reason that such transitions can be seen at all is that the vibrations undergone by the octahedron cause deviations from perfect centrosymmetry. Such singly forbidden transitions have weak molar absorbances, on the order of 1–10. The molar absorbance of the band in Figure 6.6 is ~5. The colors arising from such *d–d* transitions are of the "pastel" variety. Since tetrahedral geometry does not have a center of symmetry, tetrahedral *d–d* transitions are much stronger than octahedral ones, and give rise to deeper color.

Allowed electronic transitions have intensities of 1000–100,000. The most common example of this type of transition is in the case of charge-transfer transitions. Here the electron moves from an orbital centered mainly on the metal atom to one on the ligand (or vice versa). The energy for this type of transition is usually higher than for a *d–d* transition and therefore occurs at higher frequency. When such transitions occur in the Visible region, they are usually at the extreme violet end, and give rise to extremely deep color.

Octahedral manganese(II) complexes provide examples of doubly forbidden *d–d* transitions. There are five unpaired electrons in the ground state for most manganese(II) complexes (all t_{2g} and e_g orbitals singly occupied). There is no way of rearranging the electrons in the *d* orbitals without pairing up at least two of them. The number of unpaired electrons must therefore change during a transition, and the *d–d* transition is therefore doubly forbidden. Doubly forbidden transitions give rise to very weak color, with manganese(II) being very pale pink. Tetrahedral manganese(II) complexes give rise to somewhat stronger color, as the transitions are only spin forbidden, and are usually yellow-green.

6.B.4 Size of the Crystal Field Splitting

The size of the crystal field splitting, Δ_o, depends on several factors, which include the type of metal, the type of ligand, the charge on the metal, and the geometry of the complex. Several trends will be summarized here.

1. The size of Δ_o increases from the first to the third transition metal series. Second-row metals have Δ_o values 25–50% larger than first-row metals. Third-row metals show similar increases relative to second-row metals.
2. With certain ligands, the size of Δ_o is larger than with other ligands. The ordering of ligands in terms of the size of Δ_o induced is termed the **spectrochemical series**. The order of some of the more common ligands is

$$Br^- < Cl^- < F^- < OH^- < H_2O < NH_3 < NO_2^- < CN^-$$

3. The size of Δ_o increases sharply with increasing charge on the central metal.
4. The geometry adopted by the complex has large effects on the crystal field splitting. In tetrahedral complexes, for example, the splitting, Δ_t, is only four ninths the size of the equivalent splitting in an octahedral complex.

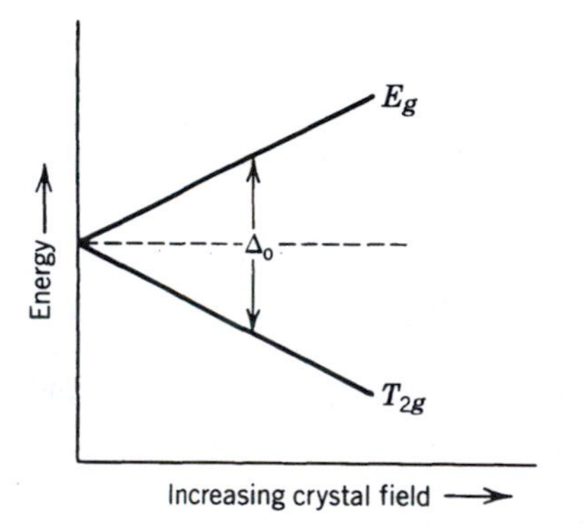

Figure 6.7. *Energy level diagram for* d^1 *octahedral complex.*

6.B.5 Energy Level Diagrams

As discussed earlier, the situation for a d^1 ion is quite simple, with the only transition possible being from the t_{2g} to the e_g orbital. The energy diagram for such a d^1 system is shown in Figure 6.7. With more than one electron in the *d* orbitals, the situation becomes much more complicated, as there are many possible ways that the electrons can occupy the orbitals. Even in the seemingly simple case of two *p* electrons, there are no fewer than 15 possible arrangements for the two electrons! The energy diagram for a d^2 ion is shown in Figure 6.8.

Each line on the energy level diagram represents the energy of a particular electronic arrangement of two *d* electrons within the five *d* orbitals. The lowest

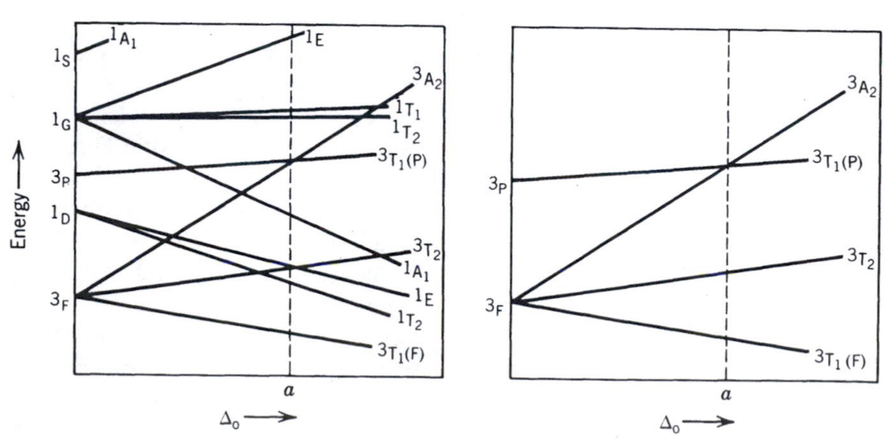

Figure 6.8. *Energy level diagrams for* d^2 *octahedral complex.*

energy line [labeled $^3T_1(F)$, read as triplet T-one-F] is termed the ground state. The superscript three for this energy state is the multiplicity of the state, and obtained by the equation[2]

$$\text{multiplicity} = n + 1$$

where n is the number of unpaired electrons. All transitions normally seen in the Visible spectrum occur from the ground state to an excited state.

Recalling the selection rules, transitions from the ground state to excited energy states are only allowed if the number of unpaired electrons does not change (e.g., the multiplicity remains constant). The energy levels for states of different multiplicity can usually be neglected, and the resulting energy level diagram simplifies considerably. The diagram on the left of Figure 6.8 shows all energy states, while the one on the right shows only the triplet states, to which transitions will not be spin forbidden.

At room temperature, only the ground state, $^3T_1(F)$, is occupied. Since there are two $^3T_1$ states, they must be distinguished from each other. The (F) notation refers to the $^3T_1$ state that arose from the splitting of the original F state. When the spectrophotometer scans the complex at different frequencies, transitions are induced from the ground state to excited states. Since the ground state is a triplet, the main transitions will be to other triplet states.

At values of Δ_o below a in Figure 6.8, the transitions would be

$$^3T_2 \leftarrow {}^3T_1(F)$$

$$^3A_2 \leftarrow {}^3T_1(F)$$

$$^3T_1(P) \leftarrow {}^3T_1(F)$$

At values of $\Delta_o > a$, the transitions would be

$$^3T_2 \leftarrow {}^3T_1(F)$$

$$^3T_1(P) \leftarrow {}^3T_1(F)$$

$$^3A_2 \leftarrow {}^3T_1(F)$$

At $\Delta_o = a$, only two transitions would be seen.

$$^3T_2 \leftarrow {}^3T_1(F)$$

$$^3T_1(P),\ {}^3A_2 \leftarrow {}^3T_1(F) \qquad \text{(degenerate)}$$

The frequency of an absorbance in the Visible spectrum may be used to calculate the energy difference between the ground state and an excited state. In this manner, we can determine "where in the energy diagram we are." Using the frequency of the transition ν, we can calculate ΔE. The energy level diagram is then searched for the location where a transition from the ground state to an excited state has the calculated value of ΔE. In practice, it can be even more difficult than this to assign the transitions, as the various energy levels can interact in different ways, leading to even more complicated energy diagrams and spectra.

REFERENCES

1. Pickral, G. M. *J. Chem. Educ.* **1983,** *60,* A 339.
2. Alternatively, the equation

$$\text{multiplicity} = 2S + 1$$

where $S = \sum m_s$ may be used.

General References

Sawyer, D. T.; Heineman, W. R.; Beeke, J. M., *Chemical Experiments for Instrumental Methods,* Wiley: New York, 1984, p. 163.

Skoog, D. A., *Principles of Instrumental Analysis,* 3rd ed., Saunders: Philadelphia, 1985, Chapter 7.

Weissberger, A., Ed., *Physical Methods of Organic Chemistry* Vol. II in *Techniques of Organic Chemistry,* Interscience: New York, 1946, Chapter 17.

Silverstein, R. M.; Bassler, C. C.; Morrill, T. C. *Spectrometric Identification of Organic Compounds,* 4th ed., Wiley: New York, 1981, Chapter 6.

Figgis, B. N., "Ligand Field Theory" in *Comprehensive Coordination Chemistry,* G. Wilkinson, Ed., Pergamon: Oxford, 1987, Vol. 1, Chapter 6, p. 213.

6.C INFRARED SPECTROSCOPY

6.C.1 Introduction

Infrared (IR) spectroscopy is the most frequently used instrumental technique in organic laboratory courses, and is extensively used in inorganic chemistry as well. The most common type of IR spectrometer is the double-beam dispersive instrument (although the FT IR spectrometer is rapidly gaining in popularity—see Section 6.C.8). A diagram of a typical IR spectrometer is shown in Figure 6.9.

A hot wire serves as the source of IR radiation, which is divided into two beams: a sample beam and a reference beam. The sample beam is reflected by mirrors so that it passes through the sample cell, while the reference beam bypasses the sample. The beams are recombined at the chopper, pass through a filter, and passed through the monochromator entrance slit. The function of the chopper is to alternately pass the sample beam or reflect the reference beam to the slit. The beam then enters the monochromator, where the various frequencies are reflected by the collimator. This renders the beam rays parallel, and they then strike the reflection grating and are dispersed by refraction. For one angle of the grating, a specific frequency of radiation is diffracted back to

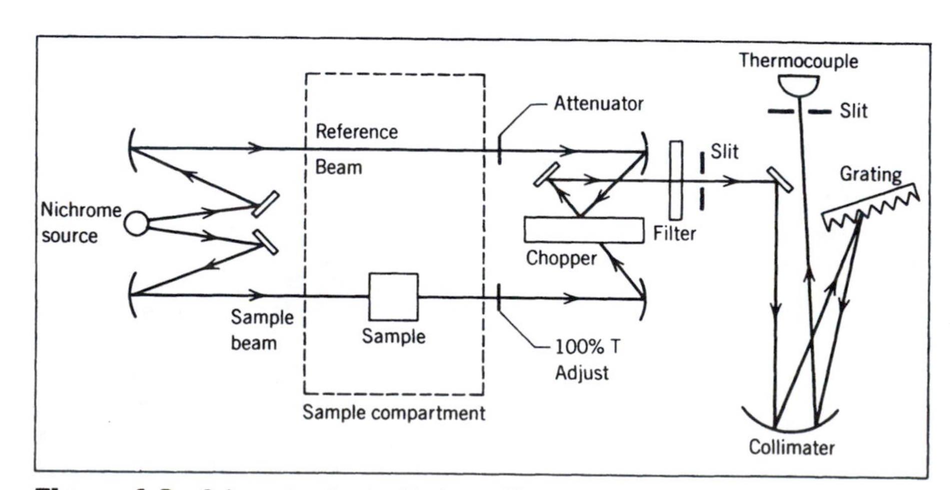

Figure 6.9. *Schematic of a double-beam IR spectrometer.*

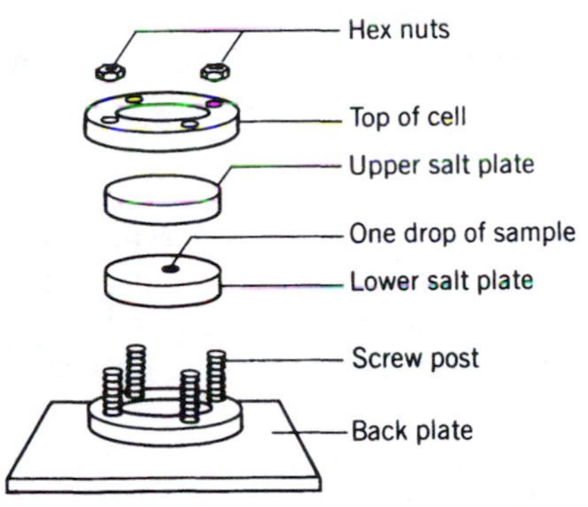

Figure 6.10. *Infrared liquid cell.*

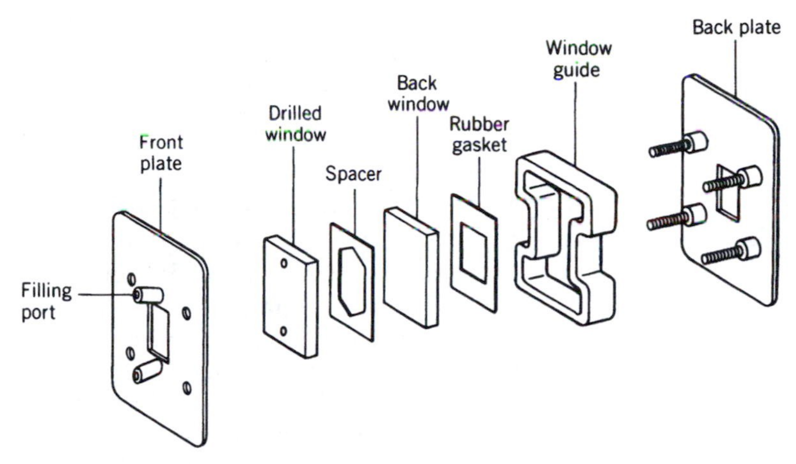

Figure 6.11. *Infrared sealed-liquid cell.*

the collimator. As the grating angle changes, different frequencies are swept. The beam is then reflected to a thermocouple detector. If no sample is in the beam, the temperature of the thermocouple will remain constant. If a sample is present, it will absorb some of the energy at various frequencies. The energy detected at the thermocouple will therefore rise and fall. The IR spectrum is a plot of the transmitted energy (transmittance) versus frequency. The fact that there are two beams allows any substance present in equal quantities in both beams to "automatically cancel." Background materials, such as water vapor or CO_2 are thereby subtracted from the spectrum, instead of providing problems by absorbing in areas of interest.

6.C.2 Sample Handling

In order to obtain an IR spectrum, the sample must be placed in a holder that does not itself absorb in the IR region. This presents somewhat of a problem, as glass or quartz absorb in the IR, rendering them unsuitable for use. Instead, cells (also called windows or plates) made from fused salts (alkali halides or silver halides) are used.

For liquids of lower volatility (bp > 100 °C), the most common type of organic sample, one or two drops of sample are placed on one salt plate, and a second salt plate is placed over the first. The two are rotated relative to each other to "smear out" the sample. The two plates are placed in a cell holder, which is in turn placed in the spectrometer. A typical sample cell is shown in Figure 6.10. For more volatile liquids, a sealed cell (Fig. 6.11) or a gas cell (Fig. 6.12) must be used. One obvious difficulty comes in cleaning the cells after use. Obviously,

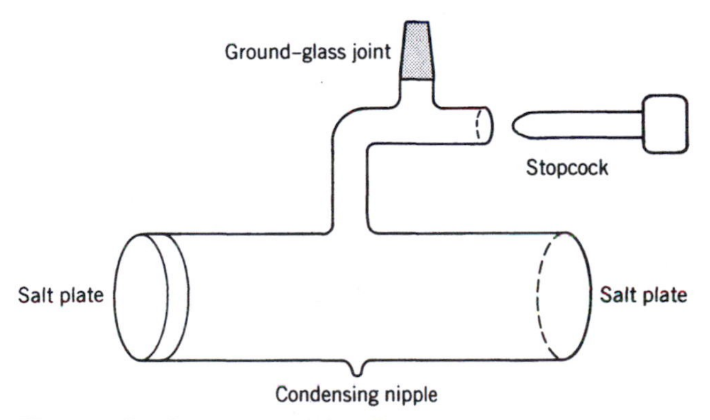

Figure 6.12. *Infrared gas cell.*

water cannot be used, or else the cell will dissolve! Instead, dry acetone is usually employed to clean cells. Very dirty NaCl cells may be cleaned by rubbing them gently on a nonfibrous cloth onto which a few drops of Brasso (a brass cleaner) were placed, followed by washing with absolute ethanol.

Gases (or volatile liquids) are handled by condensing a small amount into the nipple of a gas cell. The gas cell consists of two salt plates connected by a Pyrex tube with a path length appropriate to the spectrometer (~10 cm is normal). The tube has a stopcock fused to it for attachment to a vacuum line. A gas pressure of about 10 mm is optimal for a good IR spectrum. The cell is shown in Figure 6.12.

Solids may be handled in several ways. The most common way is to dissolve or mull them in some liquid material that will not interfere in the IR region of interest. The material usually used as a mulling agent is a mineral oil, Nujol. Nujol has absorbances in the IR region corresponding to C—H bonds. If the compound of interest has alkyl groups in its structure, Nujol will not be a good choice.

An alternate (and often better) method is to prepare a potassium bromide (KBr) disk. Approximately 5 mg of sample are finely ground in a mortar and 150 mg of finely ground KBr are added. The two materials are quickly mixed (there is some difficulty because of the hygroscopic nature of KBr) and placed in a pellet press cell. At high pressure, the KBr will flow and seal the sample in a window of salt. Since KBr is transparent to IR radiation, the IR spectrometer will detect only the sample. The cell should be disassembled and cleaned immediately after use, or else the corrosive nature of KBr to steel will result in damage to the cell.

6.C.3 Bond Vibrations

The energy emitted in the IR region is sufficient to change the vibrational state in the bonds of molecules. In order for molecules to absorb IR radiation to induce bond vibrations, there must be a change in the dipole moment of the molecule as it vibrates. This change can be either in the size of the dipole moment or in its location. Consider the molecule HF. There is a dipole in this molecule, as fluorine is more electronegative than hydrogen. The dipole moment changes as the fluorine moves away from (or toward) the hydrogen as the bond stretches. On the other hand, a homonuclear diatomic molecule (such as N_2 or O_2) does not change in its dipole as it stretches. Therefore, N_2 and O_2 do not absorb in the IR region. (This is quite convenient, as otherwise, air would interfere with the desired IR signal of the sample we wish to investigate.) Infrared vibrations are strongest when the motion involves a large change in the dipole moment of the molecule. It is for this reason that unsymmetrical C—C single bond stretches are weak, and that C—O stretches are strong.

The frequency of the stretching vibrations is easily obtained using Hooke's law,

$$\nu = \frac{1}{(2\pi c)} \sqrt{\frac{k}{\mu}}$$

where ν is the frequency of the vibration (in cm^{-1}), μ is the reduced mass of the atoms (in g) in the bond, c is the speed of light (2.998×10^{10} $cm \cdot s^{-1}$), and k is the force constant for the bond (in $dyne \cdot cm^{-1}$). The reduced mass is obtained via the equation

$$\mu = m_1 m_2/(m_1 + m_2)$$

where m_1 and m_2 are the individual masses of the atoms in the bond.

The size of the force constant, k, can be successfully related to the bond dissociation energy or the bond strength, if the molecules being compared are not too dissimilar. Thus, a strong bond is usually an indication of a large force constant. A plot of dissociation energy (or bond strength) versus force constant for a homologous series (e.g., HF, HCl, HBr, and HI) generally results in a straight line.

From Hooke's law, it is easily seen that as the masses of the atoms (m_1 and m_2) increase, the reduced mass also increases and the vibrational frequency decreases. This "heaviness" trend is a general one, seen throughout the periodic table. For example, as the mass increases down the halogen family (neglecting changes in the force constants), the stretching frequency of the hydrogen halides decreases.

H—F 4100 cm^{-1} H—Cl 3000 cm^{-1} H—Br 2650 cm^{-1} H—I 2300 cm^{-1}

This trend might well be summarized as being a vertical trend in the periodic table.

The stretching frequencies for most common functional groups have been tabulated and need not be calculated. Bonds to hydrogen are generally seen at frequencies of 2000–4000 cm^{-1}. Bonds between moderately heavy elements appear somewhat lower, from 900 to 2700 cm^{-1} (e.g., S—O, C—N, C—O, and N—O bonds). Bonds to heavier elements (e.g., C—Cl, Se—O, and P—Br) appear at low vibrational frequencies, that is, below 900 cm^{-1}. These stretching frequencies are tabulated in Figure 6.13.

Care must be exercised here, however, to distinguish between single, double, and triple bonds, as the frequency increases with bond order. This is due to the k term in Hooke's law. The force constant increases with the bond order. Thus, k is larger for triple bonds than for double bonds than for single bonds. Since k is in the numerator of Hooke's law, the frequency increases with increasing k. For this reason, a CN triple bond appears at higher frequency (2150 cm^{-1}), than a CN double bond (1650 cm^{-1}) and a CN single bond (1100 cm^{-1}).

A second factor affecting the force constant is the electronegativity difference between the atoms making up a bond. As the electronegativity difference grows larger, the bond strength increases. The force constant is related to the bond strength (so it increases) and the frequency of the vibration consequently in-

Bonds to Hydrogen	Moderate Mass Bonds	Bonds to Heavy Elements
C—H 2900 cm^{-1}	B—F 1400 cm^{-1}	C—Cl 750 cm^{-1}
Si—H 2150 cm^{-1}	C—N 1100 cm^{-1}	C—Br 650 cm^{-1}
Ge—H 2100 cm^{-1}	C—O 1100 cm^{-1}	N—Br 690 cm^{-1}
N—H 3400 cm^{-1}	N—F 1070 cm^{-1}	O—Cl 780 cm^{-1}
P—H 2300 cm^{-1}	C=C 1650 cm^{-1}	O—Br 710 cm^{-1}
As—H 2200 cm^{-1}	C=N 1650 cm^{-1}	O—I 690 cm^{-1}
O—H 3500 cm^{-1}	C=O 1700 cm^{-1}	B—Cl 950 cm^{-1}
S—H 2600 cm^{-1}	C≡C 2100 cm^{-1}	B—Br 800 cm^{-1}
Se—H 2300 cm^{-1}	C≡N 2150 cm^{-1}	S—Cl 520 cm^{-1}
F—H 4100 cm^{-1}	C≡O 2170 cm^{-1}	S—Br 400 cm^{-1}
Cl—H 3000 cm^{-1}		P—Cl 515 cm^{-1}
Br—H 2650 cm^{-1}		P—Br 390 cm^{-1}
I—H 2300 cm^{-1}		

4000 cm^{-1} 400 cm^{-1}

Figure 6.13. *Infrared correlation chart.*

creases. Thus, in the approximately equal mass bonds below, electronegativity difference is the overriding criterion.

HC 2861.6 cm^{-1} HN 3300 cm^{-1} HO 3735.2 cm^{-1} HF 4183.5 sm^{-1}

This trend might well be described as a horizontal trend, as electronegativity increases across the periodic table. These trends are all correlated in the "IR region" diagram, shown in Figure 6.14.

The motions involved in the common vibrational modes are shown in Figure 6.15. A plus sign indicates motion toward the reader, a minus sign away from the reader. In addition to the stretching vibrations discussed earlier, there are also signals associated with bending motions of bonds. Bending motions generally have much lower frequencies than stretching motions. For example, the bending vibration associated with a C—H bond occurs between 1100 and 1500 cm^{-1}, compared to the stretching frequency of 2900–3100 cm^{-1}. Some bending motions cause a drastic change in the dipole of the molecule, and give rise to strong IR peaks. The most familiar cases occur in olefins, with very strong peaks appearing between 650 and 1000 cm^{-1}, depending on the type of substitution. These correspond to the out of plane C—H bends. In acetylenes, there are strong absorptions between 610 and 700 cm^{-1} for the C—H bends (if a CH bond is present). A useful region for aromatic compounds is between 650 and 850 cm^{-1}. This region corresponds to the bending motions of the aromatic hydrogen atoms of benzene. Monosubstituted benzenes exhibit two characteristic peaks between 690 and 710 cm^{-1} and between 730 and 770 cm^{-1}. Ortho disubstituted benzenes show a single peak between 735 and 770 cm^{-1}. Meta disubstituted benzenes show two peaks between 680 and 725 cm^{-1} and between 750 and 810 cm^{-1}. Finally, para disubstituted benzenes show a single peak between 810 and 850 cm^{-1}. These frequencies are useful in identifying phenyl substituted organometallic compounds. Other bending frequencies are shown in Figure 6.16.

For molecules that are more complicated than diatomics, all atoms in the molecule undergo vibration in a synchronous manner. When a vibration involves the motion of a light group (e.g., C—H) in an otherwise heavy molecule (e.g., $CHCl_3$), very little motion of the heavier atoms occurs. The CH and CCl vibrations do not interact, and can be accurately predicted individually. If the vibration involves the motion of a group of the same mass as other groups in the molecule, the vibrations can interact, and the whole molecule will be involved in the vibration. This leads to some variation in the position of a given band. For example C═O bands occur between 1600 and 1800 cm^{-1}, depending on the nature of the remainder of the molecule.

Apart from the frequency of the peak, the IR spectrum holds a wealth of other information. The shape of the peak, for example, is instructive. Broad bands indicate intermolecular interactions. Gas-phase spectra are usually much

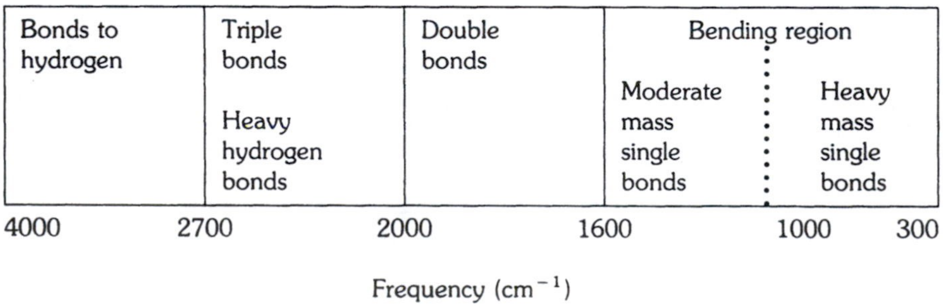

Bonds to hydrogen	Triple bonds Heavy hydrogen bonds	Double bonds	Bending region: Moderate mass single bonds	Bending region: Heavy mass single bonds
4000–2700	2700–2000	2000–1600	1600–1000	1000–300

Frequency (cm^{-1})

Figure 6.14. *Infrared region diagram.*

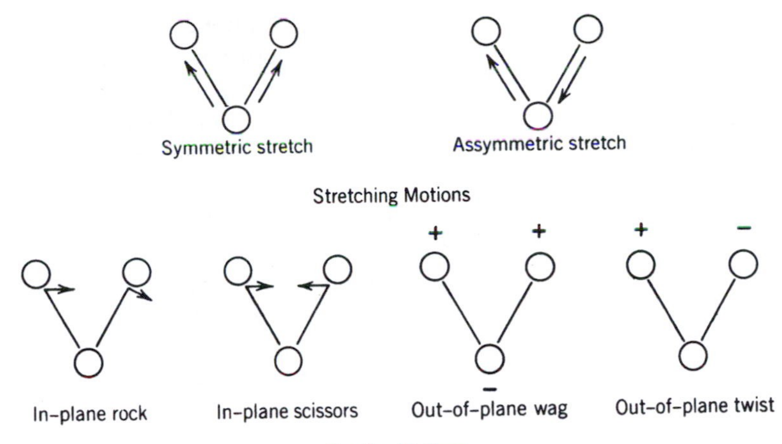

Figure 6.15. *Vibrational modes.*

sharper than liquid-phase spectra, and in fact show fine structure associated with rotational motion. In liquids, this motion is damped out because of molecular interactions and collisions. Very broad bands in IR spectra usually are indicative of hydrogen bonding, and therefore usually correspond to OH and NH groups. In solid samples, the bands are usually even broader than in liquid samples. Infrared spectra of inorganic samples, which are usually solids, are therefore often of low resolution. Sharp bands in liquid or solid spectra usually indicate ring systems, most often aromatic ring systems.

6.C.4 Vibrational Modes

Geometry plays a large role in the interpretation of IR spectra, especially in inorganic chemistry. The number of fundamental vibrations in a spectrum is determined by the number of atoms and the geometry of the molecule. Imagine making a three-dimensional graph of any molecule. To graph any of the atoms, three coordinates must be specified (such as the x, y, and z coordinates in Cartesian space). For N atoms in a molecule, there would be $3N$ coordinates needed, or a maximum of $3N$ degrees of freedom. A degree of freedom corresponds to a molecular vibration, which will be IR active provided that a change in the dipole moment occurs.

Three of these $3N$ degrees of freedom correspond to moving the molecule in the x, y, and z direction without changing any of the bond lengths or angles. For a nonlinear molecule, three more of these $3N$ degrees of freedom correspond to rotating the molecule on one of the Cartesian axes, which also does not

B—O	600–650 cm^{-1}
C—H (methyl)	1375 and 1450 cm^{-1}
C—H (methylene)	720 and 1465 cm^{-1}
C—H (olefin)	650–1000 cm^{-1} (very strong)
C—H (alkyne)	610–700 (broad)
C—H (aromatic)	See text
N—H	1590–1655 cm^{-1} (strong)
O—H	1330–1420 cm^{-1} (920 for acids, broad)
S—O	500–600 cm^{-1}
P—O	500–600 cm^{-1}
Se—O	400–450 cm^{-1}

Figure 6.16. *Infrared bending frequencies.*

change any of the bond lengths or angles. For linear molecules, there are two degrees of freedom corresponding to rotating the molecule. Since a bond length or angle must change in a molecule if an IR vibration takes place, these must be subtracted from the $3N$ degrees of freedom. Therefore, a linear molecule will have $3N - 5$ vibrations, and a nonlinear molecule will have $3N - 6$ vibrations.

There can be more bands observed in the IR spectrum than the $3N - 5$ or $3N - 6$ fundamental vibrations. The nonlinear molecule SO_2 would be expected to have three fundamental vibrations ($3 \times 3 - 6 = 3$). When the spectrum of the molecule is obtained (Fig. 6.17), seven signals are seen: 519, 606, 1151 (vs = very strong), 1361 (vs = very strong), 1871 (vw = very weak, sometimes not observed), 2305 (w = weak), and 2499 (s = strong) cm^{-1}.

Three of these frequencies (519, 1151, and 1361 cm^{-1}) indeed correspond to the three fundamental vibrations, and are called fundamental bands. The others correspond to combinations of the fundamental bands and are therefore termed combination bands. These combinations can occur as the sum of two fundamentals, as the difference of two fundamentals, or as a multiple of a strong fundamental. The band at 606 cm^{-1} is a difference band, corresponding to the difference between the 1151- and the 519-cm^{-1} fundamental bands. Note that the band (not shown in spectrum) does not occur at the exact difference, but near it. Similarly, the band at 1871 cm^{-1} is a sum band, corresponding to the sum of the 519- and 1361-cm^{-1} fundamental bands. The band at 2499 cm^{-1} is another sum band, corresponding to the sum of the 1151- and 1361-cm^{-1} fundamental bands. Finally, the band at 2305 cm^{-1} is an overtone band, corresponding to double the frequency of the strong fundamental band at 1151 cm^{-1}.

With simple molecules, the fundamental and overtone bands are fairly straightforward to assign, but with more complex molecules, this becomes far more difficult. Further complications set in because not all of the predicted bands are active in the IR spectrum. In some cases, no change in the bond–dipole moment occurs for a given vibration. The vibration, therefore, will not be IR active. In order to see some IR inactive vibrational frequencies, a second type of vibrational spectroscopy must be resorted to, called Raman spectroscopy. For a vibration to be active in the Raman spectrum, there must be a change in

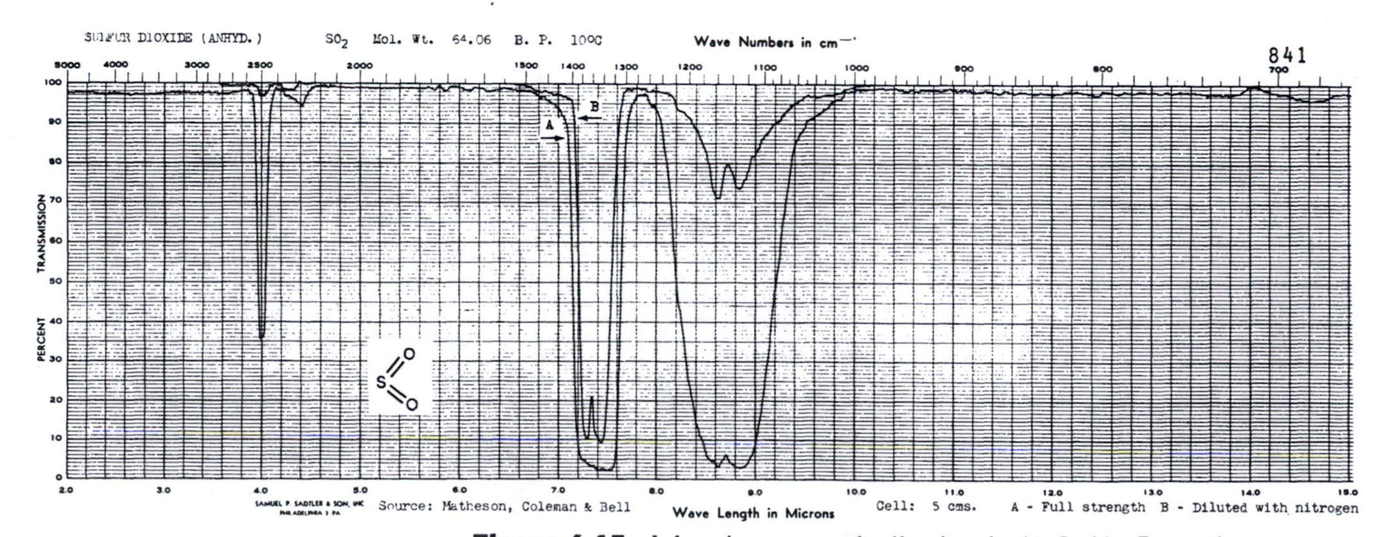

Figure 6.17. *Infrared spectrum of sulfur dioxide. (© Sadtler Research Laboratories, Division of Bio-Rad Laboratories (1961). Permission for the publication herein of Sadtler Standard Spectra® has been granted, and all rights are reserved, by Sadtler Research Laboratories, Division of Bio-Rad Laboratories, Inc.)*

polarizability during the vibration. In most cases, a vibration that is not IR active will be Raman active, and vice versa. Some vibrations are active in both types of spectroscopy. A very small number of vibrations are active in neither type. For this reason, IR and Raman spectroscopy are complementary techniques. Further information about Raman spectroscopy may be found in Refs. 1 and 2.

The geometry of a molecule (more specifically, its group symmetry) further determines the number of bands in the IR spectrum. In some cases, vibrations are degenerate, for example, more than one of the expected fundamental vibrations will occur at a given frequency. The number of fundamental absorptions in the IR spectrum therefore can be fewer (or greater—see above) than expected.

6.C.5 Inorganic Functional Groups

The horizontal and vertical trends discussed previously all apply to inorganic functional groups. One additional trend that must be considered deals with the charge on the central atom of the functional group. As a general rule, as the charge increases on the central atom, the atom becomes more electronegative, and the force constant consequently increases. Therefore, the frequency of absorption also increases.

The change in oxidation state is the most important factor in predicting the absorption frequency for an inorganic group, more important than small changes in the bond order. For example, the nitrite ion, NO_2^-, shows a strong band at approximately 1240 cm^{-1}, while the band for the nitrate ion, NO_3^-, occurs at approximatelly 1360 cm^{-1}. The N—O bond order in the nitrite ion is 1.5, whereas that in the nitrate ion is 1.33. On the basis of bond order, one would predict that the nitrite ion should occur at the higher frequency, but this is not the case. The oxidation state is the overriding criterion—the nitrogen in nitrite ion is in the III oxidation state, while that in the nitrate ion is in the V oxidation state. A similar trend is seen by comparing the frequencies of the chlorate, ClO_3^-, and perchlorate, ClO_4^-, ions. Chlorate has a major absorbance at about 960 cm^{-1}, and as expected, perchlorate appears at a higher frequency, ~1100 cm^{-1}.

As seen earlier, the vertical trend is operative for inorganic groups. A further example can be seen in comparing the sulfate, SO_4^{2-}, ion with the selenate, SeO_4^{2-}, ion. The sulfate shows a strong S—O absorbance at 1110 cm^{-1}, and the selenate, which contains the heavier selenium atom, shows a strong Se—O band at about 830 cm^{-1}.

An important point to consider about the vertical trend is that it does not depend on the masses in the bond directly, but rather on the reduced mass. When two masses are substantially different, the reduced mass is essentially equal to the lighter of the two. Further increasing the mass of the heavy atom in the bond does not appreciably change the reduced mass. For example, the Cr—O band in the chromate ion, CrO_4^{2-}, occurs at about 875 cm^{-1}. Increasing the reduced mass in the Mo—O band in molybdate, MoO_4^{2-}, shows a small decrease in the IR frequency to about 825 cm^{-1}. Increasing the mass further in tungstate, WO_4^{2-}, shows that the frequency for the W—O band has not appreciably changed, ranging from 820 to 830 cm^{-1}.

6.C.6 Synergistic Effects

When transition metals interact with ligands that possess π electrons, there are two modes of bonding in evidence. The ligand donates electrons to the empty metal σ orbitals (the $d_{x^2-y^2}$ and d_{z^2} orbitals), while the metal "back-donates" electrons from its π orbitals (d_{xy}, d_{yz}, and d_{xz}) to the empty ligand π antibonding orbitals. This type of bonding is termed synergistic bonding.

One ligand that was extensively studied in this regard is the carbonyl ligand, CO. In the free state, as carbon monoxide gas, the CO triple bond stretch occurs at 2143 cm^{-1}, while the CO double bond stretch in a ketone occurs at about 1700 cm^{-1}. In metal carbonyls, the metal would donate electrons to the π antibonding orbital of the carbonyl ligand, thereby reducing the bond order from three (in the case of carbon monoxide) to somewhat less than three. The degree of decrease in the bond order, and consequently in the stretching frequency, depends on the ability of the metal to donate electrons. Clearly, the more positively charged the metal ion, the less likely it is to donate electrons. This trend is seen in the following series:

$[V(CO)_6]^-$ 1858 cm^{-1} $[Cr(CO)_6]$ 1984 cm^{-1} $[Mn(CO)_6]^+$ 2094 cm^{-1}

Vanadium, with the lowest oxidation state (−I) is the best electron donor, and the frequency of the carbonyl band is the lowest. Managanese, having the highest oxidation state (I) is the poorest donor, resulting in a carbonyl frequency not far from carbon monoxide itself. Chromium (oxidation state 0) is intermediate.

This trend may be contrasted with metal cyanide complexes. The cyanide ion is a better σ donor, but a much poorer π acceptor than the carbonyl. This is due to the negative charge on the cyanide ion, which lowers the ion's ability to accept additional electron density donated by the metal. The ability of the metal to donate electrons is therefore not relevant in this case. However, the ability of the metal to accept electrons in σ orbitals is of paramount importance. The higher the charge on the metal, the more able it is to accept σ electrons. This is seen in the trend

$[V(CN)_6]^{5-}$ 1910 cm^{-1} $[V(CN)_6]^{4-}$ 2065 cm^{-1} $[V(CN)_6]^{3-}$ 2077 cm^{-1}

As the vanadium ion becomes more positively charged (from the I oxidation state to III), the frequency of the cyanide group increases, the opposite trend from that observed with the carbonyls.

In summary, the IR spectrum gives valuable information as to the nature of bonding and oxidation state in a compound or complex. These cases are also illustrated and discussed within specific laboratory experiments.

6.C.7 Interpretation of IR Spectra

The fundamental question to be asked is "Given an IR spectrum, how do I decide what the compound is?" Suppose you are preparing a new compound, and want to know if the preparation was successful. If you were trying to add chlorine to a boron compound, you might look at the 950-cm^{-1} region of the product and of the starting material. If your starting material did not have a peak in that area, and your product does, then you have almost certainly been successful, as B—Cl vibrations occur at approximately 950 cm^{-1}. In this case, you might conclude that the preparation was successful. In the absence of such a peak, the likelihood is that none of the desired product was obtained. Of course, if the starting material had a peak in that area, it might be difficult to determine if the desired product was obtained.

Often, the compound of interest has been prepared before by others and its IR spectrum has been published. Good sources for published spectra are the Aldrich Library of FT IR Spectra (Aldrich Chemical Company) and the Sadtler IR Spectra compendia (Sadtler). In the case of a published spectrum, all one need do is compare the experimental spectrum of the product to the published spectrum of the desired product. If they match, the preparation was obviously successful.

The most difficult case occurs when you have no idea what the product might be, and wish to characterize it using IR spectroscopy. In this case, there is a useful sequence of attack.

Step 1 Look at the overall spectral band shapes. An ultrasharp spectrum indicates an aromatic compound. Broad resonances indicate inorganic groups. Very board resonances indicate hydrogen-bonding groups.

Step 2 Check the region between 1600 and 2800 cm^{-1}. Any peak in this region is of interest and gives helpful information about the compound. For example, a peak at 1700 cm^{-1} indicates a C=O bond, and a peak at 2050 cm^{-1} indicates a C≡N triple bond in a cyanide or thiocyanide group.

Step 3 Investigate the region above 3000 cm^{-1}. Hydroxyl groups give broad and obvious bands at 3500 cm^{-1}. Amines give strong peaks at 3300 cm^{-1}. Terminal alkynes give a sharp (but weak) band at 3200 cm^{-1}. Alkenes give moderate bands at 3100 cm^{-1}.

Step 4 If the compound has aromatic functionalities, check the bending region at 650–850 cm^{-1}.

Step 5 Recognize that neutral unsaturated groups bonded to metals tend to absorb at somewhat lower frequencies than in the free compound. For example, the CO triple bond normally appears at 2100 cm^{-1}. In metal carbonyls, however, the frequency is lowered substantially because of the donation of electrons from the metal to the carbonyl antibonding orbitals. In fact, the frequency of the carbonyl stretch is a sensitive indication of the oxidation state of the metal.

Step 6 Once you have determined what type of compound you have using Steps 1–5, look up the spectra of likely possibilities, and compare your experimental spectrum to the published ones.

A more recent innovation is the computer spectral search program. One inputs to the program the IR peak wavenumbers, and the program searches its files, and tells the user which compounds are likely matches. Many advanced instruments have such search programs built in, and several "stand alone" programs are available that operate using a personal computer. These programs are geared mainly for organic chemicals, but in many cases include substantial numbers of inorganic and organometallic compounds as well. Programs are available from Aldrich Chemical Company (for organic compounds), from Sigma Chemical Company (mainly biochemicals), and from Sadtler.

6.C.8 Fourier Transform Infrared (FT IR)

Traditionally, most laboratories were equipped with dispersive IR spectrometers of the type discussed earlier. Recently, with the falling prices associated with microcomputers, the FT IR spectrometer has increased in popularity, taking a larger and larger share of the market. The cost of an FT IR spectrometer can be as low as $16,000, with $25,000 being the average price of a fully functional entry level instrument. The FT IR spectrometer offers several important advantages over the dispersive instrument. These advantages arise from two factors: an FT IR uses an interferometer instead of a monochromator, and the output data from an FT IR is stored in digital form in a computer memory, and can be manipulated.

The interferometer consists of one fixed and one movable mirror, and a beamsplitter. A source provides radiation, much like a dispersive spectrometer. One half of the radiation is reflected by the beamsplitter, strikes the fixed mirror, and is reflected back to the detector. The other one half of the radiation is transmitted by the beamsplitter to the movable mirror and is reflected back to the detector. The distance from the beamsplitter to the fixed mirror is constant, while the distance from the beamsplitter to the movable mirror is variable. If both distances are equal, the two beams of radiation interfere constructively. If the distances are different by $\lambda/2$, where λ is the wavelength of the radiation, the beams interfere destructively. The intensity of the beam at the detector therefore depends on the position of the movable mirror, and follows the equation

$$I(x) = B(\nu)\cos(2\pi x\nu)$$

where $I(x)$ is the intensity of the signal, and $B(\nu)$ is the intensity of the source radiation at frequency ν. The difference between the mirror distances is given as x. This mirror arrangement is shown in Figure 6.18. The source in an FT IR spectrometer is polychromatic, however, and there will be a cosine wave for each of the frequencies. The intensity equation therefore depends on the sum of all the individual cosine waves.

$$I(x) = B(\nu_1)\cos(2\pi x\nu_1) + B(\nu_2)\cos(2\pi x\nu_2) + \cdots$$

or

$$I(x) = \int B(\nu)\cos(2\pi x\nu)\,d\nu$$

This equation is essentially constant for all values of x, except when the two mirror distances are equal ($x = 0$) and the beams interfere constructively. At this point, there will be a "burst" at the center of the interferogram. If a compound is placed in the source path, it absorbs certain frequencies from the polychromatic source. This subtracts cosine waves from the interferogram pattern. The difference between the two interferograms (source only and sample + source) must be due to the sample itself. By performing a Fourier transform (a mathematical technique) on the interferogram, the "normal" spectrum is obtained.

The advantages of FT IR can be enormous. Since the data is stored on a computer, instead of taking one scan of the sample, multiple scans can be taken. (They can also be taken with dispersive instruments, but the price then rises to be roughly equal to that of an FT IR.)

The signal is always present in the interferogram, but the noise present is random, and will partially average out as $\sqrt{N}$, where N is the number of scans. The net gain in signal-to-noise (S/N) ratio is therefore $N/\sqrt{N} = \sqrt{N}$. For example, if 100 scans are taken, the S/N ratio will increase by a factor of 10. This is known as Fellgett's advantage. A further advantage is that an interferometer has a circular entrance opening versus the entrance slit of a dispersive instrument.

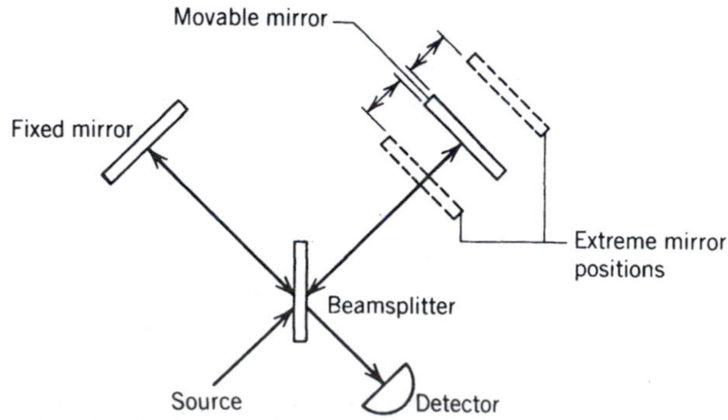

Figure 6.18. *Fourier transform infrared interferometer.*

More energy reaches the sample (by a factor of 80–200), with consequent increases in the S/N ratio. This is known as Jacquinot's advantage. In practice, however, the increases in S/N are somewhat less than theoretically predicted. The most important advantage in FT IR is that all frequencies are sampled simultaneously, unlike a monochromator where only one frequency is sampled at a time. The time necessary to take a single scan is therefore much shorter in FT IR (~1 s, compared to about 4 min for dispersive IR), and multiple scans can be accomplished in the same time as it takes a dispersive instrument to take a single scan.

This time advantage allows one to investigate problems using FT IR spectroscopy that are impossible using dispersive techniques. One such application is GC–FT IR. In this combination, the effluent from a GC is sent through a light pipe into an FT IR spectrometer, and scans are taken of the material coming by approximately every second. These scans are numbered and stored on a computer. Since the GC separates mixtures into their individual components, at any given moment, only one compound will be present in the source path in the light pipe. By summing all interferograms corresponding to a peak in the chromatogram, the FT IR spectrum of each component of the mixture can be obtained, and the component identified. This is done "on the fly," as the FT IR is fast enough at acquiring scans to be able to follow the chromatograph effluent in real time.

The main disadvantages of the FT IR instrument are functions of cost and control. Since the interferometer must be highly accurate and reproducible, it is expensive. The need for a computer further adds to the cost of the instrument.

General References

Drago, R. S., *Physical Methods in Chemistry,* 2nd ed., Saunders: Philadelphia, PA, 1977.

Nakamoto, K., *Infrared and Raman Spectra of Inorganic and Coordination Compounds,* 4th ed., Wiley–Interscience: New York, 1986.

Smith, A. L., *Applied Infrared Spectroscopy (Volume 54 in Chemical Analysis),* Wiley: New York, 1979.

George, B.; McIntyre, P., *Infrared Spectroscopy,* Analytical Chemistry by Open Learning: London, 1987.

Ebsworth, E. A. V.; Rankin, D. W. H.; Craddock, S., *Structural Methods in Inorganic Chemistry,* Blackwell: Oxford, 1987.

6.D NUCLEAR MAGNETIC RESONANCE SPECTROSCOPY

6.D.1 Introduction

Nuclear Magnetic Resonance (NMR) spectroscopy is different from IR and Visible spectroscopy in that it is the atomic nuclei that absorb energy. Nuclear magnetic resonances are reported in hertz frequency units and occur at frequencies below 1.2×10^9 Hz.

Many nuclei possess a property called spin angular momentum. The nuclei spin, leading to the generation of a magnetic field. Such nuclei are NMR active. Conversely, those that do not spin are NMR inactive. The number of protons and the isotopic mass allow one to categorize the spin quantum number (I) of the nucleus (see Table 6.3).

All nuclei with an even number of protons and an even atomic mass have spins of 0 and are NMR inactive. Examples are ^{12}C, ^{16}O, and ^{32}S. All nuclei that are not even–even are NMR active. The ease with which the NMR transitions can be observed with a particular nucleus is a function of the sensitivity of the nucleus. Sensitivities for several of the most commonly investigated nuclei are listed in Table 6.4.[1] Thus, in an experiment, the signal detected from a ^{31}P

Table 6.3 Determination of Spin Quantum Number

Number of Protons	Atomic Mass	Spin Quantum Number(I)	Example
Even	Even	0	^{12}C
Odd	Even	Integral (1, 2, 3, ...)	^{10}B (I = 3)
Even	Odd	Half-Integral ($\frac{1}{2}$, $\frac{3}{2}$, ...)	^{13}C (I = $\frac{1}{2}$)
Odd	Odd	Half-Integral ($\frac{1}{2}$, $\frac{3}{2}$, ...)	^{1}H (I = $\frac{1}{2}$)

sample will be 377 times more intense than that of a ^{13}C sample. From an NMR point of view, those nuclei having odd numbers of protons and odd atomic masses with high natural abundances are best. As is quickly seen from Table 6.4, the most commonly investigated nuclei tend to have spin quantum numbers of $\frac{1}{2}$.

In the NMR experiment, a sample is put in a tube, and placed in an external magnetic field of strength H_0 (shown in Fig. 6.19). In the magnetic field, NMR active nuclei (with $I = \frac{1}{2}$) have two different spin energy levels—one corresponding to the nucleus being aligned with the external magnetic field (called α), and one corresponding to being aligned against it (called β). The state aligned with the magnetic field is of lower energy than the state aligned against the magnetic field. If the nucleus absorbs energy equivalent to the difference between the energies of the two levels, the nucleus will make a transition to the higher energy level. This transition gives rise to the NMR signal.

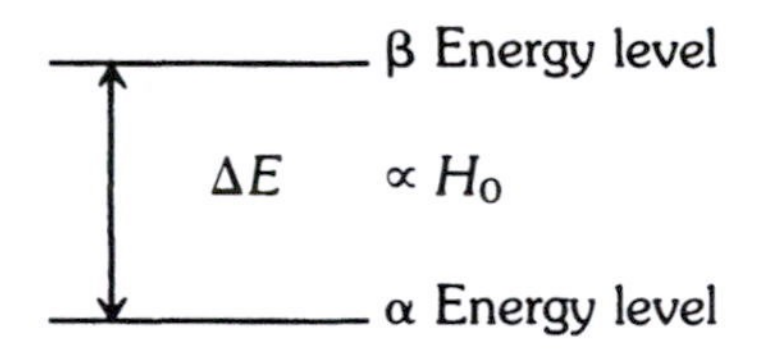

The energy difference, ΔE, between the two spin states is proportional to the NMR spectrometer magnet's field strength, H_0. In order to best see the

Table 6.4 NMR Data for Common Nuclei

Nucleus	Spin	Natural Abundance	Resonance Frequency[a]	Sensitivity[b]	Reference[c]
^{1}H	$\frac{1}{2}$	99+%	100 MHz	5556	$(CH_3)_4Si$ (TMS)
^{2}H	1	0.015	15.35	0.0082	TMS
^{11}B	$\frac{3}{2}$	80.42	32.07	754	$BF_3 \cdot O(C_2H_5)_2$
^{13}C	$\frac{1}{2}$	1.1	25.1	1	TMS
^{15}N	$\frac{1}{2}$	0.37	10.14	0.022	NH_3 (liq, 25 °C)
^{19}F	$\frac{1}{2}$	100	94.1	4611	CF_3CO_2H (TFA)
^{29}SI	$\frac{1}{2}$	4.7	19.87	2.09	TMS
^{31}P	$\frac{1}{2}$	100	40.48	377	85% H_3PO_4 (OPA)
^{57}Fe	$\frac{1}{2}$	2.19	3.24	0.0042	$Fe(CO)_5$
^{77}Se	$\frac{1}{2}$	7.58	19.07	2.98	$Se(CH_3)_2$
^{113}Cd	$\frac{1}{2}$	12.26	22.18	7.6	$Cd(ClO_4)_2 \cdot H_2O$
^{119}Sn	$\frac{1}{2}$	8.58	37.29	25.2	$Sn(CH_3)_4$
^{195}Pt	$\frac{1}{2}$	33.8	21.46	19.1	Na_2PtCl_6
^{207}Pb	$\frac{1}{2}$	22.6	20.88	11.8	$Pb(CH_3)_4$

[a] Relative to ^{1}H = 100 MHz.
[b] Relative to ^{13}C = 1.
[c] Trifluoroacetic acid = TFA and OPA = orthophosphoric acid.

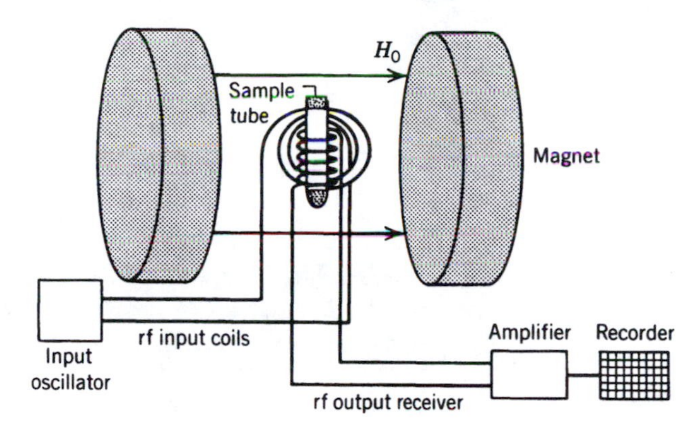

Figure 6.19. *Schematic of NMR spectrometer.*

nuclear transition, it is desirable to have the population of these two energy states be as different as possible. The ideal circumstance would be to have all nuclei occupying state α and none in state β. From the diagram above, it is clear that the easiest way to accomplish this is for ΔE to be as large as possible. Since ΔE is proportional to the magnet field strength, it is desirable to have a magnet with a large field strength for the NMR spectrometer.

This energy necessary to induce the nuclear transition is supplied by continuous energy from the input oscillator at a frequency corresponding to ΔE. This energy difference can be obtained from the equation

$$\Delta E = \mu\beta H_0/I = h\nu$$

where

μ = the magnetic moment of the nucleus (units: BM)
β = a constant, 5.049×10^{-31} J·G^{-1} BM^{-1}
H_0 = the magnet field strength (units: G)
I = the nuclear spin quantum number
h = Planck's constant, 6.626×10^{-34} J·s
ν = the frequency (units: s^{-1} or Hz)

Values of μ are given below. The energy frequency, ν, is obtained by dividing ΔE by h.

As an example, we now calculate what frequency pulse must be applied to observe ^{1}H transitions using a magnet with a field strength of 23,500 G.

$$\Delta E = [(2.7927 \text{ BM})(5.049 \times 10^{-31} \text{ J·G}^{-1} \text{ BM}^{-1})(23{,}500 \text{ G})]/\tfrac{1}{2}$$

$$\Delta E = 6.627 \times 10^{-26} \text{ J}$$

Table 6.5 Magnetic Moments of Common Nuclei

Nucleus	Magnetic Moment, μ (BM)
^{1}H	2.7927
^{13}C	0.7024
^{19}F	2.6288
^{31}P	1.1317

or the frequency is

$$\nu = 6.627 \times 10^{-26}\ \text{J}/6.626 \times 10^{-34}\ \text{J·s} = 1.000 \times 10^{8}\ \text{s}^{-1}$$

$$\nu = 1.000 \times 10^{8}\ \text{Hz} = 100.0\ \text{MHz}$$

NOTE: ***The frequency ν is called the Larmor frequency and for ^1H is in the FM radio band. For this reason, the energy is often referred to as radio-frequency (rf) energy.***

The frequency is slowly varied across the range of frequencies at which the particular nucleus absorbs (this range is quite small compared to the frequency ν). The resulting trace of signal intensity (voltage being read from the rf output) versus frequency is the NMR spectrum (see Fig. 6.20).

A radio is a good analogy to an NMR spectrometer. The radio has an oscillator that produces a certain frequency and also an adjustable magnet. When a radio is tuned, the magnetic field strength is adjusted by increasing or decreasing the number of coils on the magnet. When the proper frequency–magnetic field strength combination is reached, a radio station is tuned in. Changing the field strength again tunes in a different station. Nuclear magnetic resonance works in the same way: A frequency–magnetic field strength combination (such as the 100 MHz, 23,500 G combination above) tunes in ^1H, and a different combination would tune in a different nucleus. Different nuclei, in general, do not interfere with each other.

Equivalent results can be achieved with a radio by holding the magnetic strength constant and adjusting the frequency with a frequency synthesizer (more expensive FM receivers do this). Similarly, most NMR spectrometers tune in this manner as well. Proton spectrometers are the most common and inexpensive type of NMR instruments. There are two reasons for this:

1. Protons occur in many compounds of interest.
2. Protons are very sensitive to the NMR technique.

6.D.2 Sample Preparation

It is desirable in most cases (medical imaging being a notable exception) to have all nuclei of a similar type absorb energy at the same frequency. This is equivalent to saying that all nuclei of the same type must be in identical magnetic environments. Samples are normally prepared in 5-mm o.d. Pyrex tubes. It is clear that the sample sitting in the center of the tube would "feel" a smaller magnetic

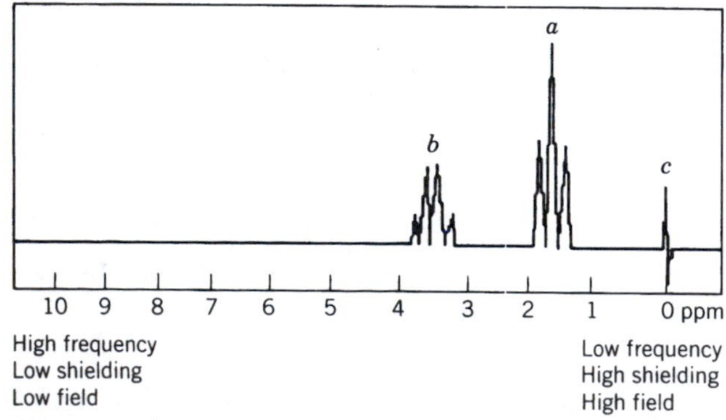

Figure 6.20. *NMR spectrum of ethyl bromide.*

field than the sample at the outside of the tube. This would result in a broad signal, an undesirable situation. To overcome this difficulty, liquid samples can be spun rapidly (~50 rps), allowing rapid rotational and translational diffusion to even out the apparent magnetic field for all nuclei. For solids, this is more problematic (as the solid will not mix even when spun), but techniques such as "magic angle spinning" and "cross polarization"[2] are available on advanced instruments to obtain more highly resolved signals (liquid samples can also benefit from cross polarization).[3] Solid signals are, in general, broad. Thus, in most cases, it is desirable to run all samples in the liquid state.

Liquid samples can be prepared either neat, or for more abundant nuclei, as ~10% solutions in a solvent that does not interfere with the signal of interest. Approximately 0.5 mL of liquid sample is necessary for a 5-mm NMR tube. For ^{1}H NMR, deuterated solvents are most commonly used, such as $CDCl_3$ and D_2O. Solid samples may be prepared by dissolving 10–25 mg of solid in 0.5 mL of solvent. More concentrated solutions may be necessary for less sensitive nuclei.

A small amount of a reference compound (TMS for ^{1}H and ^{13}C NMR, see Table 6.4 for other nuclei) is then added. In cases where the reference compound will react with the sample, the reference can be used externally. In general, all signals in the NMR spectrum are measured in fractional units of frequency from the reference, which is assigned a value of zero. The normal units used in NMR are parts per million (ppm) of the nuclear resonance frequency. For protons on a 60-MHz instrument (such as a Varian EM-360)

$$1 \text{ ppm} = 60 \text{ MHz}/1{,}000{,}000 = 60{,}000{,}000 \text{ Hz}/1{,}000{,}000 = 60 \text{ Hz}$$

The sample tube is then placed inside the sample port of the spectrometer. This places the sample between the magnet field coils. The sample is then spun, to increase the sample homogeneity to the magnetic field as discussed previously. Using the adjustment controls, the magnet field strength and homogeneity are adjusted until the signals from the reference and sample come into view and are optimized on an oscilloscope or signal meter. The resulting spectrum is then plotted, a typical result being shown in Figure 6.20.

6.D.3 Reference Materials

There are four main criteria for the selection of a reference material, such as TMS.

1. It should be readily available and stable.
2. It should give rise to a large NMR signal, so that not much reference is needed.
3. It should not appear in a part of the spectrum that will obscure peaks that we want to see in the experiment.
4. It should not react with the sample.

The common reference materials for each nucleus are listed in Table 6.4.

6.D.4 The Chemical Shift

For simplicity, we first consider the ^{1}H NMR spectrum of an organic compound, ethyl bromide. As seen from Figure 6.20, ethyl bromide and TMS give rise to three signals: *a*, *b*, and *c*. Signal *a* is larger than signal *b*. Signal *a* is split into three lines, signal *b* into four lines, and signal *c* is a single line. Also, note that signal *b* is observed with the highest shift (~3.5 ppm), signal *a* the next highest (1.7 ppm), and signal *c* the lowest (0 ppm) shift. Since this is an ^{1}H spectrum, all signals represent hydrogen.

As stated earlier, the signal at 0 ppm is the reference, TMS. The ethyl bromide must then be giving rise to two signals, *a* and *b*. Why would one compound give rise to two signals? Consider what is taking place in the spectrometer. A nucleus in a low energy spin state is excited to a higher energy spin state. In order to reach the nucleus, the excitation energy must first pass through the electron cloud surrounding the nucleus. The denser the electron cloud, the higher a field strength is needed to "break through" and excite the nucleus. Keep in mind that the high magnetic field side of the spectrum is the low parts per million side.

The three methyl hydrogen atoms are all chemically equivalent, because of free rotation about the C—C bond. They all "feel" the effect of the electronegative bromine in the same way (a small amount) and all require the same field strength for excitation. Thus, they give rise to a single signal, *a*. (Actually, the number of NMR signals depends on magnetic equivalence, not chemical equivalence. At this introductory level, these can be assumed to be synonymous.)

The two methylene hydrogen atoms are both the same distance from the bromine and are thus chemically equivalent. Since they are closer to the bromine than the methyl hydrogen atoms, the methylene hydrogens "feel" its inductive effect more strongly. The electronegative bromine pulls some of their electron density toward itself, leaving the methylene hydrogen atoms with less protection from the magnetic field. It therefore requires less field strength to excite them, and their signal, *b*, appears to the left of signal *a*. The TMS protons (signal *c*) require the highest field strength of all for excitation, since all 12 equivalent hydrogen atoms are near an electropositive silicon, which gives the hydrogens additional electron density. They are more protected than either the methyl or methylene hydrogen atoms in ethyl bromide. In general: The better protected the nucleus by electrons, the more to the right (higher magnetic field) the signal appears. An alternate way of stating this is "the more electronegative the nearby group, the lower the magnetic field the signal appears."

The distance from the reference TMS at which the signal appears is called the chemical shift, δ. A table of typical chemical shifts for organic groups appears in Table 6.6. Chemical shifts are usually measured in units of parts per million, which can be obtained from the formula

$$\delta = [(\text{frequency of signal} - \text{frequency of reference})/\text{Larmor frequency}] \times 10^6$$

The correlation of the chemical shift to the electronegativity of the substituents is clearly shown in Figure 6.21, which shows the ^{195}Pt NMR spectra of a series of platinum halide anions.[4] The reference compound for ^{195}Pt NMR is Na_2PtCl_6, which appears at 0 ppm. All the other halide anions shown are upfield from the reference (hence have negative chemical shifts). As chlorines are replaced by the less electronegative bromine, the chemical shift moves more and more to high field, as expected. It can be seen that this is a linear progression, with a shift of approximately −300 ppm with each subsequent bromine. Another interesting point is the size of the chemical shift range for ^{195}Pt. In ^{1}H NMR, the chemical shift range for most compounds is about 10 ppm. The ^{195}Pt nucleus is much more sensitive to its local magnetic environment and has a chemical shift range of approximately 4000 ppm.

An important point seen in the ^{195}Pt spectra is that different oxidation states of platinum occur in different regions of the chemical shift range. Platinum(IV) complexes appear between 0 and −2000 ppm, while platinum(II) complexes are more shielded, appearing between −1500 and −3000 ppm. A complete list of ^{13}C chemical shifts in organometallic compounds may be found in Ref. 5.

Table 6.6 ^1H Organic Functional Group Chemical Shifts

Shift[a] (ppm)	Group	Example	Type of Hydrogen
1.0	CH_3—	C**H**$_3$CH$_2$C**H**$_3$	Methyl
1.4	—CH_2—	CH$_3$C**H**$_2$CH$_3$	Methylene
1.6	—CH—	(CH$_3$)$_3$C**H**	Methyne
2.0	CH_3—CO—	C**H**$_3$COCH$_3$	Methyl α to carbonyl
2.5	—CH_2—CO—	CH$_3$C**H**$_2$COCH$_3$	Methylene α to carbonyl
	—CH_2—NR_2	CH$_3$C**H**$_2$N(CH$_3$)$_2$	Methylene α to amine
	ϕ—CH_2—	ϕ—C**H**$_2$CH$_3$	Benzyl
3.5	—CH_2—Cl	CH$_3$C**H**$_2$Cl	Methylene α to halogen
	—CH_2—OR	CH$_3$C**H**$_2$OCH$_3$	Methylene α to oxygen
5.0–6.0	—HC=CR$_2$	**H**$_2$C=CHCH$_3$	Vinyl
7.0–8.00	ϕ—H	C$_6$**H**$_5$CH$_3$	Phenyl or aromatic
9.5	—CHO	CH$_3$C**H**O	Aldehyde
10.0–12.0	—CO_2H	CH$_3$CO$_2$**H**	Acid

[a] All chemical shifts are in parts per million, downfield from TMS.

6.D.5 Integration

It was noted that signal *a* of Figure 6.20 was larger than signal *b*. If one were to measure the area under the peaks, they would find that the ratio of areas *b*:*a* was 2:3. Since signal *b* (area 2) is the CH_2 group and signal *a* (area 3) is the CH_3 group, it is obvious that the area under the signal is proportional to the number of hydrogen atoms the signal represents. Integration is a valuable tool for the interpretation of ^1H NMR spectra, but cannot be as simply interpreted for other nuclei (e.g., ^{13}C).

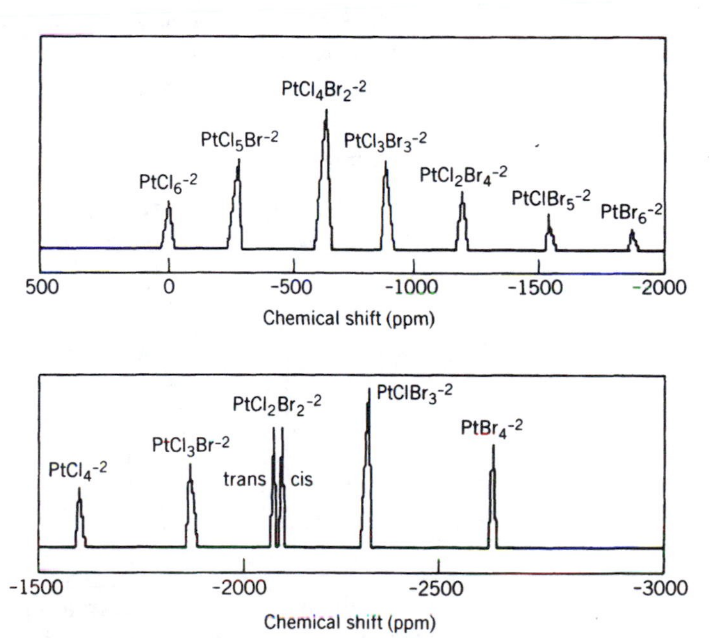

Figure 6.21. ^{195}Pt NMR spectra of platinum halide anions.

6.D.6 Spin–Spin Coupling

The question as to why the signals in Figure 6.20 are split into four lines (signal *b*), three lines (signal *a*), and only one line for TMS must now be discussed. The splittings arise because of the magnetic environment of the hydrogen atoms being observed. Consider the methyl hydrogen atoms. The methyl hydrogen atoms are nearest to the carbon they are directly bonded to. Recall that ^{12}C (99% of all carbon atoms are ^{12}C) is NMR inactive. Thus, as far as this experiment is concerned, ^{12}C is invisible. The next nucleus to consider is the other carbon (also invisible), and then the methylene hydrogen atoms. Since the experiment is a magnetic one, what the methyl hydrogen atoms will be affected by is the magnetic field of the methylene hydrogens. Each hydrogen can have two spin states: magnetic field aligned with ($S = +\frac{1}{2}$) or against ($S = -\frac{1}{2}$) the external magnetic field. There are four magnetic ways the methylene hydrogen atoms can be arranged.

Spin of H_b No. 1	Spin of H_b No. 2	Total Spin	Occurrence
$+\frac{1}{2}$	$+\frac{1}{2}$	$+1$	1
$+\frac{1}{2}$	$-\frac{1}{2}$	0	2
$-\frac{1}{2}$	$+\frac{1}{2}$	0	
$-\frac{1}{2}$	$-\frac{1}{2}$	-1	1

The methyl hydrogen atoms are affected by a total magnetic field from the methylene hydrogen atoms (total spin) of $+1$, 0, or -1. The 0 possibility occurs twice as often as the other two.

Thus, the ***a*** signal is split into three lines (corresponding to the three total magnetic spin possibilities), and the line corresponding to the total spin of 0 is twice as large as the other two. The resulting pattern is a 1:2:1 triplet.

We now try to answer why the ***b*** signal is split into four lines. Again, first in "line of sight" is the carbon the methylene atoms are bonded to (NMR inactive). Next is the other carbon (also inactive) and a bromine (inactive for reasons discussed below). Finally, there are three methyl hydrogen atoms. As before, the methylene hydrogen atoms are affected by the total magnetic field of the methyl hydrogen atoms. There are eight combination possibilities.

Spin H_a No. 1	Spin H_a No. 2	Spin H_a No. 3	Total Spin	Number of Times Occurred
$+\frac{1}{2}$	$+\frac{1}{2}$	$+\frac{1}{2}$	$+\frac{3}{2}$	1
$+\frac{1}{2}$	$-\frac{1}{2}$	$+\frac{1}{2}$	$+\frac{1}{2}$	
$-\frac{1}{2}$	$+\frac{1}{2}$	$+\frac{1}{2}$	$+\frac{1}{2}$	3
$+\frac{1}{2}$	$+\frac{1}{2}$	$-\frac{1}{2}$	$+\frac{1}{2}$	
$+\frac{1}{2}$	$-\frac{1}{2}$	$-\frac{1}{2}$	$-\frac{1}{2}$	
$-\frac{1}{2}$	$+\frac{1}{2}$	$-\frac{1}{2}$	$-\frac{1}{2}$	3
$-\frac{1}{2}$	$-\frac{1}{2}$	$+\frac{1}{2}$	$-\frac{1}{2}$	
$-\frac{1}{2}$	$-\frac{1}{2}$	$-\frac{1}{2}$	$-\frac{3}{2}$	1

Thus, the methylene hydrogen atoms see a total magnetic field (total spin) from the methyl hydrogens of $+\frac{3}{2}$, $+\frac{1}{2}$, $-\frac{1}{2}$, and $-\frac{3}{2}$. The $+\frac{1}{2}$ and $-\frac{1}{2}$ possibilities occur three times as often as the other two. The ***b*** signal is thus split into four lines (corresponding to the four total spin possibilities), and the lines corresponding to total spins of $+\frac{1}{2}$ and $-\frac{1}{2}$ are three times the size of the others. The resulting signal is a 1:3:3:1 quartet.

The **c** signal (corresponding to TMS) is unsplit, because all 12 hydrogen atoms are chemically equivalent, and there is nothing for them to couple to. Both ^{12}C (98.9% abundant) and ^{28}Si (92% abundant) are NMR inactive, hence invisible to the experiment.

Important: *Nuclei generally do not couple to other nuclei more than three atoms away. If nuclei are chemically equivalent, they do not couple.*

The number of lines in a signal (called the multiplicity) may be calculated using the equation

$$M = 2nI + 1$$

where

M = the multiplicity of the signal.

n = the number of equivalent coupled nuclei.

I = the spin of the coupled nuclei.

For nuclei with $I = \frac{1}{2}$, this equation reduces to

$$M = n + 1$$

An interesting question to consider is why the hydrogen atoms in ethyl bromide do not couple to the small fraction of the carbon atoms that are ^{13}C. In fact they do couple, but since there are so few ^{13}C nuclei in the sample, the coupling is minor. The signal from the hydrogen atoms is indeed split, but the "splitting lines" are very small, and usually disappear into the noise. Coupling of hydrogen atoms to ^{13}C can be observed in ^{1}H FT–NMR spectra (although the signals are still weak, the S/N ratio is high enough so that they can be seen). Generally, though, one turns to the ^{13}C FT–NMR spectrum to see these couplings. Observation of ^{13}C—^{13}C couplings is very difficult, as the odds of having two ^{13}C nuclei next to each other is quite small.

Coupling is seen, of course, in nonproton spectra as well. Figure 6.22 shows the ^{77}Se NMR spectrum of bis(trifluoromethyl) selenide, $(CF_3)_2Se$. The ^{19}F nucleus has a spin quantum number of $\frac{1}{2}$, the same as ^{1}H, and is 100% naturally abundant. Fluorine therefore will give rise to similar splitting patterns. The only difference will be the size of the coupling constant. In the case of $(CF_3)_2Se$, there are six equivalent fluorine atoms coupled to the ^{77}Se nucleus. Substituting into the multiplicity equation,

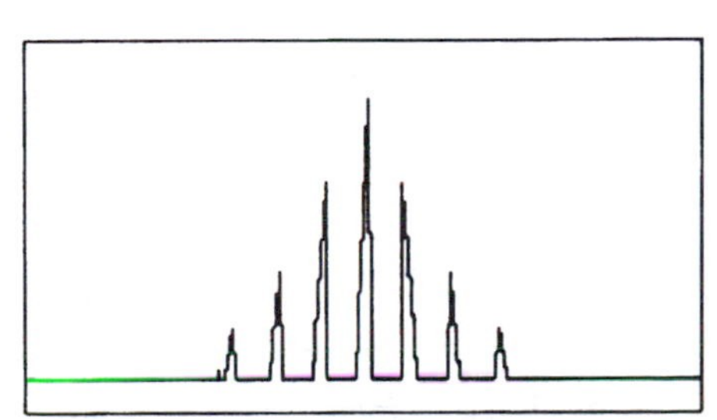

Figure 6.22. *^{77}Se NMR spectrum of $(CF_3)_2Se$.*

$$M = 2(6)(\tfrac{1}{2}) + 1 = 7$$

Therefore, the signal should be a septet, which is observed.

6.D.7 ^{13}C-NMR Spectroscopy

Nuclei with $I = \frac{1}{2}$ are collectively referred to as dipolar nuclei. Some common examples are ^{13}C, ^{19}F, ^{31}P, ^{77}Se, and ^{195}Pt. Along with proton NMR spectroscopy, the NMR of these additional nuclei has become a common tool for characterizing and identifying inorganic and organometallic compounds. Of these, ^{13}C NMR is the most common. It has been used extensively for the study of compounds having direct metal-to-carbon bonds. Valuable information may also be obtained on mechanisms of fluxional behavior.

As can be seen in Table 6.4, all other nuclei are less sensitive than ^{1}H, and many are even less sensitive than ^{13}C. The common type of ^{1}H NMR spectrometer is called a continuous wave (CW) NMR. Theoretically, it is possible to detect any fairly sensitive nucleus (e.g., ^{11}B, ^{19}F, or ^{31}P) on a CW NMR. Probes for CW instruments are commercially available for ^{19}F and ^{31}P. Usually, however, one uses a far more sophisticated instrument called a Fourier transform NMR (FT NMR) spectrometer.

In FT NMR spectroscopy, the sample receives a powerful pulse of rf radiation. This pulse contains a broad range of frequencies and causes all NMR active nuclei to resonate at their appropriate Larmor frequencies. The detector of the FT NMR spectrometer follows the change in magnetization as the excited nuclei relax back to their original energy state. The time required for the nuclei to relax back to their original energy states is called the spin–lattice relaxation time, T_1.

A plot of the magnetization versus time is called a free induction decay (FID). By Fourier transforming the FID, the "normal" NMR spectrum is obtained. (Compare this with Fourier transforming the interferogram to obtain the "normal" IR spectrum in Section 6.C.8.) The FT NMR spectrum enjoys many of the same advantages seen earlier for the FT IR spectrum. Since all frequencies are excited simultaneously, the experiment is much faster than in CW NMR, where the frequencies are slowly, individually scanned. This speed allows multiple scans to be taken, and the technique is consequently often called multiple pulse NMR. By taking multiple scans of a sample, storing the signals on a computer, and coadding them, the signal increases and the noise averages down. The S/N ratio thus increases (as it did in FT IR) by $\sqrt{N}$, where N is the number of scans. The speed at which one can scan is limited only by the requirement that the nuclei must relax to their original energy state after the pulse is applied. In practice, a waiting time of about $5T_1$ is employed between pulses. Typical values of T_1 for nuclei with $I = \frac{1}{2}$ are on the order of 2 s. The multiple pulse technique allows the chemist to use NMR to investigate nuclei of low sensitivity (sensitivities less than or equal to that of ^{13}C), which would be impossibly difficult with CW NMR.

The low sensitivity of the carbon nucleus and low natural abundance of the ^{13}C isotope (1.1%) gives rise to a S/N ratio that is 5556 times less than that of the proton. In routine ^{13}C NMR, the spectra are obtained with complete proton decoupling. This process increases the sensitivity of the signals and reduces the complexities of the spectrum as well. An additional gain in sensitivity is achieved from the nuclear Overhauser effect (NOE). The NOE involves saturation of one signal in the spectrum and observation of the changes of the intensities of other signal. In $^{13}C—^{1}H$ systems, the intramolecular dipole–dipole relaxation is responsible for this effect. Thus, the sensitivity of a signal may be enhanced by a factor of 2.98 (called nuclear Overhauser enhancement factor).

To a good approximation, three items influence the ^{13}C chemical shifts. They are

- A paramagnetic shielding term, mainly caused by mixing higher electronic states with the ground state.
- A diamagnetic shielding term arising from electron density at the nucleus.
- An anisotropy term that includes the contribution of electron currents from neighboring atoms.

Chemical shift values are given in parts per million downfield from the TMS reference signal.

There are two main problems associated with the measurement of ^{13}C NMR spectra of diamagnetic organometallic compounds. The first is the lengthy data collection time. This is a result of the long spin–lattice relaxation times for those carbon atoms that are not directly bonded to protons. For example, carbonyl compounds do not exhibit strong signals. This problem can be partially alleviated by adding a paramagnetic compound [such as $Cr(acac)_3$, prepared in Experiment 22A] to the solution.

The second problem is associated with high natural abundances of quadrupolar metal nuclei, which interfere with the spectra of organometallic compounds by broadening the signals. Quadrupolar nuclei are those with $I > \frac{1}{2}$. For example, the natural abundances of the ^{55}Mn nuclei ($I = \frac{5}{2}$) and ^{59}Co ($I = \frac{7}{2}$) are 100%. Thus, carbonyl complexes of these two metals would be expected to exhibit broadening of the carbonyl signals. In some cases (as in cyclopentadienyl complexes of Mn), however, sharp signals are observed nonetheless.

One of the most common inorganic applications of ^{13}C NMR spectroscopy is the structural investigation of organometallic complexes, and the influence of other ligands on overall structural changes in these systems. Carbon-13 NMR spectroscopy provides the following information:

- The dynamic processes in metal carbonyl systems in solution can be investigated.
- Molecular fluxionality (see Experiment 45) can be studied.
- The chemical shift of carbon is very much dependent on the chemical environment. For example, different chemical shifts are observed for the carbonyl signal in metal carbonyls depending on the metal involved. The shift also depends on the ligands present in the compound. An empirical relationship can be found between the carbonyl stretching frequencies and ^{13}C NMR signals. Table 6.7 lists organic ^{13}C chemical shifts of common functional groups.
- While $^{13}C—^{1}H$ coupling constants are rarely determined, off-resonance decoupling experiments provide useful information about the number of protons directly bonded to a carbon atom.
- Relaxation data provide information on the mobility of carbon atoms in a compound.

Inorganic ^{13}C chemical shifts vary over a wide range, depending on the metal, ligand, solvent, and other factors. A complete overview of applications of ^{13}C

Table 6.7 ^{13}C Organic Functional Group Chemical Shifts

Shift[a]	Group	Example	Type of Carbon
0–40	$\mathbf{C}H_3$—	$\mathbf{C}H_3\mathbf{C}H_3$	Methyl
10–50	—$\mathbf{C}H_2$—	$CH_3\mathbf{C}H_2CH_3$	Methylene
15–50	—$\mathbf{C}$H—	$(CH_3)_3\mathbf{C}H$	Methyne
10–65	—$\mathbf{C}$—X	$CH_3\mathbf{C}H_2Br$	Carbon attached to halogen, or amine
100–170	—$\mathbf{C}$=	$\mathbf{C}H_2{=}CHCH_3$	Vinyl
100–170	ϕ	$\mathbf{C}_6H_6$	Phenyl
120–130	—$\mathbf{C}$N	$CH_3\mathbf{C}N$	Nitrile
150–220	—$\mathbf{C}$=O	$CH_3\mathbf{C}OCH_3$	Ketone

[a] All chemical shifts are in parts per million, downfield from TMS.

NMR to inorganic chemistry and listings of chemical shifts for various organometallic compounds may be found in Ref. 6.

6.D.8 Quadrupolar Nuclei

We noted that the hydrogen nuclei in Figure 6.20 were not split by the bromine. Bromine has two NMR active isotopes, ^{79}Br and ^{81}Br, both in ~50% natural abundance. Both have spin quantum numbers of $\frac{3}{2}$.

Hydrogen, with a spin of $\frac{1}{2}$, has two spin states: $+\frac{1}{2}$ and $-\frac{1}{2}$. In general, any nucleus of spin I will have $2I + 1$ spin states: $I, I - 1, I - 2, \ldots, -I$. Thus for bromine, the spin states are $+\frac{3}{2}$, $+\frac{1}{2}$, $-\frac{1}{2}$, and $-\frac{3}{2}$. Nuclei with spins $>\frac{1}{2}$ are called quadrupoles. Nuclei with spins of $\frac{1}{2}$ are called dipoles. Quadrupolar nuclei tend to move quickly (have fast relaxation times), and average out in the NMR timeframe.[7] In this case, the bromine nucleus is relaxing so quickly that the hydrogen only sees the average spin state (0). As far as the ^{1}H spectrum is concerned, therefore, the bromine is rendered NMR inactive by its quadrupolar relaxation, and the hydrogen signal appears to be unsplit. Since bromine is actually NMR active, one can observe its resonance at the appropriate frequencies for ^{79}Br and ^{81}Br NMR spectroscopy. Other quadrupolar nuclei do not relax as quickly as bromine and can therefore spin couple. The ^{11}B nucleus, for example, has $I = \frac{3}{2}$ and a fairly long relaxation time. Hydrogen atoms next to a single ^{11}B may therefore be split into four lines of equal size, corresponding to the four possible spin states of ^{11}B ($+\frac{3}{2}$, $+\frac{1}{2}$, $-\frac{1}{2}$, $-\frac{3}{2}$). The relatively high natural abundance ^{11}B nuclei can also couple to each other if they are in different magnetic environments.[8]

Spectra between these two extremes (bromine—no coupling, some ^{11}B compounds—total coupling) are well known, and discussed extensively in the chemical literature.[9]

6.D.9 The Coupling Constant

When an NMR signal is split through coupling, the separation between the lines is a useful piece of chemical information, called the coupling constant. Nuclei "see" each other through their electron clouds (i.e., through their bonds). The more electrons present between the nuclei, the better they see each other and the more separated the lines occur. For example, in the ^{13}C spectrum, the ^{13}C—^{13}C coupling constant in ethane (single bond) is 35 Hz, in ethylene (double bond) is 67 Hz, and in acetylene (triple bond) is 171 Hz. The symbol given to the coupling constant is J.

A more sophisticated use of coupling constants is seen in the ^{11}B—^{11}B coupling constants of the boron hydrides. In the case of diborane(6), B_2H_6, the borons are connected through two BHB bridging three-center bonds. There is no electron density directly between the boron nuclei, therefore the coupling constant is quite small ($J_{BB} < 1$ Hz).[10] In B_3H_7CO, the three boron nuclei are connected via a BBB three-center bond (the bond order is therefore $\frac{2}{3}$), and a coupling constant of $J_{BB} = 11$ Hz is observed.[11] In tetraborane(10), B_4H_{10}, boron atoms 1 and 3 are connected by a "normal" two-center B—B bond (bond order is 1), with a coupling constant of $J_{BB} = 25$ Hz.[12] Boron atoms 1 and 2 (and also 1,4; 2,3; and 3,4) are connected through hydrogen bridge bonds, and as seen in diborane(6), the coupling constant is small, on the order of 1 Hz. Structures of the three compounds are shown in Figure 6.23. It is easily seen that as the bond order between the boron nuclei increases, the coupling constant also increases. This is essentially the only manner in which the nature of bonding in boron hydrides can be established experimentally. Coupling is summarized in Table 6.8.

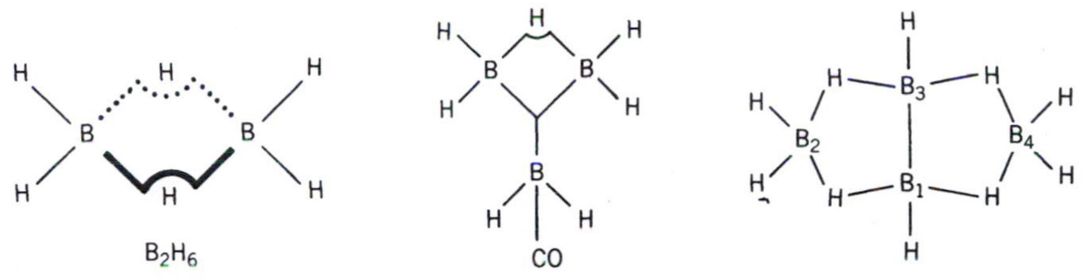

B_2H_6 B_3H_7CO B_4H_{10}

Figure 6.23. *Structures of several boron hydrides.*

6.D.10 Interpretation of Inorganic Spectra

Generally, inorganic NMR spectra are no more difficult to interpret than ^1H spectra, except when quadrupoles are involved. Figure 6.24 shows the ^{11}B NMR spectrum of B_4H_{10}. There are two resonances, a triplet [6.9 ppm upfield from $BF_3{\cdot}O(C_2H_5)_2$] and a doublet (40 ppm upfield). In the structure of B_4H_{10} (Fig. 6.23), we can readily see that boron atoms 1 and 3 are chemically equivalent, as are boron atoms 2 and 4.

The resonances are split because of coupling, so one must consider what types of coupling may be present. The boron nuclei have two different types of hydrogen bonded to them: terminal (normal) hydrogen atoms and bridging hydrogen atoms. We have already seen that bridging hydrogen atoms are bonded through three-center bonds, so their coupling constants should be very small. The only significant $^{11}B—^1H$ coupling will therefore be to the terminal hydrogen atoms. There will be no significant $^{11}B—^{11}B$ coupling, as we already saw that coupling through hydrogen bridge bonds is small (which eliminates coupling between boron atoms 1 and 2, 2 and 3, 3 and 4, and 1 and 4). Furthermore, boron atoms 1 and 3 are equivalent, as are boron atoms 2 and 4, so only a $^{10}B—^{11}B$ coupling can be obtained (often not observed due to the low abundance of ^{10}B). We conclude that the only significant coupling is between the boron atoms and their terminal hydrogen atoms.

Boron atoms 1 and 3 have one terminal hydrogen, so we would expect the ^{11}B signal to be split into a doublet. This is obviously the signal at 40 ppm. Boron atoms 2 and 4 have two terminal hydrogen atoms, so we would expect the ^{11}B signal to be split into a triplet. This is obviously the signal at 6.9 ppm. The signal at 6.9 ppm also shows some weak residual coupling, which appears

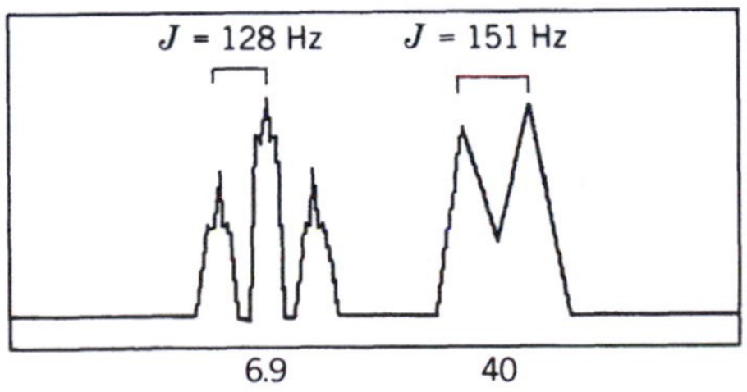

Table 6.8 Summary of Coupling

Splitting	A signal will be split by any neighbors that are NMR active (except for quickly relaxing quadrupoles). The maximum number of lines (M) in the resulting signal is easily obtained from the formula $$M = 2\,nI + 1$$ where n is the number of equivalent nuclei that the observed nucleus is coupled to, and I is the spin of the coupled nuclei.
Size of peaks	The relative size of each peak in the split signal can be obtained from the appropriate Pascal's triangle.
Separation of lines	The separation of peaks in a split signal depends on the electron density between the coupled nuclei. Typical values may be found in Ref. 1.

J = 128 Hz J = 151 Hz

6.9 40

Chemical shift (ppm) from $BF_3{\cdot}O(C_2H_5)_2$

Low field High field

Figure 6.24. *^{11}B NMR spectrum of B_4H_{10}.*

to be splitting each line in the triplet into another triplet. This is due to very weak coupling to the two bridge hydrogen atoms. This type of coupling is so weak that it is not seen at all in the 40 ppm signal.

The hydrogens on boron atoms 2 and 4 have somewhat smaller coupling constants to the boron (128 Hz) than do the hydrogens on boron atoms 1 and 3 (151 Hz). This is not surprising, as similar differences in coupling constants are seen when comparing CH_2 group couplings with CH group couplings in organic NMR.

6.D.11 Spectral Collapse

Note that the spectrum for B_4H_{10} (Fig. 6.24) is very broad compared to an organic NMR spectrum. This occurs because ^{11}B is a quadrupolar nucleus. It was seen earlier in the case of ethyl bromide that the coupling to bromine was totally gone (collapsed or decoupled) resulting from the fast relaxation time of the bromine nucleus. The ^{11}B relaxation times are intermediate between the fast relaxation time of bromine and the slow relaxation times of dipolar nuclei (such as ^1H). This leads to spectra that are more or less collapsed, depending on the specific relaxation time of ^{11}B in the compound. A convenient set of terms to predict the degree of spectral collapse is

$$C = \text{degree of collapse} = 2\pi J T_1$$

For large values of C (10 or larger, if J is in units of hertz and T_1 in seconds), the spectrum is sharp and well defined. As C decreases, the spectrum begins to broaden out, and collapse. At a value of $C = 1$ or less, the spectrum has totally collapsed into a singlet (the coupling has disappeared), which gets sharper as C gets smaller. This is shown in Figure 6.25.

A large problem that is readily seen is that as the spectrum collapses, the resonance gets narrower (the total quartet is broader than the doublet of doublets, which is broader than the doublet, which is broader than the singlet, etc.). This being the case, the separation between the lines in the resonance also gets narrower as the spectrum collapses. If one measures the coupling constant from these various spectra, it appears to be getting smaller, even though it is actually T_1 that is changing. In spectra of quadrupolar nuclei, therefore, great care must be taken in obtaining coupling constants to ensure that they are accurate, and one must often resort to computer simulation of the spectrum.[12,13]

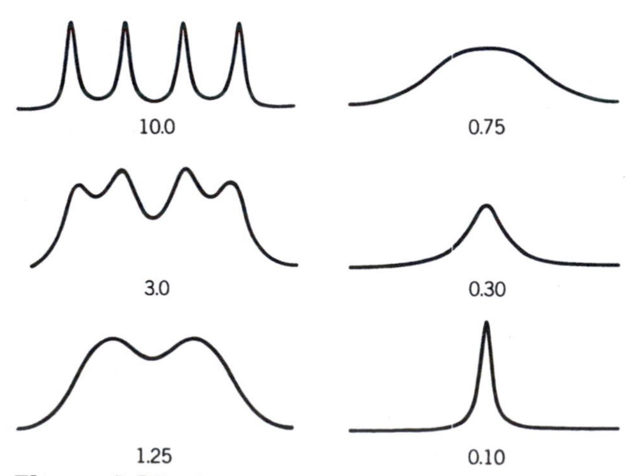

Figure 6.25. *Spectral collapse.*

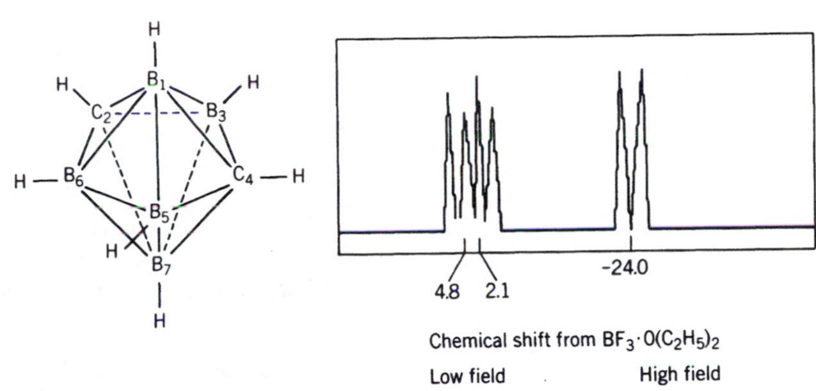

Figure 6.26. *Structure and* ^{11}B *NMR spectrum of* $C_2B_5H_7$.

6.D.12 Decoupling

In many cases, an NMR spectrum is so complicated that it is difficult to interpret. In these cases, it may be desirable to remove the coupling of (decouple) one nucleus or another. In the case of a boron hydride, for example, one can remove the coupling of the hydrogen and observe the ^{11}B spectrum. The decoupling phenomenon is accomplished by irradiating the sample with energy (a few watts is sufficient) at the frequency of the nucleus (^{1}H in this case) to be decoupled. This causes the nucleus to change spin states rapidly in the NMR timeframe, similar to a fast quadrupolar nucleus. If the irradiation frequency is one specific frequency, any given signal in the NMR can be selectively decoupled. If the irradiation frequency is actually a broad band of frequencies, all of the coupled nuclei (all of the hydrogen atoms in this case) can be decoupled.

Figure 6.26 shows the structure and ^{11}B spectrum of 2,4-dicarba-*closo*-heptaborane(7), $C_2B_5H_7$.[14] It is easily seen that there are three different types of boron: apical (B_1 and B_7), equatorial (B_5 and B_6), and one unique equatorial boron (B_3). Each boron has a single hydrogen bonded to it, so each boron signal will appear as a doublet. The boron nuclei do not couple among themselves in this case. The spectrum should therefore consist of three doublets. It is clear that two of the doublets are overlapping in the 3 ppm region (a minus sign indicates that the signal is upfield from the reference compound). By decoupling the ^{1}H nuclei, the spectrum becomes much simpler, and easier to assign. The decoupled spectrum is shown in Figure 6.27.

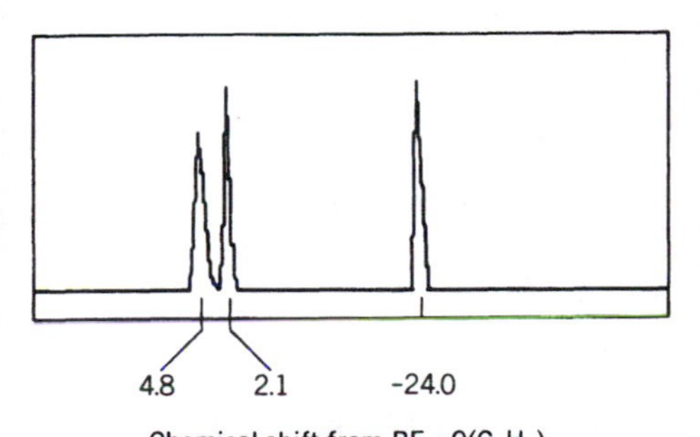

Figure 6.27. ^{1}H *Decoupled* ^{11}B *NMR spectrum of* $C_2B_5H_7$.

From the decoupled spectrum, it is easily seen that the first and third peaks in the coupled spectrum are one doublet, and the second and fourth peaks in the coupled spectrum are a second doublet. The peaks can now be assigned. The high field doublet is assigned to the apical boron atoms, and the two low field doublets are assigned to the equatorial boron atoms. The doublet at 4.8 ppm can be assigned to B_3 for two reasons.

1. The peak is smaller. There is only one boron of this type, and two of the other, B_5 and B_6.
2. B_3 is next to two carbon atoms, which are more electronegative than boron, and should therefore be to lower field than B_5 and B_6, which are next to only one carbon.

6.D.13 Identification of Isomers

Nuclear magnetic resonance spectroscopy is an excellent tool for distinguishing among geometrical isomers. It is far better than IR spectroscopy for this purpose. Consider the NMR spectra of pentaborane(9), B_5H_9, and its monochlorination

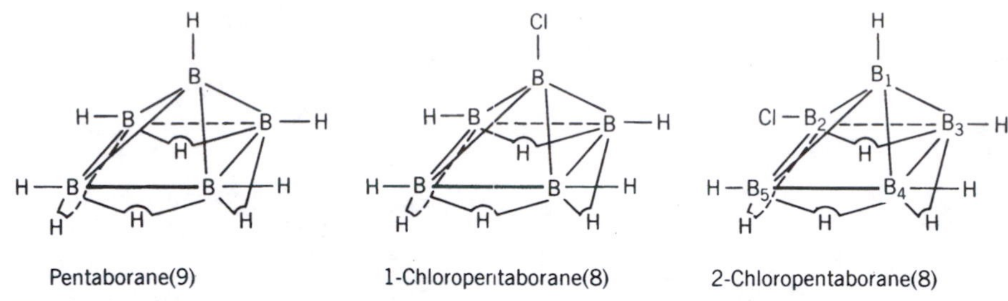

Figure 6.28. *Structures of pentaborane(9) and chlorination derivatives.*

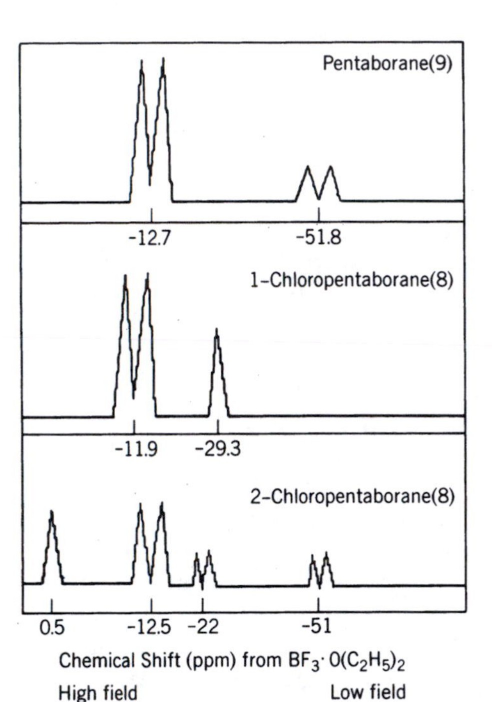

Figure 6.29. *^{11}B NMR spectra of pentaborane(8) and chloro derivatives.*

products 1-chloropentaborane(8) and 2-chloropentaborane(8).[14] The structures of these compounds are shown in Figure 6.28 and their spectra in Figure 6.29.

From the structure of pentaborane(9), it is readily seen that there are two different types of boron: the single apical boron atom and the four equatorial boron atoms. The four equatorial boron atoms are equivalent, so two signals are expected in the NMR spectrum. Each boron is directly bonded to a single hydrogen, therefore the two signals should be doublets. The larger doublet (−12.7 ppm) is from the four equatorial boron atoms and the smaller doublet (−51.8 ppm) from the single apical boron.

When pentaborane(9) is chlorinated, if the chlorine substitutes on the apical boron, all four equatorial boron atoms remain equivalent. Since the equatorial boron atoms are still bonded to hydrogen, their signal remains a doublet, although it is shifted slightly downfield (to −11.9 ppm) because of the presence of the chlorine atom some distance away. The apical boron resonance is no longer split, since the chlorine does not couple (chlorine is another fast quadrupole, similar to bromine). The resonance is therefore a singlet. The signal is shifted significantly downfield, due to the presence of the electronegative chlorine (from −51.8 to −29.3 ppm).

If the chlorine substituent is on one of the equatorial boron atoms, the four equatorial atoms are no longer equivalent. It is easily seen that boron atoms 3 and 5 are equivalent, while all other boron atoms are unique. The chlorine substituted boron (no. 2) is readily identified, as it is the only boron without a hydrogen bonded to it. The signal is therefore a singlet and should be at very low field (0.5 ppm). The large doublet at −12.5 ppm corresponds to boron atoms 3 and 5, for both size and chemical shift reasons. The last of the equatorial boron atoms, boron 4, appears as a doublet (as it is bonded to one hydrogen) at −22 ppm. The apical boron, boron 1, appears as a doublet at −51 ppm. Boron atoms 4 and 1 may be distinguished from each other by comparing the spectrum of pentaborane(9) with that of 2-chloropentaborane(8). The position of the apical boron is essentially unchanged, moving from −51.8 to −51 ppm. The spectra of the two chlorinated pentaboranes bear no similarity to each other, and are easily assigned to the proper isomer.

REFERENCES

1. Brevard, C.; Granger, P., *Handbook of High Resolution Multinuclear NMR*, Wiley: New York, 1981.

2. Pines, A.; Gibby, M. G.; Waugh, J. S. *J. Chem. Phys.* **1973,** *59*, 569.

3. Bertrand, R. D; Moniz, W. B.; Garroway, A. N.; Chingas, G. C. *J. Am. Chem. Soc.* **1978,** *100*, 5227.

4. Kerrison, S. J. S.; Sadler, P. J. *J. Magn. Reson.* **1978,** *31*, 321.

5. Mann, B. E.; Taylor, B. F., *^{13}C NMR Data for Organometallic Compounds*, Academic Press: London, 1981.

6. (a) Chisholm, M. H.; Godleski, S., "Applications of Carbon-13 NMR in Inorganic Chemistry" in *Progress in Inorganic Chemistry*, S. J. Lippard,

Ed., Wiley: New York, 1976, Vol. 20, p. 299. (b) Jolly, P. W.; Mynott, R., "The Application of ^{13}C-NMR to Organo-Transition Metal Complexes" in *Advances in Organometallic Chemistry*, Academic Press: New York, 1981, Vol. 19, p. 257.

7. Schlichter, C. P., *Principles of Magnetic Resonance*, 2 ed., Springer–Verlag: New York, 1978.

8. For example, Stampf, E. J.; Garber, A. R.; Odom, J. D.; Ellis, P. D. *Inorg. Chem.* **1975**, *14*, 2446.

9. For example, Gunther, H.; Moskau, D.; Bast, P; Schmalz, D. *Angew. Chem. Int. Ed.* **1987**, *26*, 1212.

10. Odom, J. D.; Ellis, P. D.; Lowman, D. W.; Gross, M. H. *Inorg. Chem.* **1973**, *12*, 95.

11. Stampf, E. J.; Garber, A. R.; Odom, J. D.; Ellis, P. D. *J. Am. Chem. Soc.* **1976**, *98*, 6550.

12. Szafran, Z., *A Theoretical and Experimental Study of Boranes*, Ph.D. Thesis, University of South Carolina, Columbia, 1981.

13. Weiss, R.; Grimes, R. *J. Am. Chem. Soc.* **1978**, *100*, 1401.

14. Information taken from, and figures redrawn from: Smith, W. L. *J. Chem. Educ.* **1977**, *54*, 469.

General References for NMR

Derome, A. E., *Modern NMR Techniques for Chemical Research*, Pergamon: Oxford, 1987.

Chakravorty, A., "High-Resolution Nuclear Magnetic Resonance," in *Spectroscopy in Inorganic Chemistry*, Vol. 1, C. N. R. Rao and J. R. Ferraro, Eds., Academic Press: New York, 1970.

Dechter, J. J., "NMR of Metal Nuclides, Part 1" in *Progress in Inorganic Chemistry*, S. J. Lippard, Ed., Interscience: New York, 1982, Vol. 29, p. 285.

Dechter, J. J., "NMR of Metal Nuclides, Part 2" in *Progress in Inorganic Chemistry*, S. J. Lippard, Ed., Interscience: New York, 1985, Vol. 33, p. 393.

Chisholm, M. H.; Godleski, S., "Application of Carbon-13 NMR in Inorganic Chemistry" in *Progress in Inorganic Chemistry*, S. J. Lippard, Ed., Interscience: New York, 1976, Vol. 20, p. 299.

Ebsworth, E. A. V.; Rankin, D. W. H.; Craddock, S., *Structural Methods in Inorganic Chemistry*, Blackwell: Oxford, 1987.

Levy, G. C., *Topics in Carbon-13 NMR Spectroscopy*, Wiley: New York, 1976. Volume 2 contains a long section on "^{13}C NMR Studies of Organometallic and Transition Metal Complex Compounds" on pp. 213–341.

6.E ATOMIC ABSORPTION SPECTROSCOPY

6.E.1 Introduction

Atomic absorption spectroscopy (AAS), for quantitative analysis purposes, dates from 1955. Atomic absorption is based on neutral or ground-state atoms of an element absorbing electromagnetic radiation over a series of very narrow wavelengths. A simplified system is shown in Figure 6.30.

In AAS, radiation sources [hollow-cathode (HC) lamps] are used that are specific for each element. It is possible to have dual or triple lamp sources in selected cases. The sample to be analyzed is converted into an atomic vapor, generally accomplished by spraying a solution into a flame.

6.E.2 Theory

When a solution of the species to be analyzed is aspirated into a flame (or graphite furnace), the heat of the flame first causes the solvent to evaporate. Species to

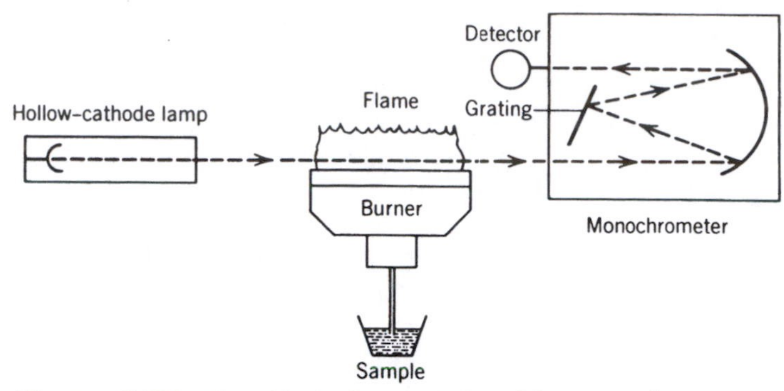

Figure 6.30. *Simplified schematic of an AA spectrophotometer.*

be analyzed are generally in the form of salts. Using $MgCl_2 \cdot H_2O$ as an example,

$$MgCl_2 \cdot H_2O \rightarrow MgCl_2\ (s) \qquad \text{(desolvation, } \sim 2000\ ^\circ\text{C)} \tag{6.1}$$

The microcrystals remaining are then partially or wholly dissociated into single atoms in the gas phase.

$$MgCl_2\ (s) \rightarrow MgCl_2\ (g) \qquad \text{(vaporization)} \tag{6.2}$$

$$MgCl_2\ (g) \rightarrow Mg^0\ (g) + 2Cl^0\ (g) \qquad \text{(atomization)} \tag{6.3}$$

Some of these gaseous atoms absorb radiant energy at precisely the same wavelengths of those emitted by the hollow-cathode light source. As a consequence of this absorption, these atoms become excited to a higher electronic state.

$$Mg^0\ (g) + h\nu \rightarrow Mg^* \qquad \text{(excitation)} \tag{6.4}$$

Because of this absorption, the intensity (power) of the radiation from the hollow-cathode lamp source is decreased. The measurement of this decrease due to absorption is the basis for AA analysis.

It is important to realize that the selected wavelength spread is very narrow, both from the emission line of the source and the absorption line of the element in the flame or furnace. For this reason, there is essentially no chance of interference by absorption of spectral lines of other elements. Most elements can be detected by AA analysis (Fig. 6.31). The sensitivity for most elements is ~1 ppm (1 $\mu g \cdot mL^{-1}$). Accuracy is usually in the range of $\pm 2\%$.

6.E.3 The Instrument

The Flame

Most inexpensive AA instruments employ an acetylene–air flame system, although the N_2O–air combination is also commonly used.

The Nebulizer Burner System

The burner system (Fig. 6.32) converts the test substance in the sample to atomic vapor. Ideal droplets (water solvent) are ~20 nm in diameter. Those larger than 20 nm are lost in the spray chamber and are collected by drainage. A further portion fail to desolvate completely before leaving the flame. This loss, by the above two processes, accounts for 90% of the aspirated volume. On the average, the aspiration rate is ~5 $mL \cdot min^{-1}$.

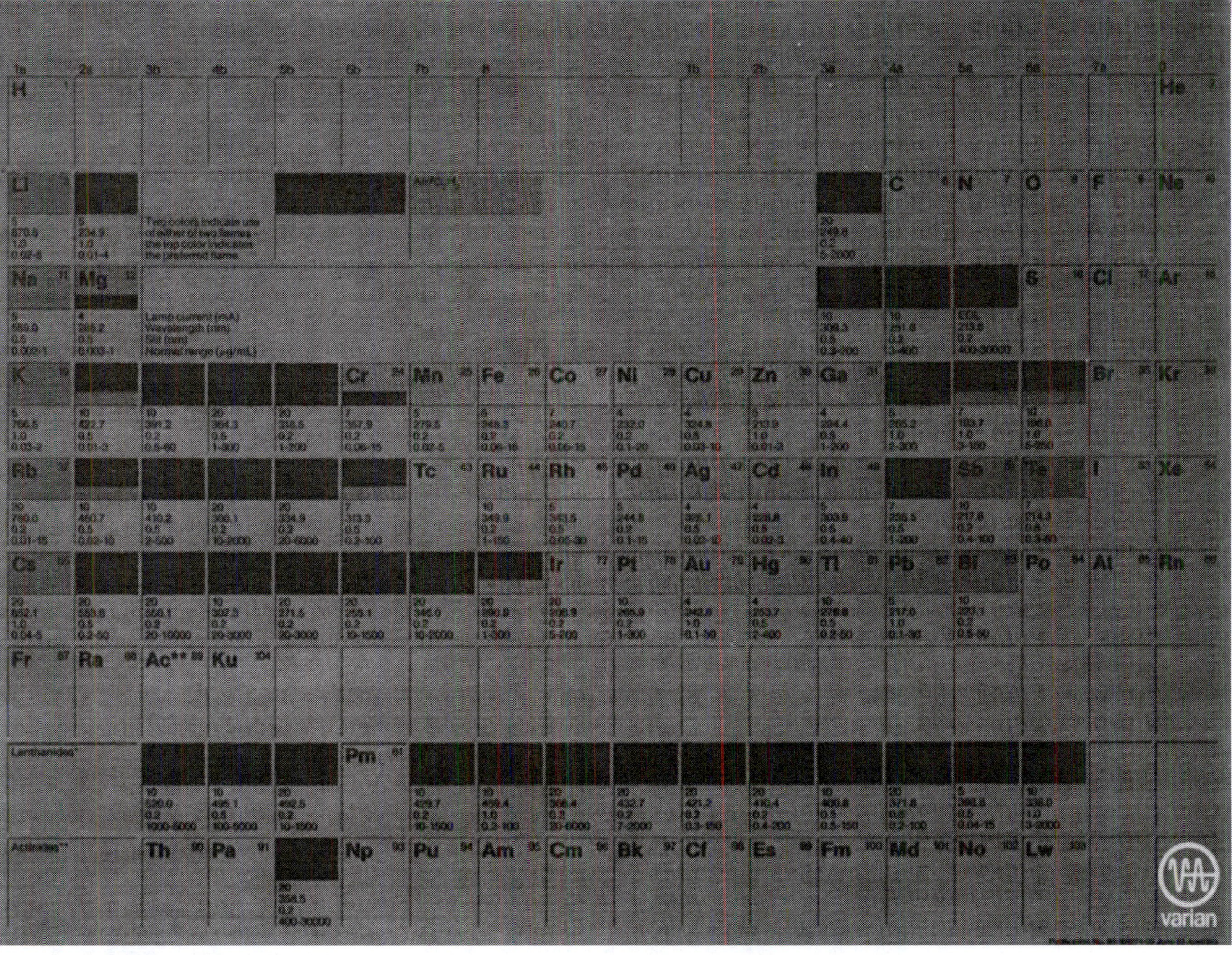

Figure 6.31. *Operating conditions for flame AA spectroscopy (dark areas–N_2O/C_2H_2; light areas–air/C_2H_2) (Courtesy of Varian Techtron Pty Limited, Mulgrave, Australia.)*

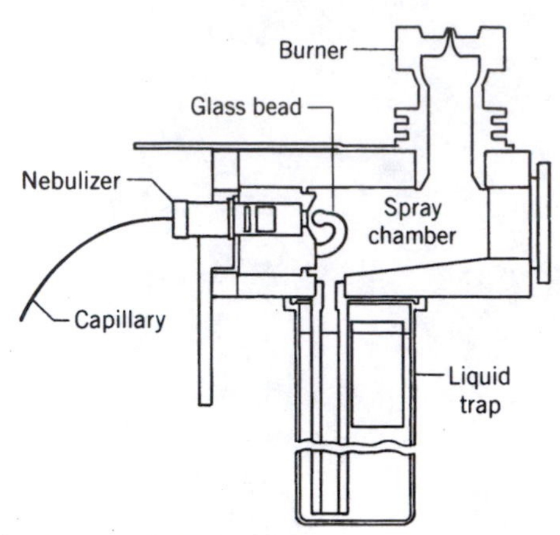

Figure 6.32. *Cross section of spray chamber–burner system. (Courtesy of Varian Techtron Pty Limited, Mulgrave, Australia.)*

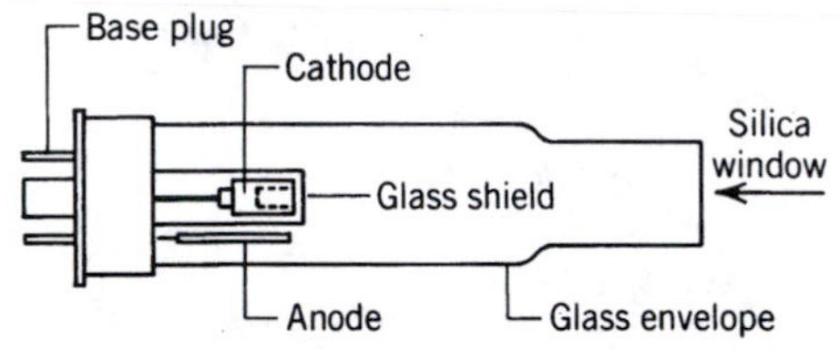

Figure 6.33. *Varian-Techtron hollow-cathode lamp. (Courtesy of Varian Techtron Pty Limited, Mulgrave, Australia.)*

Figure 6.34. *Spectral output of a nickel hollow-cathode lamp. (Courtesy of Varian Techtron Pty Limited, Mulgrave, Australia.)*

The Light Source

The hollow-cathode lamp (Fig. 6.33) is evacuated to a pressure of 1–5 mm but contains a small amount of inert gas (He, Ne, or Ar). The cathode is composed of the same metal as the element to be determined in the analysis. The anode is composed of W, Ni, or Zr.

A voltage of about 400 V charges the atoms of the inert gas. One sees a low pressure discharge glow, similar to that observed in a Crooke's tube. This is a result of positive ions being formed by impact with high speed electrons.

$$He + e^- \rightarrow He^{2+} + 3e^-$$

The positive gas ions bombard the cathode. This ionic bombardment sputters metal atoms into the tube atmosphere, where they become excited by collision (with e^- and ions), which results in the emission of light of a wavelength characteristic of the cathode metal, that is, an emission spectrum is generated. The output of a nickel hollow-cathode lamp is shown in Figure 6.34.

The Monochromator

As seen from Figure 6.34, the hollow-cathode lamp generally emits more than one emission line. The required spectral line can be isolated by means of a relatively low dispersion monochromator. The most intense line is chosen to provide maximum sensitivity. The exit and entrance slit width of the monochromator need only be sufficiently narrow to isolate the particular line being used in the analysis. An acceptable spectral bandwidth ranges from 0.1–1.0 nm.

Although the absorption line of an element to be measured is narrow, it is still broad compared to the emission line used. This is advantageous, in that the absorption line can be measured at its peak maximum (Fig. 6.35).

The Detector

The most common detector used in AA instruments is a photomultiplier tube. This tube has the capacity to convert light energy into an electrical current. This electrical impulse is then relayed to the output system of the instrument.

6.E.4 Measurement of Concentration

The ratio of radiant power of the incident beam, P_0 (from the HC lamp), and of the transmitted beam, P_t (after passing through the flame), is measured. The amount of light absorbed depends on the number of absorbing atoms in the light path. Provided the flame is hot enough to convert a chemical compound to free atoms, the light absorbed is almost independent of the flame temperature and of the absorption wavelength.

If the flame conditions and rate of aspiration of the sample into the flame are kept approximately constant, the absorbance, $\log(P_0/P_t)$, is then directly proportional to the concentration of the given metal atom in the sample. Thus, the Beer–Lambert law is followed:

$$\log(P_0/P_t) = A = \epsilon bc$$

where

A = absorbance
ϵ = molar absorptivity
b = path length in centimeters
c = concentration in moles per liter

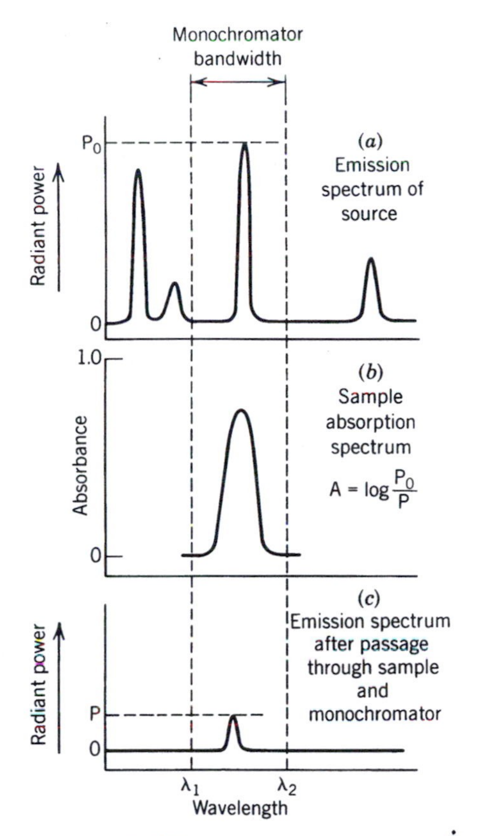

Figure 6.35. *Absorption of a resonance line by atoms.* (From Principles of Instrumental Analysis, *3rd ed., by D. A. Skoog, copyright © 1985 by Saunders College Publishing Co., a division of Holt, Rinehart and Winston, Inc., reprinted by permission of the publisher.)*

This relationship is a fundamental law governing all types of electromagnetic radiation. It is important to note that the equation states that absorbance is directly proportional to the concentration of the absorbing species when the path length is fixed. Conversely, the absorbance is directly proportional to the light path length when the concentration is constant. Generally, the instrument is operated at a fixed light path length, and thus absorbance relates directly to concentration.

For quantitative analysis, the standard solutions and the sample solutions should be as similar as possible. A calibration curve is obtained, using a series of solutions of known concentrations over the range of interest (see Fig. 6.36). A new curve should be plotted each day for analytical work, since it is often difficult to exactly reproduce flame and aspiration conditions, thus leading to a change in the slope of the curve.

6.E.5 Other Considerations

The AA method is quite insensitive to the matrix, that is, the solvent system and other ions. For example, Mg^{2+} aspirated in distilled water gives the same results as in native or seawater.

If one does observe interference, the sample can be "doped." For example, in the analysis of Ca^{2+}, the presence of sulfate or phosphate ion in the matrix can affect the linearity of the working curve. In this case, the sample solution can be "doped" with lanthanum ion, La^{3+}, which scavenges the sulfate or phosphate ions and releases Ca^{2+} to behave in the normal manner.

One disadvantage of flame AAS is the low spray efficiency (10%). It is therefore necessary to have a good size sample and aspirate at approximately 2–5 $mL \cdot min^{-1}$. However, the efficiency can be increased using one of several techniques.

1. Use of a prehot mode.
2. Multiple pass of the signal through the flame.
3. Use of a higher temperature fuel.
4. Use of an organic phase, such as a 50:50 solution of methyl isobutyl ketone–isopropyl alcohol. This phase gives a finer spray and can increase the efficiency by a factor of 10.

Nonflame methods are also available. The most important of these is the graphite furnace technique. Figure 6.37 shows a cross-sectional view of a commercial electrothermal atomizer.

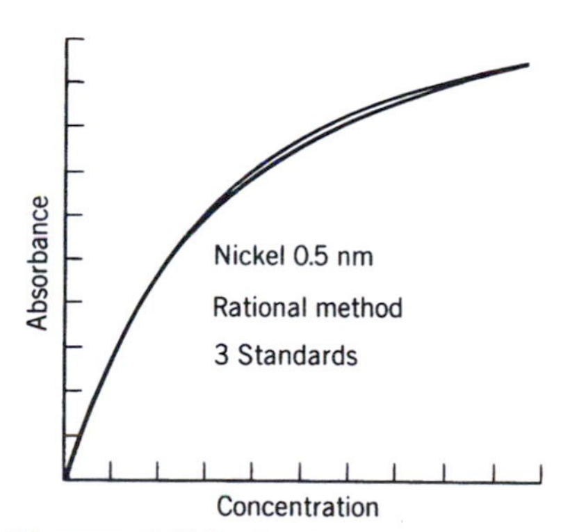

Figure 6.36. *Parabolic rational function correction to the calibration curve. (Courtesy of Varian Techtron Pty Limited, Mulgrave, Australia.)*

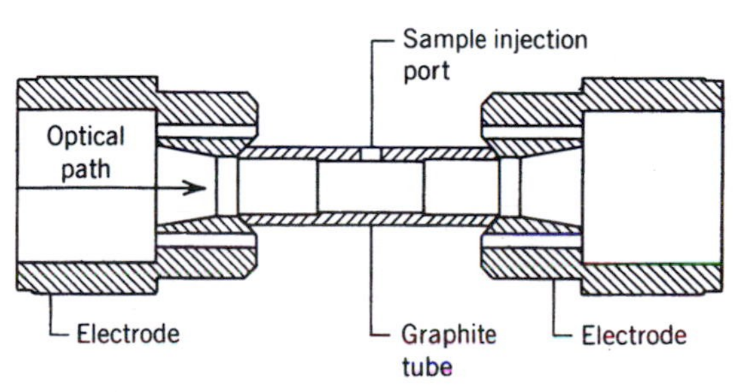

Figure 6.37. *Graphite tube atomizer. (Courtesy of Varian Techtron Pty Limited, Mulgrave, Australia.)*

General References

Walsh, A. *Anal. Chem.* **1974,** *46,* 698A.

Kahn, H. *J. Chem. Educ.* **1966,** *43,* A7.

Varmer, A., *Handbook of Atomic Absorption Analysis,* CRC Press: Boca Raton, FL, Vols. I and II, 1982.

Christian, C. D.; O'Reilly, J. E., *Instrumentation Analysis,* Allyn & Bacon: Boston, 1986, Chapter 10.

Skoog, D. A., *Principles of Instrumental Analysis,* 3rd ed., Saunders: Philadelphia, PA, 1985, Chapter 9.

Bennett, P. A.; Rothery, E., *Introducing Atomic Absorption Analysis,* Varian Techtron Pty Ltd.: Mulgrave, Australia, 1983.

Welz, B., *Atomic Absorption Spectrometry,* 2nd ed., VCH: Weinheim, Federal Republic of Germany, 1985.

Chapter 7
Chemistry of the Main Group Elements

Experiment 1 Preparation of Sodium Amide

INTRODUCTION

Water and ammonia are both examples of liquids of low molecular weight that have high heats of vaporization and high boiling points due to hydrogen bonding. Ammonia is a liquid in the temperature range −78 to −33 °C under atmospheric pressure, it conducts well, and is an excellent solvent. The N—H bonds in ammonia are less polar than the O—H bonds in water. The lower boiling point suggests that hydrogen bonding is much weaker in ammonia than in water. The dielectric constant and the self-ionization constant are both lower, indicating the liquid ammonia will be a poorer solvent than water for ionic compounds.

Most salts are less soluble in liquid ammonia than in water. There are, however, a few very interesting exceptions to this generalization. Ammonium salts, iodides, and nitrates are quite soluble in liquid ammonia. The familiar solubility rules for water are somewhat different in ammonia. When silver nitrate is added to a solution of barium chloride in water, silver chloride precipitates. In liquid ammonia, silver chloride and barium nitrate react to form a precipitate of barium chloride. This illustrates that silver chloride, as well as barium nitrate, are soluble in ammonia but barium chloride is not; whereas silver nitrate and barium chloride are soluble in water, but silver chloride is not.

The self-ionization of liquid ammonia is written as follows:

$$2NH_3 = NH_4^+ + NH_2^-$$

Thus, an acid (in ammonia) is defined as a substance that will increase the concentration of NH_4^+, and a base as one that will increase the concentration of NH_2^-. The most common base in liquid ammonia solution is potassium amide, KNH_2, which is more soluble than sodium amide. Acid–base titrations may be carried out between KNH_2 and ammonium compounds. The ammonia system is helpful in detecting very weakly acidic function in molecules. Thus, urea, $CO(NH_2)_2$, which is a weak base in water, acts as a weak acid in liquid ammonia.

The most striking property of liquid ammonia is its ability to dissolve all of the alkali and alkaline earth metals except beryllium. The ammoniated cations are metastable solutions from which the alkali metals can be recovered unchanged. Dilute solutions are deep blue and paramagnetic, because of the presence of solvated electrons.

$$Na \rightarrow Na^+_{(ammoniated)} + e^-_{(ammoniated)}$$

These solutions are strongly conducting and have reducing properties.

Concentrated solutions of alkali metals have a metallic bronze color, and are diamagnetic, indicating that electron pairing has occurred. The blue solutions are stable for a few weeks, slowly decomposing to form hydrogen gas and the metal amide.

$$M + NH_3 \rightarrow MNH_2 + \tfrac{1}{2}H_2$$

This laboratory is designed to introduce the techniques of working with liquid ammonia and to prepare a sodium amide salt.

Prior Reading and Techniques

Section 1.B.4: Compressed Gas Cylinders and Lecture Bottles

Section 5.C: Vacuum and Inert Atmosphere Techniques

EXPERIMENTAL SECTION

Safety Recommendations

Ammonia (CAS No. 7664-41-7): This compound is a pungent, nonbreathable gas. IHL-HMN LCLo: 5000 ppm/5M. It is harmful if swallowed, inhaled, or absorbed through the skin. Ammonia will form explosive compounds with many heavy metals and with the halogens.

Iron(III) nitrate nonahydrate (CAS No. 7782-61-8): This compound is harmful if swallowed or inhaled. ORL-RAT LD50: 3250 mg/kg.

Sodium (CAS No. 7440-23-5): This element is harmful if swallowed, inhaled, or absorbed through the skin. Contact of sodium with water can result in an explosion. It is extremely destructive to the skin and tissues. **Only handle while wearing gloves.** Exposure to air forms sodium oxide. Store the metal under mineral oil or kerosene.

−78 °C Slush: The 2-propanol (isopropyl alcohol)–dry ice slush will cause severe cold-burns if contacted with the skin.

Chemical Data

Compound	FW	Amount	mmol	mp (°C)	bp (°C)	Density
Na	22.99	100 mg	4.35	97.8		0.968
NH_3	17.03	Excess		−78	−33	

Required Equipment

N_2 gas source, 10-mL test tube with stopcock fitted side arm (Schlenk tube if available), mercury bubbler, dry ice–2-propanol slush in Dewar flask.

Time Required for Experiment: 2.5h.

EXPERIMENTAL PROCEDURE[1]

> **NOTE: *Carry out all reactions involving liquid ammonia in the HOOD.***

Secure a *tared* 10-mL test tube with a stopcock-fitted side arm to a ring stand. Add a crystal of $Fe(NO_3)_3 \cdot 9H_2O$ (which acts as a catalyst) to the tube and connect the side arm to a source of N_2 gas. Connect the tube opening to a mercury bubbler (see Fig. 7.1). The bubbler acts as an outlet for ammonia vapors, and also prevents air from entering into the system when all the ammonia is vaporized.

Flush the tube with dry N_2 gas for 15 min. Close the side arm stopcock, disconnect the line to the nitrogen source, and attach a line to a lecture bottle of ammonia, using rubber tubing. *Reopen the stopcock.* Cautiously, open the valve of the ammonia tank so that ammonia gas slowly bleeds into the tube. **Make sure that ammonia can freely escape through the mercury bubbler.** After a few minutes of flushing with ammonia, cool the tube in a −78 °C bath (2-propanol–dry ice) and collect 2–4 mL of liquid ammonia in the tube.

Close the valve on the ammonia tank, close the stopcock, and disconnect the ammonia lecture bottle. Quickly reattach the system to the nitrogen source. Some liquid ammonia will evaporate and escape through the mercury bubbler at this point—this is normal.

Place a pea-sized piece of sodium metal in kerosene in a crystallizing dish. Cut the sodium metal into small pieces, keeping them immersed under kerosene at all times. If a sodium press is available, sodium ribbon may be used with excellent results. Freshly exposed sodium surface is necessary for success with this experiment.

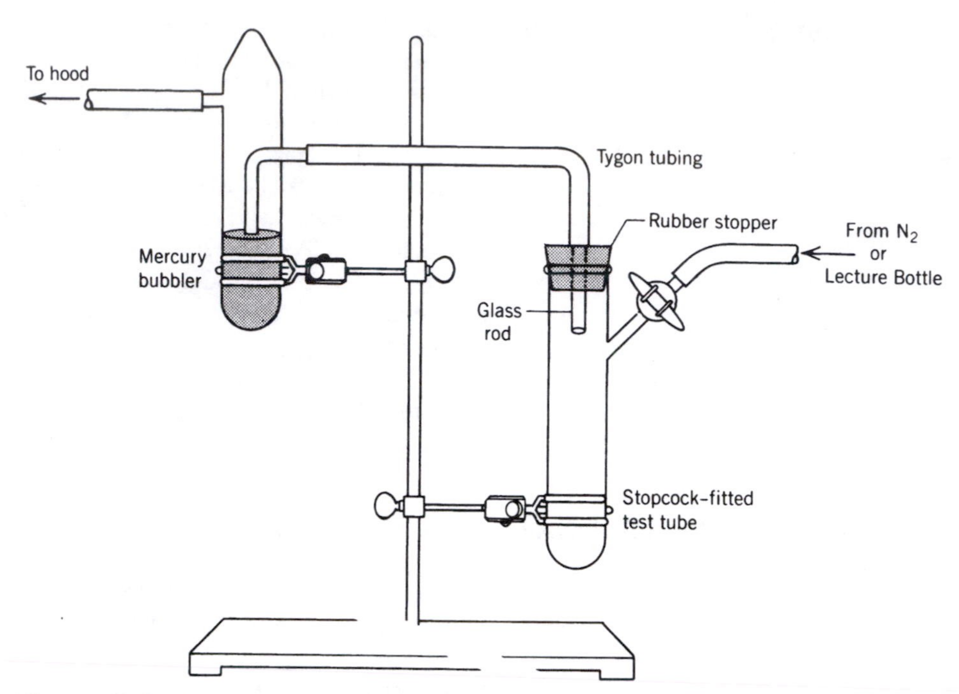

Figure 7.1. *Apparatus for Experiment 1.*

> **NOTE:** ***Do not expose the sodium to the atmosphere for long periods of time. It reacts with air, forming sodium oxide.***

Open the stopcock and resume the nitrogen flow through the system. With forceps, transfer ~100 mg of sodium (weigh on a rough balance) to the test tube by disconnecting the system from the mercury bubbler **momentarily.** The solution should become a deep blue upon the addition of the sodium. After ~30 min, the distinctive blue color of the solution should disappear, as sodium amide forms.

Isolation of Product

Remove the −78 °C bath, and gently flush N_2 gas over the solution to expel all ammonia as the system warms to room temperature. **It is very important that no air enters into the flask** or else the sodium amide may be partially oxidized.

> **NOTE:** ***Oxidized product is known to behave as an explosive, detonating by friction or heat. The oxidized material will be yellow or brown in color.***

Determining the Purity of Product

The purity of the product may be checked by adding excess standardized 0.01*M* HCl to a known weight of the product, and back-titrating the remaining HCl formed with standard 0.01*M* NaOH using phenolphthalein as an acid–base indicator. The principal impurity in the product is NaOH, which is formed as a result of the reaction of sodium amide with moisture.

> **NOTE:** ***Each millimole of sodium amide consumes 2 mmol of the acid and each millimole of NaOH reacts with 1 mmol of the acid.***

The reactions are

$$NaNH_2 + 2HCl \rightarrow NaCl + NH_4Cl$$

$$NaOH + HCl \rightarrow NaCl + H_2O$$

Use of Sodium Amide

In organic chemistry, sodium amide is useful as a reducing agent, as a dehydrohalgenating agent, and for other more specialized uses.[2] There are several interesting reactions that have been published utilizing this reagent, two of which are summarized here.

1. The heating of sodium amide and 3-cyclohexyl-2-bromopropene in mineral oil produces 3-cyclohexylpropyne (a reductive dehydrohalogenation) in good yield.[3]
2. The treatment of sodium amide in ammonia solution with chromium(III) chloride followed by a workup with concentrated nitric acid affords a high yield of $[Cr(NH_3)_6](NO_3)_3$.[4]

QUESTIONS

1. The reactions of sodium with liquid ammonia and with liquid water are quite different. Describe both and explain the difference.
2. Chlorides are generally quite soluble in water, whereas iodides are quite soluble in ammonia. Explain.
3. Detail three uses for sodium amide in organic chemistry not described in this experiment.
4. A colleague who teaches organic chemistry once said "Never send a nucleophile to do a base's job". Is sodium amide an exception to this rule?
5. While the alkali metal cations are well known, alkali metal anions can also exist in solution. Describe these anions, their preparations, and how their solutions were characterized. What work appears in the literature in the period from 1976–present? A useful starting reference is Dye, J. L. *J. Chem. Educ.* **1977,** *54*, 332.

REFERENCES

1. Greenlee, K. W.; Henne, A. L. *Inorg. Syn.*, **1946,** *2,* 128.
2. Fieser, L. F.; Fieser, M., *Reagents for Organic Synthesis,* Vol. 1, Wiley: New York, 1967, p. 1034.
3. Lespieau, R.; Bourguel, M., *Organic Syntheses* Collective Volume 1, 2nd ed., Wiley: New York, 1964, p. 191.
4. Angelici, R. J., *Synthesis and Technique in Inorganic Chemistry,* 2nd ed., Saunders: Philadelphia, 1977, p. 43.

GENERAL REFERENCE

Jolly, W. L., "Metal-ammonia Solutions" in *Progress in Inorganic Chemistry,* F. A. Cotton, Ed., Vol. 1, Interscience: New York, 1959, p. 235.

Experiment 2 Synthesis and Thermal Analysis of the Group 2 (IIA) Metal Oxalate Hydrates

INTRODUCTION

One of the most common ways of determining the stability of a compound is by measuring its physical response to the application of heat. There are many types of analytical thermal analysis techniques, but the two most common are **thermogravimetric analysis (TGA)** and **differential scanning calorimetry (DSC).**

In TGA, the sample being analyzed is heated following a preset temperature program (ramping the temperature by 10–20 °C per minute is most common), and the *loss of weight* of the compound is recorded. In DSC, it is the *absorption or release of energy* that is recorded over the same temperature range. Section 5.B gives a brief introduction to thermal analysis techniques and instrumentation.

The Group 2 (IIA) metal ions form insoluble oxalates from neutral or weakly acidic solution. These oxalates precipitate as white crystalline hydrated compounds of specific composition, which are ideal candidates for study by thermal methods.[1,2] Periodic regularities in the thermal properties of these salts are observed as one proceeds down the metal family (see Questions 4 and 5).

In this laboratory, the oxalates of Group 2 (IIA) metal ion oxalates are prepared via precipitation from solution at approximately pH 5, following the general reaction

$$M^{2+}\ (aq) + C_2O_4^{2-}\ (aq) = MC_2O_4\ (s)$$

where $M^{2+} = Mg^{2+}, Ca^{2+}, Sr^{2+}$, and Ba^{2+}.

THE METHOD OF HOMOGENEOUS PRECIPITATION

Since the Group 2 (IIA) oxalates are the salts of the weak acid, oxalic acid, their solubility will increase with increasing hydrogen ion concentration. The oxalates can therefore be precipitated by making the solution more alkaline, for example, by raising the pH. We wish to produce the precipitate in the form of large, individual crystals. The method of **homogeneous precipitation** is used to accomplish this goal. In this technique, the precipitating reagent is not added initially, but rather, forms slowly *within* the solution. In this way, supersaturation is minimized and local buildups of concentration of precipitating reagent are avoided.

Urea, when hydrolyzed, forms ammonia.

$$(H_2N)_2C{=}O + H_2O = 2NH_3 + CO_2$$

The formation of ammonia slowly raises the pH of the solution to approximately 5, because of its own hydrolysis. This pH is sufficient to generate free oxalate and precipitate the metal oxalates.

$$NH_3 + H_2O = NH_4^+ + OH^-$$

The degree of hydrolysis of urea depends on the temperature, so that essentially any desired final pH can be achieved by the method of homogeneous precipitation, using careful temperature control.

Prior Reading and Techniques

Section 5.B: Thermal Analysis

Section 5.D.3: Isolation of Crystalline Products (Suction Filtration)

Related Experiments

Magnesium and Calcium: Experiment 3

Thermal Analysis: Experiment 12

EXPERIMENTAL SECTION

Safety Recommendations

Magnesium oxide (CAS No. 1309-48-4): This compound is not normally considered dangerous, however, the usual safety precautions (see Section 1.A.3) should be taken.

Calcium carbonate (CAS No. 471-34-1): This compound is not generally considered to be dangerous, however, the usual safety precautions (see Section 1.A.3) should be taken. ORL-RAT LD50: 6450 mg/kg.

Strontium carbonate: No safety data is available for this compound. However, it would be prudent to use this compound with great care, as strontium is known to cause heavy metal poisoning. Do not ingest or breathe the dust from this compound. Avoid contact with the skin. Wash repeatedly with water if skin is contacted.

Barium carbonate (CAS No. 513-77-9): This compound may be fatal if inhaled, swallowed, or absorbed through the skin. Barium compounds are known to cause heavy metal poisoning. Wash repeatedly with water if skin is contacted. ORL-HMN LDLo: 17 mg/kg, ORL-RAT LD50: 418 mg/kg.

Ammonium oxalate monohydrate (CAS No. 6009-70-7): Ammonium oxalate is harmful if swallowed, inhaled, or absorbed through the skin. It is classified as a mild poison. No LD50 data is available. Wash repeatedly with water if skin is contacted.

Urea (CAS No. 57-13-6): Urea is not generally considered dangerous and is classified as a diuretic. ORL-RAT LD50: 8471 mg/kg. The usual safety precautions (see Section 1.A.3) should be taken.

CHEMICAL DATA

Compound	FW	Amount	mmol	mp (°C)	Density
MgO	40.31	40 mg	1.0		3.580
$CaCO_3$	100.09	25 mg	0.25	825	2.830
$SrCO_3$	147.63	25 mg	0.17	1100[a]	3.700
$BaCO_3$	197.35	25 mg	0.13	1300[a]	4.430
$(NH_4)_2C_2O_4 \cdot H_2O$	142.11	Saturated solution			1.500
Urea	60.06	1.5 g	25.0	133	1.335

[a] Decomposes.

Required Equipment

Magnetic stirring bar, 25-mL beaker, 10-mL graduated cylinder, microwatch glass, magnetic stirring hot plate, Hirsch funnel, water trap, clay tile, or filter paper.

Time Required for Experiment: 2.5 h.

EXPERIMENTAL PROCEDURE[1,2]

Weigh 25 mg of the desired metal carbonate (in the case of magnesium, use 40 mg of MgO instead of the carbonate) into a 25-mL beaker. Add 2.0 mL of deionized water, a magnetic stirring bar, and cover the beaker with a small watch glass. Add 6*M* HCl dropwise to the solid with stirring and gently warm the solution on a magnetic stirring hot plate until all the solid has dissolved.

Using deionized water, dilute the solution to 10.0 mL for all metals except magnesium. Add one drop of 1% methyl red indicator. At this stage the solution should be acidic and develop a light red color. Add 1.5 mL of saturated ammonium oxalate solution and 1.5 g (25 mmol) of solid urea to the solution. In the case of magnesium, the amount of urea added should be 4.5 g.

Isolation of Product

With stirring, gently boil the solution until the color changes from red to yellow. If necessary, add water to compensate for loss of water due to evaporation (urea may precipitate out from the concentrated solution). Colorless crystals of metal oxalate should begin to precipitate at this point. If no precipitate forms, add a few drops of 6*M* ammonia to neutralize any excess acid. Cool the solution to room temperature. Using a Hirsch funnel and water trap, collect the product crystals by suction filtration. Wash the product with cold water until it is free from chloride ion (test with $AgNO_3$ solution; the product is chlorine-free when a few drops of the filtrate do not show any turbidity with a drop of 1% $AgNO_3$ solution). Dry the product on a clay tile or on filter paper. Calculate the percentage yield of each product.

Characterization of Product

Obtain a TGA or DSC thermogram of each metal oxalate hydrate, as directed by your laboratory instructor. Oxalate hydrates (with some exceptions) decompose in three steps.

$$MC_2O_4 \cdot nH_2O = MC_2O_4 + nH_2O$$

$$MC_2O_4 = MCO_3 + CO$$

$$MCO_3 = MO + CO_2$$

Determine the temperature for each of these decompositions, and calculate how accurately the thermal technique used measures the mass lost in each step.

NOTE: *The final decomposition for the barium product occurs above 1200 °C, which may be beyond the temperature capacity of the thermal instrument.*

Calculate the number of waters of hydration for each oxalate hydrate on the basis of your thermogram.

If desired, the magnesium and calcium oxalates prepared in this experiment can be analyzed using AAS in Experiment 3.

QUESTIONS

1. Describe the reaction of calcium carbonate with HCl. Write the chemical equation.
2. What kind of indicator is methyl red? What is the range of pH change for this indicator?

3. An excess of ammonium oxalate must be avoided in preparing magnesium oxalate hydrate. Why?
4. Correlate the temperatures at which the waters of hydration are lost. What periodic trend does this illustrate?
5. Correlate the temperature at which the carbonate decomposes to form an oxide. What periodic trend does this illustrate?
6. The temperature at which water is lost from a hydrated compound is indicative of the manner in which it is bound. In the Group 2 (IIA) oxalates, in what manner is the water bound?
7. In copper sulfate, there are clearly two modes of binding of hydrated water (see Section 5.B). What are they? Perform a literature search to find the structure of $CuSO_4 \cdot 5H_2O$. A convenient place to start is *Comprehensive Coordination Chemistry*, Vol. 5, under "Copper." (*Hint:* Sulfate is an *oxygen* ligand.) Who first determined this structure and how?

REFERENCES

1. Erdey, L.; Liptay, G.; Svehla, G.; Paulik, F. *Talanta* **1962**, *9*, 489.
2. Hill, J. O.; Magee, R. J. *J. Chem. Educ.* **1988**, *65*, 1024.

Experiment 3 Atomic Absorption Analysis of Magnesium and Calcium

INTRODUCTION

Atomic absorption spectroscopy (AAS) is the instrumental measure of the amount of radiation absorbed by unexcited atoms in the gaseous state. The absorption spectrum of an element in its gaseous atomic form consists of well defined, narrow lines arising from the electronic transitions of the valence electrons. For metals, the energies of these transitions generally correspond to wavelengths in the UV and Visible regions. A wavelength must be selected for each element where the element absorbs strongly, and where no other element interferes. For calcium, the usual wavelength is 422.7 nm, and for magnesium it is 285.2 nm.

Atomic absorption spectroscopy has been used for the determination of more than 70 elements. Applications in industry include clinical and biological samples, forensic materials, foods, beverages, water and effluents, soil analysis, mineral analysis, petroleum products, pharmaceuticals, and cosmetics. In this experiment, synthetic water samples will be analyzed for calcium and magnesium content. The water samples will be typical of hard water found in the midwestern United States.

Prior Reading and Technique

Section 6.E: Atomic Absorption Spectroscopy

Related Experiment

Hydrates: Synthesis and Thermal Analysis of the Group 2 (IIA) Metal Oxalate, Experiment 2

EXPERIMENTAL SECTION

Safety Recommendations

Calcium carbonate (CAS No. 471-34-1): This compound is not generally considered to be dangerous, however, the usual safety precautions (see Section 1.A.3) should be taken. ORL-RAT LD50: 6450 mg/kg.

Magnesium (CAS No. 7439-85-4): This element is harmful if swallowed. The normal precautions (Section 1.A.3) should be observed.

CHEMICAL DATA

Compound	FW	Amount (g)	mmol	mp (°C)	Density
$CaCO_3$	100.09	1.249	12.48	825	2.83
Mg	24.31	1.000	41.14	648	1.74

Required Equipment

Two 1-L volumetric flasks, two 250-mL volumetric flasks, sixteen 100-mL volumetric flasks, 5 ppm Cu^{2+} standard solution.

Time Required for Experiment: 3 h.

PREPARATION OF STOCK SOLUTIONS AND CALIBRATION STANDARDS

NOTE: ***Be sure to mark all flasks carefully as to the metal they contain and their concentration. Sufficient stock solutions are prepared to supply a full laboratory of students.***

Prepare a 500 ppm stock solution of calcium ion, Ca^{2+}, by dissolving 1.249 g of calcium carbonate in 50 mL of distilled water in a 1-L volumetric flask. Add enough concentrated HCl (**Caution:** *Corrosive!*) to just complete the dissolution of the calcium carbonate. Add distilled water to the mark on the volumetric flask, to make 1.00 L of stock solution.

Prepare a 1000 ppm stock solution of magnesium ion, Mg^{2+}, by dissolving 1.000 g of magnesium ribbon (remove oxide film with sandpaper) in enough 6*M* HCl in a 1-L volumetric flask to effect dissolution. Add 1% (v/v) HCl to the mark on the volumetric flask, to make 1.00 L of solution.

Prepare a 50 ppm Ca^{2+} solution by diluting 25 mL of the 500 ppm stock solution to 250 mL in a 250-mL volumetric flask. Prepare a series of calibration solutions, as follows, using 100-mL volumetric flasks.

1. 2 ppm: Dilute 4 mL of 50 ppm solution to 100 mL
2. 4 ppm: Dilute 8 mL of 50 ppm solution to 100 mL
3. 6 ppm: Dilute 12 mL of 50 ppm solution to 100 mL
4. 8 ppm: Dilute 16 mL of 50 ppm solution to 100 mL
5. 10 ppm: Dilute 20 mL of 50 ppm solution to 100 mL
6. 12 ppm: Dilute 24 mL of 50 ppm solution to 100 mL

Prepare a 20 ppm Mg^{2+} solution by diluting 5 mL of the 1000 ppm stock solution to 250 mL. Prepare a series of calibration solutions, as follows, using 100-mL volumetric flasks.

1. 0.4 ppm: Dilute 2 mL of 20 ppm solution to 100 mL
2. 0.8 ppm: Dilute 4 mL of 20 ppm solution to 100 mL
3. 1.2 ppm: Dilute 6 mL of 20 ppm solution to 100 mL
4. 1.6 ppm: Dilute 8 mL of 20 ppm solution to 100 mL
5. 2.0 ppm: Dilute 10 mL of 20 ppm solution to 100 mL
6. 2.4 ppm: Dilute 12 mL of 20 ppm solution to 100 mL

Obtain an unknown hard water sample from your laboratory instructor. (The oxalates prepared in Experiment 2 may be used for this purpose: Dissolve 18 mg of magnesium oxalate hydrate in a 250-mL volumetric flask, using 1 mL of 12*M* HCl. Fill to the mark with water.) Prepare a series of dilutions of the unknown.

1. Dilute 5 mL of the unknown to 100 mL
2. Dilute 2 mL of the unknown to 100 mL
3. Dilute 1 mL of the unknown to 100 mL
4. Dilute 0.5 mL of the unknown to 100 mL

Calibration of the AA Unit

NOTE: *The following instructions are based on a Perkin–Elmer 2280 AA. Other units will have similar operation. The instrument should be turned on at least 30 min before use.*

Select the Cu lamp and properly insert it in the hold position. Make sure that water is in the waste loop. The instrument dials should be set in the following manner: **Signal:** Lamp 1 **Gain:** Fully counterclockwise **Lamp 1:** Fully counterclockwise. Turn the **Lamp 1** dial clockwise until **Lamp/Energy** reads 15. Set the **Slit Width** at 0.7 nm. Turn the **Signal** dial to the Abs. setting.

Set the **Wavelength.** For Cu^{2+}, the wavelength should be 324.5 nm. For Ca^{2+}, the wavelength should be 422.7 nm. For Mg^{2+}, the wavelength should be 285 nm.

Turn the **Gain** dial clockwise until **Lamp/Energy** reads 75. Adjust the **Lamp/Energy** reading to a maximum using the **Wavelength** dial and the horizontal and vertical controls on the lamp. Return the **Lamp/Energy** reading to 75 after each adjustment, using the **Gain** dial.

Turn on the vent, and ignite the flame by (a) turning the air pump on, (b) setting the acetylene tank pressure to 12 psi, (c) turning the knob on the instrument to *air* (40 on gauge), and (d) turning the acetylene toggle switch to the up position, and pushing the ignition button.

Aspirate water, then zero the instrument by pressing the **AZ** button. Aspirate the 5 ppm Cu^{2+} solution. The absorbance of this standard should have a value of between 0.18 and 0.24. If the desired reading with Cu^{2+} is obtained, shut off the flame by closing the acetylene toggle switch, shutting off the air at the instrument, and turning off the power.

Replace the Cu lamp with a Ca–Mg lamp, and repeat the previous steps to adjust the lamp and ignite the flame.

EXPERIMENTAL PROCEDURE

Beginning with the most dilute solution, obtain the absorption (abs) reading for each of the six Ca^{2+} calibration solutions. Aspirate distilled water between each Ca^{2+} measurement. Then record the absorption of the unknown Ca^{2+} solution.

Turn the dial from "abs" to "conc." Select two of the known solutions (preferably a high and low concentration). Remember to aspirate distilled water between each solution. While aspirating the lower dilution, punch in your known solution concentration. For example, enter 2.0 and press "S1" twice. Switch from "conc" to "abs" and record the absorbance. Repeat the procedure for as many standards as the unit allows.

Press 5.0 (s) and button "t." Aspirate the unknown sample and press "readout." The concentration of the unknown can now be read directly. Record this value. (If the oxalates from Experiment 2 were used, calculate the percentage of Mg or Ca in the oxalate.)

Repeat this procedure for Mg^{2+}, remembering to set the wavelength to 285 nm. Recalibrate the instrument as before.

> **NOTE:** ***If the instrument does not allow for direct reading of unknown concentrations, make a graph of concentration (x axis) versus absorbance (y axis) for the known solutions. Fit the points with the best possible straight line. The concentration of the unknown may be obtained by reading from the absorbance axis until the line is reached, and down to the concentration value.***

To shut down the instrument, close down the acetylene tank, and then the air. Turn off the compressor and the power switch.

QUESTIONS

1. What is meant by the term "sputter" as it relates to the operation of a hollow-cathode tube in AA analysis?
2. What graphic approach can be used in the quantitative AAS determination of a specific species? Give an illustration.
3. What two basic phenomena account for the inefficiency of the nebulizer-flame burner system used in AAS?
4. Why must the light beam (from the hollow-cathode source) be chopped?
5. In an H_2–O_2 flame, the absorption peak for iron was found to decrease in intensity in the presence of large concentrations of sulfate ion. Suggest an explanation for this observation. Suggest a method for overcoming this potential interference of sulfate in a quantitative determination of iron.
6. Lead and mercury are well-known environmental heavy metal poisons. Both are commonly analyzed using AAS, although the methods for each differ. Perform a literature search about the history of AAS, and how it has been used to analyze these elements. A good review article to start from is Walsh, A. *Anal. Chem.* **1974,** *46*, 698A.

GENERAL REFERENCES

Christian, C. D.; O'Reilly, J. E., *Instrumentation Analysis,* Allyn and Bacon: Boston, 1986, Chapter 10.

Skoog, D. A., *Principles of Instrumental Analysis,* 3rd ed., Saunders: Philadelphia, PA, 1985, Chapter 9.

Varmer, A., *Handbook of Atomic Absorption Analysis,* CRC Press: Boca Raton, FL, Vols. I and II, 1982.

Bennett, P. A.; Rothery, E., *Introducing Atomic Absorption Analysis,* Varian Techtron Pty Ltd.: Mulgrave, Australia, 1983.

Experiment 4 Preparation of Trialkoxyboranes

Part A. **Preparation of Tri-*n*-propylborate**

Part B. **Preparation of a Poly(vinylalcohol)–Borate Copolymer**

INTRODUCTION

Boric acid will react with organic alcohols to yield trialkoxyboranes (also named as trialkylborates) according to the reaction scheme

$$B(OH)_3 + 3ROH \rightarrow B(OR)_3 + 3H_2O$$

Alternatively, in base solution, one can form the tetraalkoxyborate anion

$$B(OH)_3 + 4NaOR \rightarrow B(OR)_4^- + 3H_2O$$

Trialkoxyboranes are extensively used in industry for a huge variety of purposes, a small number of which are listed here.

Photography. Mixtures of titanium(IV), trialkoxyborane, and poly(vinylalcohol) form photosensitive solutions. The solution will first absorb oxygen and undergo photoreduction to form a dark blue Ti(III) complex, which will reoxidize in the presence of air to form colorless Ti(IV).[1]

Polymer Chemistry. Trialkoxyboranes are extensively used with the Ziegler–Natta catalytic system, $TiCl_3$ and $(CH_3CH_2)_3Al$, as polymerization catalysts for various olefins.[2] Alternatively, trialkoxyboranes can be incorporated into polymers with poly(vinylalcohol) (see Part 4.B) or siloxanes.[3]

Biochemistry. The lighter trialkoxyboranes are useful as chemosterilants of various houseflies and screwworm flies, useful in proportion to their boron content. Flies fed these trialkoxyboranes reproduced at a rate only 1% of normal. These compounds are also useful as plant growth regulators.[4]

Glasses and Glazing. Glass surfaces treated with aluminum butoxide, tetraalkoxytitanium(IV) and trialkoxyboranes form glazes that are impervious to alkali materials in the glass. These are used to coat fluorescent light bulbs.[5]

Trialkoxyboranes are air-stable boron compounds, although the lighter ones are susceptible to hydrolysis to reform boric acid. The B—O bond is especially stable, as the lone pairs of electrons on the oxygen atoms can be donated into the empty *p* orbital of the boron, drastically reducing the Lewis acidity of the compound.

Prior Reading and Techniques

Section 2.F: Reflux and Distillation
Section 5.I.5: Drying of the Wet Organic Layer
Section 6.C: Infrared Spectroscopy
Section 6.D: Nuclear Magnetic Resonance Spectroscopy

Related Experiments

Boron Chemistry: Experiments 5–7
Industrial Chemistry: Experiments 8, 12, and 34

EXPERIMENTAL SECTION

Safety Recommendations

Boric acid (CAS No. 10043-35-3): While not generally considered dangerous, the compound is toxic if ingested. ORL-RAT LD50: 5.14 g/kg. Death has occurred from ingestion of 5–20 g in adults.

n-Propanol (CAS No. 71-23-8): This compound is not generally considered dangerous. It is a volatile, flammable liquid. ORL-WMN LDLo: 5700 mg/kg, ORL-RAT LD50: 1870 mg/kg.

Poly(vinylalcohol) (CAS No. 9002-89-5): This compound is classified as a possible carcinogen. ORL-MUS LD50: 14,700 mg/kg.

Toluene (CAS No. 108-88-3): This compound is harmful if swallowed, inhaled, or absorbed through the skin. ORL-HMN LDLo: 50 mg/kg, ORL-RAT LD50: 5000 mg/kg.

CHEMICAL DATA

Compound	FW	Amount	mmol	mp (°C)	bp (°C)	Density
$B(OH)_3$	61.83	1 g	16.2	171	300[a]	2.46
$CH_3(CH_2)_2OH$	60.10	4 mL	53.5	−127	97	0.804

[a] Decomposes.

Part A. Preparation of Tri-*n*-propylborate

Required Equipment

Water condenser, 10-mL graduated cylinder, 10-mL distilling flask, one-hole stopper, boiling stone, wire gauze square, iron ring, 5-mL round-bottom flask, Hickman still, thermometer, air condenser, Keck clip, Pasteur pipet, thermometer, microburner.

Time Required for Experiment: 2 h.

EXPERIMENTAL PROCEDURE

Place 1.0 g (16.2 mmol) of boric acid in a 10-mL distilling flask containing a boiling stone. Using a 10-mL graduated cylinder, add 4 mL (53.5 mmol) of dry *n*-propanol, and 2 mL of toluene (which forms an azeotrope with the water, carrying it to the side arm). Attach a short piece of Tygon tubing to the side arm, clamping it off at the end (see Fig. 7.2). Attach a water condenser to the flask using a one-hole stopper. The distilling flask should be mounted over a wire gauze square, held on a iron ring, with the side arm slanted downwards as shown in the figure.

Heat the solution with a microburner until reflux is achieved. Toluene and water will collect in the side arm, with the denser water moving toward the bottom and the toluene recycling to the distillation pot. Continue the reflux until no further water collects in the side arm (~30 min, it may be necessary to drain the side arm periodically). After the reflux period, drain the side arm.

Remove the condenser, and transfer the remaining solution to a 5-mL round-bottom flask containing a boiling stone, using a Pasteur pipet. Attach a Hickman still fitted with an air condenser to the flask with a Keck clip. Clamp a thermometer inside the Hickman still so that its bulb is at the same level as the still's collar.

Isolation of Product

Heat the solution to reflux with the microburner, using the same wire gauze arrangement as above. Collect any remaining toluene and alcohol in the still's collar. Remove the collected distillate with a Pasteur pipet, collecting the product that distills up to a temperature of 120 °C. Discontinue the heating and allow the flask to return to room temperature. The product, which remains in the flask, is of sufficient purity for characterization and further work. Calculate a percentage yield of crude product.

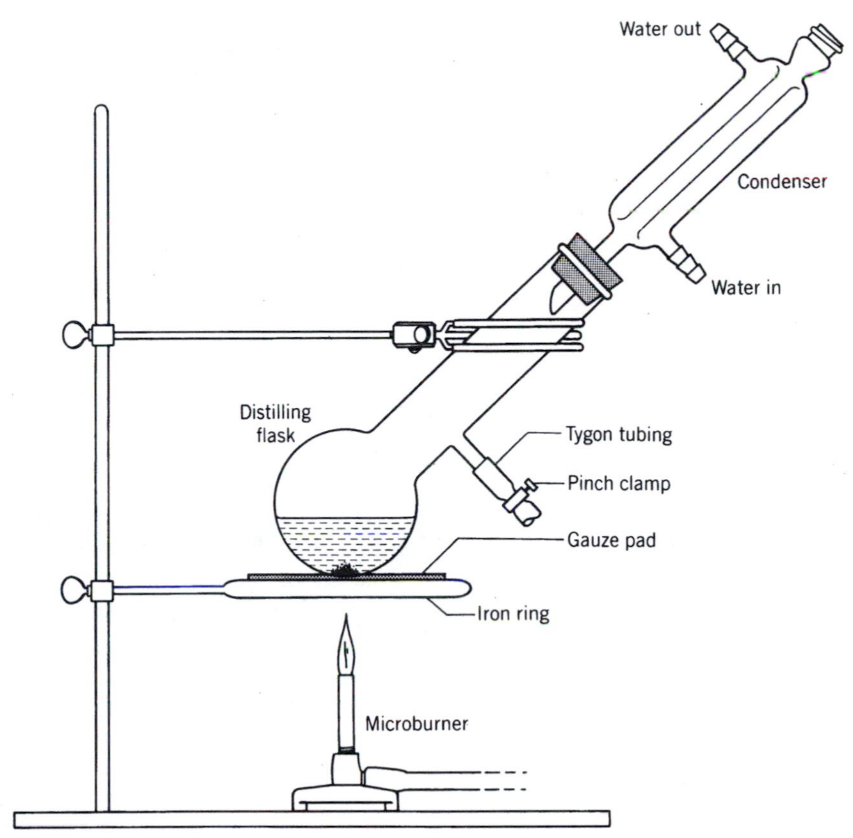

Figure 7.2. *Apparatus for Experiment 4.*

Purification of Product

The tri-*n*-propylborate may be purified by distillation at atmospheric pressure by continuing the distillation in the previous step, and collecting the product that distills between 175 and 185 °C. Some decomposition of the ester usually occurs during this process, resulting in a fairly low yield.

Alternatively, the ester may be transferred to a vacuum distillation apparatus, and distilled in far greater yield at a reduced pressure of approximately 100 torr.

Characterization of Product

Obtain the IR spectrum of the product (neat), noting any residual presence of water. Determine the frequency of the B—O and O—C bands. Compare the spectrum with *n*-propanol and published sources[6] for the borate. If desired, the ^{1}H NMR spectrum can be obtained in $CDCl_3$, and compared with the literature.[7]

Part B. Preparation of a Poly(vinylalcohol)–Borate Copolymer

Required Equipment

Glass rod, 25- and 100-mL beakers, 10-mL graduated cylinder, Pasteur pipet.

Time Required for Experiment: 30 min.

EXPERIMENTAL PROCEDURE[8]

Prepare an approximately 4% (by weight) solution of the tri-*n*-propylborate (prepared in Part 4.A) by dissolving 200 μL of the borate in 5 mL of water (graduated cylinder) in a 25-mL beaker. Alternatively, 200 mg of boric acid may be used to replace the the alkoxyborane.

In a separate 100-mL beaker, prepare a 4% (by weight) solution of poly(vinylalcohol) by slowly dissolving 1.2 g poly(vinylalcohol) in 30 mL of

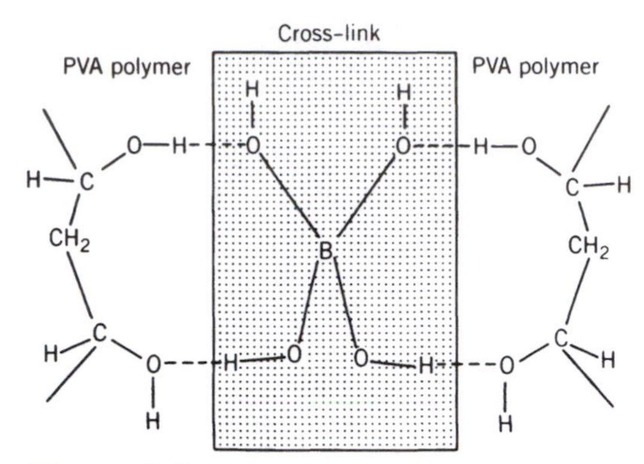

Figure 7.3. *Borate cross-linking of polymers.*

stirred boiling water. Allow the solution to cool to room temperature. Add a few drops of phenolphthalein solution, to act as a base indicator.

Pour the borate solution into the poly(vinylalcohol) solution. Stir the resulting solution with a glass rod. Dropwise, using a Pasteur pipet, add 6*M* NaOH. As the base reacts with the mixture, polymerization occurs immediately, and microfibrils of polymer can be seen. A slimy copolymer results from this addition. As much NaOH may be added as needed to obtain a polymer of the desired rigidity (no more than 1.5 mL). The structure of the copolymer is shown in Figure 7.3.

QUESTIONS

1. What function do boron esters serve in polymers?
2. While most three-coordinate boron compounds react vigorously with water, borate esters react quite slowly, if at all. Why is the B—O bond so stable?
3. There are many types of borates. Describe the structure of three different types.
4. Borates are extensively used in the glassmaking industry, as are silicates and aluminates. Perform a literature search and describe the use of these materials in the manufacture of glass. What attractive properties do they confer to the glass? Suggest reasons why these compounds should be related.

REFERENCES

1. Wade, R. C., US Patent 4,173,521, 1980; *Chem. Abstr. 92,* abs No. 155914z, 1980.
2. Yamaguchi, K.; Kano, N.; Tanaka, T.; Tanaka, E.; Suzuki, M.; Murakami, A. Japan Kokai 76,76,189, 1976; *Chem. Abstr. 85,* abs No. 124703f, 1976.
3. Reikhsfel'd, V. O.; Khanzhodzhaeva, D. A.; *Vysokomol. Soedin., Ser. A.* **1967,** *9, 638; Chem. Abstr. 67,* abs No. 33595j, 1967.
4. For example, Settepani, J. A.; Crystal, M. M.; Borkovec, A. B. *J. Econ. Entomol.* **1969,** *62,* 375; *Chem. Abstr. 70,* abs No. 105408n, 1969.
5. Hammer, E. E.; Martyny, W. C., US Patent 3,847,643, 1982; *Chem. Abstr. 82,* abs No. 63489e, 1975.
6. Pouchert, C., Ed., *Aldrich Library of FT–IR Spectra,* Aldrich Chemical Co.: Milwaukee, WI, Vol. II, Spectrum No. 1132C.
7. Pouchert, C., Ed., *Aldrich Library of NMR Spectra,* Aldrich Chemical Co.: Milwaukee, WI, Vol. II, Spectrum No. 1010C.
8. Casassa, E. Z.; Sarquis, A. M.; Van Dyke, C. H. *J. Chem. Educ.* **1986,** *63,* 57.

Experiment 5 Synthesis of Tetrafluoroberyllate and Tetrafluoroborate Complexes

Experiment 5A: Synthesis of Ammonium Tetrafluoroberyllate, $(NH_4)_2[BeF_4]$

Experiment 5B: Synthesis of Ammonium Tetrafluoroborate, $NH_4[BF_4]$

CAUTION: ***Beryllium compounds are highly toxic. Be sure to read the safety section before attempting to work with beryllium.***

INTRODUCTION

The Group 13 (IIIA) elements and beryllium in Group 2 (IIA) often form covalent compounds that are electron deficient. This deficiency occurs as there are fewer valence electrons than there are valence orbitals. The element boron, for example, has a valence shell electron configuration of $2s^22p^1$. One of the *s* electrons is easily promoted to a $2p$ level, thereby allowing the formation of three sp^2 hybridized bonds. In most of its simple compounds, boron is three coordinate, and trigonal planar, as in BCl_3 and $B(OH)_3$. The Lewis structure of these simple boron compounds shows that the central boron has only six valence electrons, two short of a complete valence shell. Such three-coordinate boron compounds are therefore very susceptible to attack by nucleophiles, and act as Lewis acids.

In the case of beryllium, the valence shell configuration is $2s^2$. One of the *s* electrons could easily be promoted to a $2p$ level, forming two *sp* hybridized orbitals. This coordination is, in fact, not generally seen. The central beryllium is four electrons short of a complete valence shell, and any two-coordinate beryllium compounds formed would be powerful Lewis acids, and would immediately undergo further reaction with nucleophiles. Instead, most "simple" beryllium compounds (e.g., $BeCl_2$) are polymeric, although still Lewis acids.

One way of alleviating the electron deficiency of boron and beryllium compounds is by forming complexes. The product of the reaction between a boron Lewis acid and a nucleophile (a Lewis base) is an adduct complex, in which the boron is four coordinate, tetrahedral, and negatively charged.

$$BF_3 + F^- = [BF_4]^-$$

Similarly, the product between a beryllium Lewis acid and a nucleophile is usually also a four-coordinate adduct.

$$BeF_2 + 2F^- = [BeF_4]^{2-}$$

These anions have complete valence shells of electrons, and are, in general, quite stable.

Ammonium tetrafluoroberyllate is prepared[1] using an ion exchange reaction between ammonium fluoride and beryllium hydroxide.[2] The fluoride ion acts as a Lewis base and complexes with the Lewis acid $Be(OH)_2$.

$$4NH_4F + Be(OH)_2 = (NH_4)_2[BeF_4] + 2NH_3(aq) + 2H_2O$$

Beryllium hydroxide is unique among *s* element hydroxides in that it is amphoteric. The hydroxide is not stable in acidic media, reacting to form polymerized hydrolyzed species, such as $[(H_2O)_3BeOBe(H_2O)_3]^{2+}$. Such species react easily with excess fluoride ion to form the very stable $[BeF_4]^{2-}$ adduct.

Ammonium tetrafluoroborate, NH_4BF_4, is prepared[3] using a double decomposition reaction.

$$2NH_4[HF_2] + B(OH)_3 \rightarrow NH_4[BF_4] + NH_3 + 3H_2O$$

Ammonium tetrafluoroborate is commonly used as the starting material in the production of boron trifluoride, BF_3. It is difficult to prepare $NH_4[BF_4]$ in the same manner as $(NH_4)_2[BeF_4]$ in Part 5.A, since the use of NH_4F requires acidification with H_2SO_4. This operation produces a precipitate of $(NH_4)_2SO_4$, which is difficult to separate from the product. Instead, the more powerful complexing agent, ammonium hydrogen difluoride is used.

Prior Reading and Techniques

Section 5.D.3: Isolation of Crystalline Products (Suction Filtration)

Section 5.D.4: The Craig Tube Method

Section 5.F.2: Evaporation Techniques

Section 6.D: Nuclear Magnetic Resonance Spectroscopy

Related Experiments

Boron Chemistry: Experiments 6 and 7

Main Group Complexes: Experiments 10 and 18

EXPERIMENTAL SECTION

Part A. Synthesis of Ammonium Tetrafluoroberyllate, $(NH_4)_2[BeF_4]$

Safety Recommendations

Beryllium hydroxide (CAS No. 13327-32-7) and **Ammonium tetrafluoroberyllate: ALL BERYLLIUM COMPOUNDS ARE EXTREMELY TOXIC.** Avoid contact with the skin. **NEVER** handle these compounds without wearing gloves. Do not breathe the dust. Use only in an efficient fume hood. In case of contact, flush affected area with large amounts of water. Specific TLV data for these compounds is not available.

Ammonium fluoride (CAS No. 12125-01-8): This compound is harmful if swallowed, inhaled, or absorbed through the skin. In case of contact, flush affected area with large quantities of water. Under fire conditions, fumes of HF and NH_3 are emitted.

CHEMICAL DATA

Compound	FW	Amount	mmol	mp (°C)	Density
$Be(OH)_2$	43.04	100 mg	2.32	138	1.92
NH_4F	37.04	350 mg	9.45	Decompose	1.015

Required Equipment

Magnetic stirring hot plate, 10-mL Erlenmeyer flask, magnetic stirring bar, side arm flask, Hirsch funnel, clay tile or filter paper.

Time Required for Experiment: 1 h.

EXPERIMENTAL PROCEDURE[1]

> **NOTE: *The entire procedure should be carried out in an efficient fume hood. Gloves should be worn throughout the experiment.* Be sure to read the safety recommendations before attempting to do this laboratory.**

Dissolve 350 mg (9.45 mmol) of NH_4F in 2 mL of water in a 10-mL Erlenmeyer flask equipped with a stirring bar. With stirring, using a sand bath set on top of a magnetic stirring hot plate, heat the solution to just below boiling. Slowly, add 100 mg (2.32 mmol) of $Be(OH)_2$ to the stirred, hot solution.

When addition is complete, heat the solution, with stirring, to reduce its volume by 50%. Allow the solution to cool to room temperature, and then further cool it over cracked ice. If precipitation does not begin, reduce the volume again and repeat the procedure.

Collect the small, colorless needles and prisms by suction filtration using a Hirsch funnel, and wash with two 0.5-mL portions of 95% ethanol. Dry the product on a clay tile or on filter paper and calculate a percentage yield.

Characterization of Product

Dissolve ~50 mg of ammonium tetrafluoroberyllate in 1 mL of D_2O to obtain a saturated solution. Transfer the solution (Pasteur pipet) to a 5-mm (o.d.) NMR tube, and add two drops of concentrated trifluoroacetic acid to serve as a fluorine reference. Obtain the ^{19}F NMR spectrum. What is the multiplicity of the signal? Since all fluorines in the complex are equivalent, what is the source of the multiplicity?

> **NOTE: *The product is frequently contaminated with $NH_4BeF_3 \cdot H_2O$. Does the ^{19}F NMR show this?***

Part B: Synthesis of Ammonium Tetrafluoroborate, $[NH_4BF_4]$[3]

Safety Recommendations

Ammonium bifluoride (CAS No. 1341-49-7): This compound is a toxic solid, and is extremely destructive to tissue and skin. Inhalation is to be avoided. In case of contact, flush area with large quantities of water. The compound will also react with glass, so only plastic utensils should be used.

Boric acid (CAS No. 10043-35-3): While not generally considered dangerous, the compound is toxic if ingested. Oral LD50 in rats is 5.14 g/kg. Death has occurred from ingestion of 5–20 g in adults.

Ammonium tetrafluoroborate (CAS No. 13826-83-0): This compound is a toxic solid, and is extremely destructive to tissue and skin. Inhalation is to be avoided. In case of contact, flush area with large quantities of water.

CHEMICAL DATA

Compound	FW	Amount	mmol	mp (°C)	Density
$B(OH)_3$	61.83	130 mg	2.10	171	2.46
NH_4HF_2	57.04	330 mg	5.79	125	1.500

Required Equipment
Plastic centrifuge tube, magnetic stirring hot plate, magnetic stirring bar (cannot be made of glass), Craig tube, centrifuge, watch glass, oven.

Time Required for Experiment: 1.5 h.

EXPERIMENTAL PROCEDURE

> **NOTE: *Only handle ammonium bifluoride while wearing gloves.***

In a plastic centrifuge tube equipped with a spin vane, add 330 mg (5.79 mmol) of NH_4HF_2 to 1 mL of water. With stirring, add 130 mg (2.10 mmol) of powdered $B(OH)_3$. Continue stirring until the boric acid dissolves. Transfer the solution to the bottom section of a Craig tube using a Pasteur pipet. In the **HOOD,** clamp the Craig tube in a warm sand bath set on top of a magnetic stirring hot plate. Add a boiling stone and concentrate the solution to dryness under a gentle stream of air or nitrogen.

> **NOTE: *During the heating, ammonia and hydrogen fluoride may be released, as well as water. Use of a hood is therefore required during this operation.***

Purification of Product
Dissolve the residual solid by adding a few drops (no more than three should be necessary) of near-boiling water. Allow the solution to cool slowly to room temperature, and then cool over cracked ice. Insert the top of the Craig tube, and centrifuge. Collect the white, crystalline product on a tared watch glass. Place the watch glass in an oven set at approximately 100 °C for 1 h to dry the product. Calculate the percentage yield.

Characterization of Product
Prepare an NMR tube sample of the product as in Part 5.A. Acquire the ^{19}F NMR spectrum. Explain the multiplicity of the signal. If facilities permit, obtain the ^{11}B NMR spectrum. Why is there a difference from $BF_3{\cdot}O(CH_2CH_3)_2$?

QUESTIONS

1. Draw Lewis dot structures for BeF_4^{2-} and BF_4^-.
2. The ^{19}F NMR chemical shifts for the two complexes (in ppm from trifluoroacetic acid) are quite different. Explain why.
3. ^{19}F NMR is nearly as easy to perform as ^1H NMR spectroscopy. What properties make a given nucleus particularly amenable to NMR? Classify the following nuclei as difficult or amenable to NMR.

$$^{13}C,\ ^{31}P,\ ^{77}Se,\ ^{11}B,\ ^{10}B,\ ^{9}Be,\ ^{32}S$$

   Explain your classifications.
4. Boron trifluoride is an electron deficient molecule. It is not, however, a particularly strong Lewis acid (e.g., much weaker than BCl_3). Explain why, in terms of electronic interactions between the boron and fluorine. Would this type of interaction be possible for BF_4^-? Explain.
5. Beryllium nearly always forms electron deficient covalent molecules similar to boron. The other Group 2 (IIA) elements nearly never do. Explain why.

6. To what non-Group 2 (IIA) element does beryllium show the closest chemical relationship? Explain.
7. The ^{19}F spectrum of commercial $(NH_4)_2BeF_4$ shows *two* quartets, one for the compound itself and a smaller one for the main contaminant, $NH_4BeF_3 \cdot H_2O$. The ^{9}Be NMR spectrum, however, shows a single signal, which is a quintet. Account for the multiplicity of the signals, and explain why only one signal is observed in the ^{9}Be spectrum. Use this as a springboard for a literature search of chemical shift ranges for the nuclei listed in Question 3. (A useful reference is Kotz, J. C.; Schaeffer, R. and Clouse, A. *Inorg. Chem.* **1967,** *6,* 620.)

REFERENCES

1. Experiment adapted from: Brauer, G., *Handbook of Preparative Inorganic Chemistry,* Academic Press: New York, 1963.
2. $Be(OH)_2$ may be purchased from ICN K&K Laboratories, 4911 Commerce Parkway, Cleveland, OH 44128.
3. Experiment adapted from: Booth, H. S.; Rehmar, S. *Inorg. Syn.* **1946,** *2,* 23.

Experiment 6 Synthesis of Dichlorophenylborane

INTRODUCTION

Boron has an extensive organic chemistry, which represents "*one of the most sturdy bridges between the areas of inorganic and organic chemistry, i.e. organometallic chemistry.*" [1] Thousands of organoboron compounds are known, the first having been synthesized more than 100 years ago. Organoboron compounds, BR_3, are isoelectronic with the equivalent carbocations CR_3^+, and thereby provide an interesting comparison with various aspects of organic chemistry.

Organoboron compounds are synthesized via the normal organometallic preparative routes, that is, by organic group transfer (transmetallation). Many compounds have been used as organic group-transfer agents, such as organomercury compounds, organotin compounds, organolithium compounds, and organoaluminum compounds. Organomercury compounds have more recently fallen into disfavor because of their high toxicity.

Various boron starting materials have been used as well, including boric acid, trimethyl borate, and boron trihalides. It is desirable to choose a boron starting material whose bonds are not too stable, or else very low yields are obtained. Therefore BCl_3 is an ideal choice, as the B—Cl bonds are fairly reactive, and BCl_3 is not particularly difficult to work with, if moisture is excluded.

The reaction performed in this experiment is a direct organic transfer from tetraphenyltin to boron trichloride.

$$(C_6H_5)_4Sn + 2BCl_3 \rightarrow 2C_6H_5BCl_2 + (C_6H_5)_2SnCl_2$$

If stoichiometric amounts of reagents are used, and if the temperature is kept low, no contamination from chlorodiphenylborane, $(C_6H_5)_2BCl$, is observed.

Prior Reading and Techniques

Section 1.B.4: Compressed Gas Cylinders and Lecture Bottles

Section 5.C: Vacuum and Inert Atmosphere Techniques

Section 6.D: Nuclear Magnetic Resonance Spectroscopy

Related Experiments

Boron Chemistry: Experiments 4, 5, and 7

Organometallic Chemistry of Main Group Elements: Experiments 7, 11, and 15

Tin Chemistry: Experiments 9, 10, and 15

EXPERIMENTAL SECTION

Safety Recommendations

Boron trichloride (CAS No. 10294-34-5): This compound is a gas at room temperature (bp 12.5 °C), which reacts vigorously with moisture, forming borates and HCl gas. It is extremely destructive to the respiratory tract and must not be inhaled. Use only in an efficient **HOOD**.

Tetraphenyltin (CAS No. 595-90-4): This compound's toxicity data is not known. It would be prudent to handle it with care, as many tin compounds are toxic. Do not contact with the skin.

CHEMICAL DATA

Compound	FW	Amount	mmol	mp (°C)	bp (°C)	Density
BCl_3	117.15	100 mg	0.85	−107	12.5	
$(C_6H_5)_4Sn$	427.11	185 mg	0.43	224	>420	

Required Equipment

Vacuum manifold, 3 stopcock tubes, liquid N_2, BCl_3 lecture bottle and stand.

Time Required for Experiment: 3 h.

EXPERIMENTAL PROCEDURE

Attach a lecture bottle of BCl_3 to a vacuum system. A thick-walled rubber hose, wired to the lecture bottle at one end, and to a ground-glass joint at the other, should be used to make the attachment. With the lecture bottle valve closed, evacuate the rubber hose and joint. Open a path so as to allow 0.85 mmol of BCl_3 to expand into a previously calibrated U-trap (or series of U-traps)[2] open to a mercury manometer. *Make sure that this path is not also open to the pump!* Since the volume (V) of the U-trap is known, as is the ambient temperature (T), and the desired number of moles of BCl_3 ($n = 8.5 \times 10^{-4}$), the necessary pressure can be measured using the ideal gas law, $PV = nRT$. Allow the BCl_3 to expand at ambient temperature until this pressure is reached, and then shut off the lecture bottle. Close off the U-trap, and condense the BCl_3 by placing a liquid nitrogen Dewar around the trap. Close off the stopcock leading to the manometer, so as not to condense mercury in with the BCl_3. The BCl_3 remaining elsewhere in the vacuum system may be condensed into a second stopcock tube for later use.

Add 185 mg (0.43 mmol) of tetraphenyltin to an empty stopcock tube.

> **NOTE: *Regardless of the amount of BCl_3 condensed, the amount of BCl_3 and tetraphenyltin should be maintained at a 2:1 mole ratio.***

Attach the tube to the vacuum line, place a liquid N_2 Dewar around the tube, and evaculate the air. Close off the stopcock, isolating the tube, and allow the tetraphenyltin to warm to room temperature. Refreeze the tube and pump off the air once again. This process is called **freeze–thaw degassing** and should be repeated until no air is present.

Replace the liquid N_2 filled Dewar around the stopcock tube, remove the Dewar around the BCl_3, and open a path from the BCl_3 to the stopcock tube. *Be sure that this path is not open to the pump.* Condense the BCl_3 onto the tetraphenyltin. Close the tube stopcock, remove the Dewar, and allow the tube to warm slowly to room temperature, rotating and gently shaking the tube periodically. If the reaction in the tube appears to be too vigorous, cool the stopcock tube briefly with liquid nitrogen. Allow the tube to sit at room temperature for 30 min.

Separation of Product (Vacuum Fractional Distillation)

Open the stopcock tube to the vacuum system along a path that goes through a −78 °C bath (isopropyl alcohol–dry ice), and then a liquid N_2 bath cooled series of U-traps under dynamic vacuum. The product, dichlorophenylborane will collect in the −78 °C bath, while unreacted BCl_3 will collect in the nitrogen bath. Unreacted tetraphenyltin and any side product $(C_6H_5)_2SnCl_2$ will remain in the stopcock tube. Condense the collected product into a clean stopcock tube. The product is stable to air for ~2 weeks, and then will slowly colorize and decompose.

Characterization of Product

Acquire an IR spectrum of the product by condensing ~10-torr pressure of it into an IR gas cell (less pressure is necessary if an FT IR spectrometer is used). Obtain a ^1H NMR spectrum, and if equipment is available, also record the ^{11}B and ^{13}C NMR spectra of the product, all in $CDCl_3$. The normal reference standards are TMS (^1H, ^{13}C) and $BF_3{\cdot}O(CH_2CH_3)_2$ (^{11}B).

Additional Work

If desired, an adduct of dichlorophenylborane can be prepared with trimethylphosphine (**Caution:** *Toxic!*) by simply co-condensing a 1 : 1 molar mixture of the organoborane and phosphine in an evacuated stopcock tube at −196 °C, and allowing the mixture to warm slowly to room temperature. Pumping off any residual phosphine yields the pure adduct in quantitative yield. A preparation for the cyclopropyl analog of this compound is found in the reference cited in Question 5.

Obtain the ^{13}C, ^1H, and ^{11}B NMR spectra of the adduct, and compare the chemical shifts to those of the parent organoborane. What effect on the boron electron density does conversion from a trigonal to a tetrahedral complex have? A similar change in chemical shift may be observed by comparing the ^{31}P NMR spectrum of the adduct to that of the parent phosphine.

QUESTIONS

1. Organoboron compounds are generally quite susceptible to attack by base, water, or air. Explain.
2. In what way could an organomercury compound be synthesized?
3. Phenyldifluoroborane can be prepared from the reaction of tetraphenyltin and boron trifluoride, but the reaction proceeds in extremely low yield. Suggest a reason for this.
4. What geometric orientation would you expect the BCl_2 group to have relative to the phenyl ring in order to maximize electron donation through the π system?
5. Discuss the ability of phenyl, vinyl, and cyclopropyl groups to act as π-electron donors to the electron poor BCl_2 moiety. Draw a diagram showing

the electron donation in the case of vinyldichloroborane. What effect would you expect this to have on the ^{13}C chemical shift of the B-bonded carbon, and the carbon atoms α to it, compared to ethylene itself? How can a cyclopropyl group act as a π-electron donor at all? A useful starting reference is Odom, J. D.; Szafran, Z.; Johnson, S. A.; Li, Y. S.; Durig, J. R. *J. Am. Chem. Soc.* **1980,** *102,* 7173.

REFERENCES

1. Odom, J. D., "Non-cyclic Three and Four Coordinated Boron Compounds" *Comprehensive Organometallic Chemistry,* G. Wilkinson, Ed., Pergamon Press: Oxford, UK, 1982, Vol. 1, Chapter 5.1, p. 253.
2. Shriver, D. F.; Drezdzon, M. A., *The Manipulation of Air Sensitive Compounds,* 2nd ed., Wiley: New York, 1986.

Experiment 7 Synthesis and Reactions of Carboranes

Part A: Preparation of Potassium Dodecahydro-7,8-dicarba-*nido*-undecaborate(1-), a Carborane Anion

Part B: Preparation of 3-(η^5-Cyclopentadienyl)-1,2-dicarba-3-cobalta-*closo*-dodecaborane(11), a Metal Carborane

INTRODUCTION

Among the most interesting electron-deficient compounds formed by boron are the boron hydrides, also known as boranes, which were first prepared by Alfred Stock in 1912. The simplest isolable borane is diborane(6), B_2H_6. Diborane(6) has two fewer electrons than ethane (being isoelectronic with $C_2H_6^{2+}$), and if a Lewis dot structure for the molecule is attempted, one winds up two electrons short. Most boron hydrides (of which hundreds are known) are similarly electron deficient. The bonding in these compounds was first explained by William Lipscomb,[1] as consisting of both two-center "normal" bonds, where two electrons are shared by two atoms, and three-center bonds, where the two electrons are shared by three atoms. These three-center bonds are relatively unstable and are very reactive. As a result, most boranes are quite unstable, decomposing or exploding on contact with air, heat, or water. The most accessible of the boranes is decaborane(14), $B_{10}H_{14}$, which was once considered for use as a rocket fuel, and is consequently readily available. The larger boron hydrides are known as polyhedral boranes.

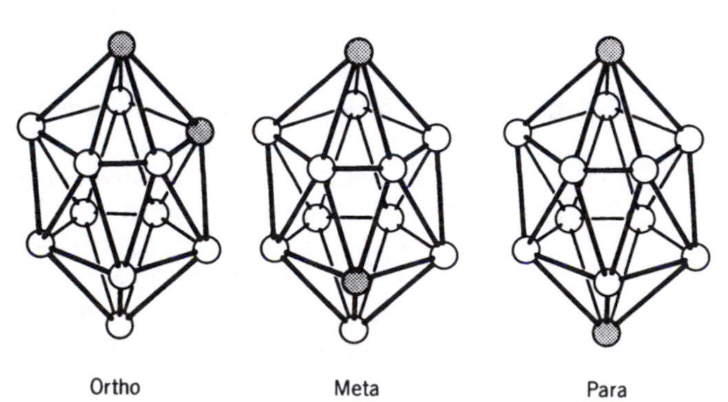

Figure 7.4. Ortho-, meta-, *and* para-*carborane. (There is one hydrogen atom per boron or carbon atom, omitted for clarity. Boron in white, carbon in black.)*

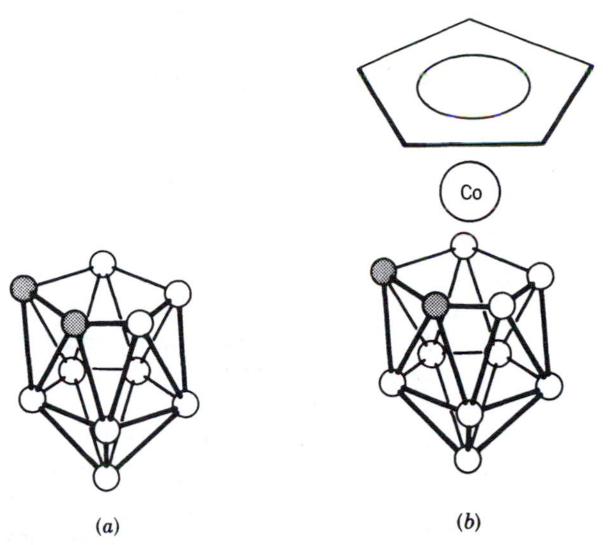

Figure 7.5. (a) *The dodecahydro-7,8-dicarba-nido-undecaborate(1-) anion.* (b) *3-(η^5-cyclopentadienyl)-1,2-dicarba-3-cobalta-closo-dodecaborane(11). [There is one hydrogen atom per boron or carbon atom, omitted for clarity. The 12th hydrogen bridges the open face in Part* (a). *Boron in white, carbon in black.]*

If a B^- unit in a polyhedral borane is formally replaced by an isoelectronic carbon atom, one obtains the formula of a carborane. Most carboranes have two carbon atoms, but some are known to contain 1, 3, or 4 atoms. The development of carborane chemistry was somewhat hampered by the difficulty in synthesizing carboranes, as most routes involve the use of boranes, and proceed in low yield. Despite this fact, carboranes are very stable and have an extensive and interesting chemistry.[2] The most familiar carboranes are the dicarba-*closo*-decaborane(12) isomers, of which there are three. These are known by the much simpler names of *ortho-*, *meta-*, and *para*-carborane, and are shown in Figure 7.4. The ortho, meta, and para prefixes refer to the position of the carbon atoms within the polyhedral framework.

When *o*-carborane reacts with a strong base, a degradation of the icosahedral cage occurs, with the loss of a boron atom. The product is a carborane anion, $C_2B_9H_{12}^-$. Treatment of the anion with various metal reagents gives rise to a family of very stable metallocarboranes, wherein the carborane anion functions much like an aromatic ring does in ferrocene (see Experiment 40). In Part 7.A, potassium dodecahydro-7,8-dicarba-*nido*-undecaborate(1 −) (Fig. 7.5*a*) will be synthesized[3] via the base degradation of *o*-carborane.

$$1,2\text{-}C_2B_{10}H_{12} + 3CH_3OH + KOH \rightarrow K[7,8\text{-}C_2B_9H_{12}] + B(OCH_3)_3 + H_2O + H_2$$

The product will be used to synthesize a metallocarborane, 3-(η^5-cyclopentadienyl)-1,2-dicarba-3-cobalta-*closo*-dodecaborane(11), in Part 7.B. The product is shown in Figure 7.5*b*.

Prior Reading and Techniques

Section 2.F: Reflux and Distillation
Section 5.C.2: Purging with an Inert Gas
Section 6.D: Nuclear Magnetic Resonance Spectroscopy

Related Experiments

Boron Chemistry: Experiments 5 and 6

Cobalt Complexes: Experiments 17, 26, 27, 30, 35, and 47

Organometallic Chemistry of the Main Group Elements: Experiments 6, 11, and 15

EXPERIMENTAL PROCEDURE

Part A: Preparation of Potassium Dodecahydro-7,8-dicarba-*nido*-undecaborate(1-)

Safety Recommendations

o-Carborane {**1,2-dicarba-*closo*-dodecaborane(12)**} (CAS No. 16872-09-6): Toxicity hazards for this compound have not been established. Many boranes, however, are quite toxic, so it would be prudent to handle this compound with care, including wearing gloves and working in a well-ventilated area.

Methanol (CAS No. 67-56-1): This compound can be fatal if swallowed, and is harmful if inhaled or absorbed through the skin. ORL-HMN LDLo: 143 mg/kg, ORL-RAT LD50: 5628 mg/kg. This compound is flammable and highly volatile.

Potassium hydroxide (CAS No. 1310-58-3): This compound is highly corrosive and very hygroscopic. Ingestion will produce violent pain in the throat. ORL-RAT LD50: 1.23 g/kg. If contacted with the skin, wash with large quantities of water.

CHEMICAL DATA

Compound	FW	Amount	mmol	mp (°C)	Density
o-Carborane	144.2	217 mg	1.50		
KOH	56.11	280 mg	5.00		

Required Equipment

10-mL side arm flask, magnetic stirring bar, magnetic stirring hot plate, water condenser, Keck clip, calcium chloride drying tube, mineral oil bubbler, source of nitrogen, calibrated Pasteur pipet.

Time Required for Experiment: 3 h.

EXPERIMENTAL PROCEDURE

NOTE: ***All glass equipment used in this experiment should be scrupulously dry.***

The experimental apparatus is shown in Figure 7.6. Attach a water condenser to a 10-mL side arm flask containing a magnetic stirring bar using a Keck clip. Connect a calcium chloride drying tube to the top of the condenser, and attach the tube to a mineral oil bubbler. Connect the side arm of the flask to a source of nitrogen, and flush the reaction apparatus for 10 min. Maintain a positive flow of nitrogen throughout the addition of reagents. Add 217 mg (1.5 mmol) of *o*-carborane and 280 mg (5.00 mmol) of KOH to the flask (**NOTE:** *Caustic!*).

Remove the drying tube momentarily, and using an calibrated Pasteur pipet,

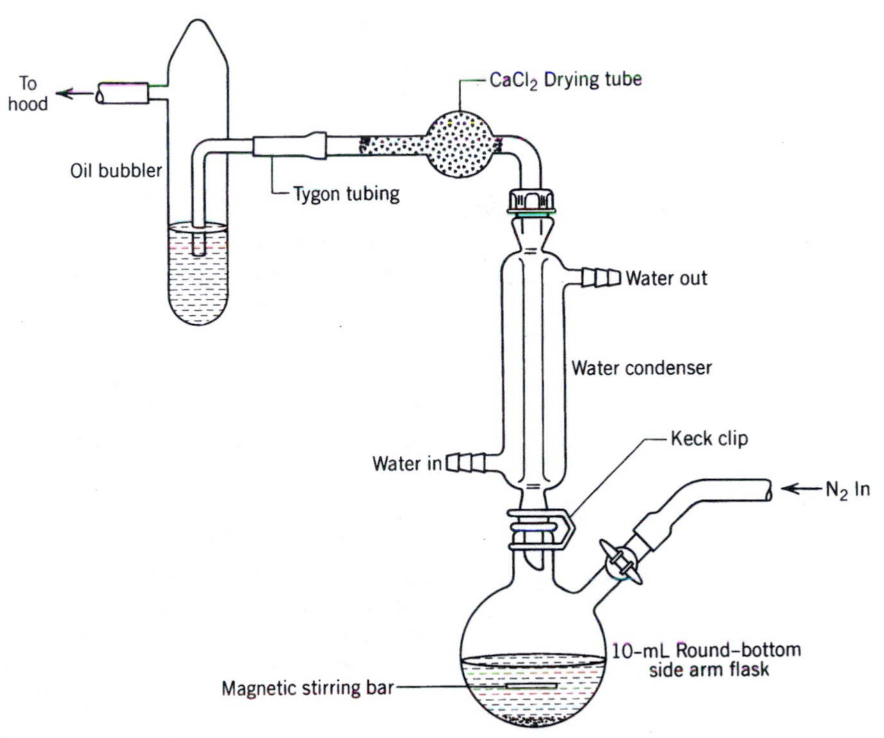

Figure 7.6. *Apparatus for Experiment 7.*

add 1.5 mL of methanol at once through the top of the condenser. Close the side arm stopcock and shut off the nitrogen flow. Slowly stir the solution. Hydrogen gas will be evolved, as evidenced by the bubbling of the solution. When gas evolution ceases at room temperature (~30 min), reflux the reaction mixture for about 1 h, or until gas evolution ceases. The resulting solution of K[7,8-$C_2B_9H_{12}$] is used, as is, for Part 7.B, and is reasonably air stable.

Part B: Preparation of 3-(η^5-Cyclopentadienyl)-1,2-dicarba-3-cobalta-*closo*-dodecaborane(11)

PREPARATION OF CYCLOPENTADIENE

Safety Recommendations

Dicyclopentadiene (CAS No. 77-73-6): This compound is harmful if swallowed, inhaled, or absorbed through the skin. ORL-RAT LD50: 353 mg/kg.

Silicone Oil (CAS No.: none): No toxicity data is available for this compound, but it is normally not considered dangerous. The usual precautions should be taken (Section 1.A.3).

CHEMICAL DATA

Compound	FW	Amount	mmol	mp (°C)	bp (°C)	Density
Dicyclopentadiene	132.21	2 mL	16.2	−1	170	1.071

Required Equipment

Magnetic stirring hot plate, 25-mL round-bottom side arm flask with stopcock, distillation head, water condenser, vacuum adapter, 10-mL round-bottom flask, sand bath, ice–water bath, magnetic stirring bar, calibrated Pasteur pipet, source of nitrogen, source of vacuum.

Time Required for Experiment: 1.5 h.

EXPERIMENTAL PROCEDURE

NOTE: *This part of the experiment should be carried out in the* HOOD, *because of the strong odor of the dicyclopentadiene and cyclopentadiene.*

Freshly prepared cyclopentadiene is necessary to accomplish this synthesis. Assemble the inert atmosphere apparatus shown in Figure 9.5. Add 1 mL of silicone oil (Nujol or mineral oil may be used instead) to the 10-mL round-bottom side arm flask using a calibrated Pasteur pipet.

Purge the apparatus with N_2 gas for 15 min and maintain a slow nitrogen flow throughout the experiment (20–30 bubbles of N_2 gas per minute as measured with a bubbler). Heat the oil to about 60 °C, and using a calibrated Pasteur pipet, dropwise, add 2 mL (16.2 mmol) of dicyclopentadiene through the thermometer port. Replace the thermometer and collect the fraction distilling at 42–45 °C in the receiving flask. The vacuum adaptor may be connected to a vacuum system, if desired, and the cyclopentadiene distilled under slightly reduced pressure. Otherwise, use the vacuum adaptor line as an exhaust to a **HOOD.**

NOTE: *While the distillation is in progress, proceed to the next part of this experiment.*

Placing an ice–water bath around the receiving flask will assist the collection, by preventing loss of cyclopentadiene due to evaporation. The cyclopentadiene prepared must be used immediately, or else it will undergo a Diels–Alder reaction, reforming dicyclopentadiene. Alternatively, it must be stored at a temperature of −78 °C or below.

PREPARATION OF 3-(η^5-CYCLOPENTADIENYL)-1,2-DICARBA-3-COBALTA-*CLOSO*-DODECABORANE(11)

Additional Safety Recommendations

Cobalt(II) chloride hexahydrate (CAS No. 7791-13-1): This compound is harmful if swallowed, inhaled, or absorbed through the skin. ORL-RAT LD50: 766 mg/kg.

CHEMICAL DATA

Compound	FW	Amount	mmol	mp (°C)	Density
$CoCl_2 \cdot 6H_2O$	237.93	178 mg	0.75		1.920
KOH	56.11	280 mg	5.00		

Required Equipment

Magnetic stirring hot plate, 10-mL side arm flask, magnetic stirring bar, ice–water bath, calibrated Pasteur pipet, automatic delivery pipet, Pasteur pipet, Hirsch funnel, extra filter flask.

Time Required for Experiment: Two 3-h laboratory periods, two overnight waiting periods, plus one additional hour.

EXPERIMENTAL PROCEDURE

Flush a second side arm flask containing a magnetic stirring bar with N_2 gas. Maintain the nitrogen flow throughout the addition of reagents. Add 1.5 mL of warm methanol (calibrated Pasteur pipet), followed by the addition of 178 mg (0.75 mmol) of cobalt(II) chloride hexahydrate. With stirring, slowly add 280 mg (5.0 mmol) of KOH flakes. (**NOTE:** *Caustic !*). Stir the mixture thoroughly. If the mixture is very thick, add an additional 1 mL of methanol. Occasionally shake the flask so as to thoroughly mix the contents. At this point, the flask will contain a gray-blue precipitate. Cool the flask in an ice–water bath.

Using an automatic delivery pipet, add 130 μL (1.5 mmol) of freshly distilled cyclopentadiene (prepared previously) to the solution prepared in Part 7.A. Make sure that the addition is made under an N_2 atmosphere. Using a Pasteur pipet, add this mixture slowly dropwise to the rapidly stirring gray-blue precipitate prepared above. The addition should not take >5 min. Continue to stir the mixture until the end of the laboratory period under a gentle N_2 purge. Allow the solution in the flask to stir overnight.

Connect the flask, which now contains a thick slurry of gray-brown precipitate, to a vacuum line and remove any excess cyclopentadiene under dynamic vacuum. Collect the excess in a liquid N_2 trap. Stop the vacuum and repressurize the contents of the flask with N_2. Add 2 mL of deionized water (calibrated Pasteur pipet), and suction filter the flask contents using a Hirsch funnel. Wash the filter cake with three 0.5-mL portions of water, followed by three 0.5-mL portions of 10% HCl, which dissolves any unreacted cobalt hydroxide. Wash the remaining yellow-black precipitate with two additional 0.5-mL portions of water. Allow the precipitate to dry as much as possible on the filter under suction.

Transfer the Hirsch funnel to a clean, dry, filter flask; dissolve the filter cake in 1 mL of acetone, filter it under suction and collect the filtrate in the clean flask. The filtrate is yellow-green in color. The residue left on the filter is black and contains metallic cobalt.

Isolation of Product

Transfer the filtrate (Pasteur pipet) to a stopcock-fitted side arm flask. Add 1 mL of toluene to the filtrate, which should immediately turn yellow. Concentrate the solution to 1 mL under vacuum (2–3 min). Flush the flask with N_2 and dissolve any solid formed by warming the solution in a water bath at 60 °C. Add 1 mL of dry hexane to complete the precipitation. Allow the flask to stand overnight. Collect the yellow precipitate by suction filtration using a Hirsch funnel and dry the product under suction.

Characterization of Product

Determine the percentage yield and the melting point. Obtain an IR spectrum of the product as a KBr pellet. In what ways do the bands from *o*-carborane and cyclopentadiene change upon formation of the metallocarborane?

Obtain a ^1H NMR spectrum in acetone-d_6. Two signals should be evident, one for the cyclopentadienyl protons, and one (broad) for the C—H groups in the carborane. Why is it broad? If equipment is available, the ^{11}B NMR spectrum should also be obtained, in acetone-d_6. Five doublets are observed over a wide chemical shift range from +6.6 to −21.8 ppm versus $BF_3 \cdot O(CH_2CH_3)_2$.[3] Why are the individual boron atoms shielded so differently? Why are the signals doublets?

The metallocarborane also exhibits strong absorbances in both the UV and Visible regions. These spectra were reported using dichloromethane as the solvent.[3]

The magnetic moment of this complex should be obtained, if possible. Based upon your result, what is the oxidation state of the cobalt?

QUESTIONS

1. In base attack on *o*-carborane, why is it always the boron atom closest to the carbon atoms that is lost?
2. The bonding in boron hydrides is quite unusual, being described by a semi-topological bonding scheme known as "styx notation." Describe this type of bonding and solve for the structure of $B_3H_8^-$. Two answers are obtained.
3. It has been said that Alfred Stock gave his life for boron chemistry. Write a brief biographical sketch of this scientist's life and death.
4. Carboranes are known to thermally isomerize from the 1,2- to the 1,7-isomer. By what mechanism does this isomerization occur? Can the 1,12-isomer be prepared in the same manner?
5. Metal complexes resembling ferrocene, but with carborane ligands in place of the cyclopentadiene groups have been widely reported. Perform a literature search for this most interesting class of compounds. What general types of "sandwiches" were prepared? (For a starting point, see the references listed in this experiment.)

REFERENCES

1. Lipscomb, W. M., *Boron Hydrides,* Benjamin: New York, 1963.
2. Onak, T., "Polyhedral Organoboranes" in *Comprehensive Organometallic Chemistry,* G. Wilkinson, Ed., Pergamon Press: Oxford, UK, 1982, Vol. 1, Chapter 5.4, p. 411.
3. Plesek, J.; Hermanek, S.; Stibr, B. *Inorg. Syn.* **1983,** *22,* 231.

GENERAL REFERENCES

Onak, T., "Carboranes and Organo-substituted Boron Hydrides" in *Advances in Organometallic Chemistry,* F. G. A. Stone and R. West, Eds., Academic Press: New York, 1965, Vol. 3, p. 263.

Todd, L. J., "Transition Metal-Carborane Complexes" in *Advances in Organometallic Chemistry,* F. G. A. Stone and R. West, Eds., Academic Press: New York, 1970, Vol. 8, p. 87.

Grimes, R. N., "Metallacarboranes and Metallaboranes" in *Comprehensive Organometallic Chemistry,* G. Wilkinson, Ed., Pergamon Press: Oxford, UK, 1982, Vol. 1, Chapter 5.5, p. 459.

Experiment 8 Silicone Polymers: Preparation of Bouncing Putty

INTRODUCTION

Silicones, or more specifically organopolysiloxanes, are polymeric materials that contain silicon, oxygen, and organic groups. The methyl silicones (also called dimethylpolysiloxanes) are perhaps the most important members of this class of materials. A typical structure of a methyl silicone is the silcone oil shown in Figure 7.7.

Figure 7.7. *Structure of a typical methyl silicone.*

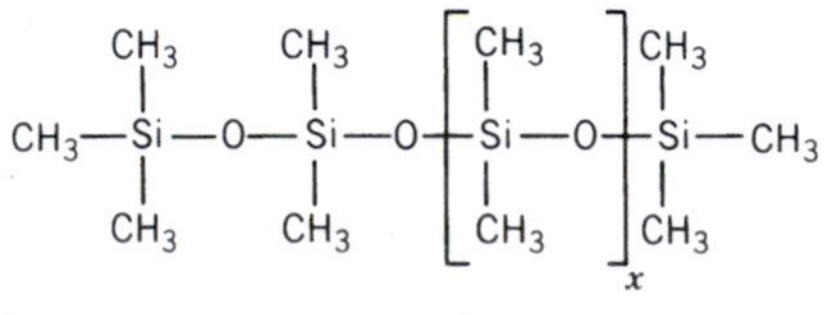

$$CH_3-Si(CH_3)_2-O-Si(CH_3)_2-O-[Si(CH_3)_2-O]_x-Si(CH_3)_2-CH_3$$

The formation of these polymers is controlled by the type of Si—O linkage. There are three types available, shown in Figure 7.8. The nonorganic groups are known as **functionalities**. The R unit may be various organo groups, such as methyl, ethyl, phenyl, or vinyl. The ratio of the difunctional to monofunctional groups control the length of the polymer chain (i.e., x in Fig. 7.7). These materialls are named as derivatives of disiloxane, $H_3SiOSiH_3$. Thus $(CH_3)_3Si—O—Si(CH_3)_3$ is named hexamethyldisiloxane. The polymers are often referred to as diorganopolysiloxanes. Trifunctional units are employed in the formation of cross-linked structures known as silicone resins.

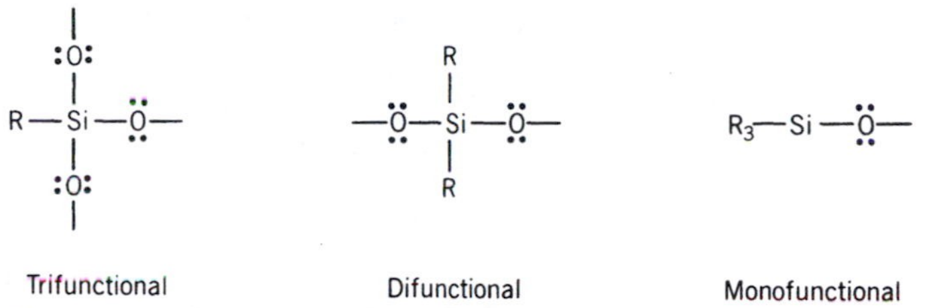

Trifunctional Difunctional Monofunctional

Figure 7.8. *Functionality of Si-O linkage.*

The term "silicone" was first used by Friedrich Wöhler (better known for disproving the theory of organic vitalism) in 1857. The term was used to describe chemicals that had an empirical formula of R_2SiO, in an analogy to the organic compounds called ketones, R_2CO, where R is an organic group. Silicones bear little resemblance to ketones, however, and the differences between these two types of compounds also illustrate a basic difference between silicon and carbon chemistry. Carbon shows a strong tendency to form π bonds, and ketones have the structure RR′C═O, with a double bond between the carbon and oxygen. They are generally simple molecular compounds. Silicon, on the other hand, does not show this tendency, and $p\pi$–$p\pi$ double bonds between silicon and itself or other elements are relatively unstable, not being discovered until the mid-1970s.

A second type of π bonding is possible, however. Silicon, unlike carbon, possesses empty, low energy *d* orbitals, which can effectively overlap with filled $p\pi$ orbitals on other elements, such as nitrogen, oxygen, or fluorine. Evidence for this $p\pi$–$d\pi$ type of bonding is given by the unusually short bond lengths, even after accounting for differences in electronegativity, and by unusual bond angles in many compounds. While other explanations have been offered for these phenomena, the simple fact is that silicon forms extremely stable "single" bonds to oxygen (466 $kJ \cdot mol^{-1}$), and forms them under a wide variety of situations.

The most common starting material for the formation of the methyl silicone polymers is dichlorodimethylsilane, $(CH_3)_2SiCl_2$. This compound is, in turn, made in huge industrial quantities by the action of CH_3Cl on silicon powder in the presence of a copper catalyst at 250–300°C. The resulting mixture of compounds includes $(CH_3)_2SiCl_2$, $(CH_3)_3SiCl$, CH_3SiCl_3, $SiCl_4$, and other less abundant products. The mixture is separated by very careful fractional distillation. Dichlorodimethylsilane is a useful starting material for two reasons:

1. The Si—Cl bonds are easily hydrolyzed, making the compound very reactive.
2. The compound is bifunctional, since there are two chlorines. The chain can therefore propagate in two directions, resulting in high molecular weight polymers.

The organopolysiloxanes are prepared by the hydrolysis of the selected chlorosilane. Thus, hydrolysis of $(CH_3)_2SiCl_2$ gives rise to the corresponding silanol, $(CH_3)_2Si(OH)_2$, and hydrogen chloride. The outstanding characteristic of the silanols is the ease in which they condense to yield siloxane polymers, as shown in the following reaction sequence.

$$n(CH_3)_2SiCl_2 + nH_2O \rightarrow n(CH_3)_2Si(OH)_2 + 2nHCl$$

$$n(CH_3)_2Si(OH)_2 \rightarrow [(CH_3)_2SiO]_n + nHO\text{—}Si(CH_3)_2\text{—}[O\text{—}Si(CH_3)_x\text{—}]Si(CH_3)_2\text{—}OH$$

The polysiloxanes formed are a mixture of cyclic compounds (where $n = 3$, 4, 5, etc.) and open chained compounds having hydroxyl end groups. In the industrial preparation of silicones, the cyclic species are obtained in good yield by carrying the hydrolysis out in dilute ether solution. They are isolated and purified by fractional distillation. The cyclic compound (the tetramer is most often used) is then polymerized to the linear polymer by a process called **equilibration.** Equilibration of cyclic siloxanes is the process by which the Si—O— linkages are continuously broken and reformed until the system reaches an equilibrium condition at the most thermodynamically stable state. Heat alone, or more commonly acid or base catalysis, is used in the process.

The chain length of the polymer can be controlled by adding $(CH_3)_3SiCl$, a monofunctional compound called an endblocker, to the reaction mixture. This compound terminates the chains with an $—OSi(CH_3)_3$ group (see Fig. 7.7). The absence of an endblocker can produce chains of high molecular weight, often referred to as silicone gums.

Silicones can be prepared having a wide range of viscosities, lubricating properties, and reactivities. They see extensive use in industrial chemistry in automobile polishes, cosmetics, water repellants, high-temperature electrical insulation, gaskets, release agents, antifoam agents, high-temperature paints, glass cloth laminates, elastomers (rubbers), greases, and other general polymers. The framework of all the polymers is the very stable —Si—O—Si— sequence. This gives silicones good thermal stability at high temperatures (above 500 °F) as well as flexibility at low temperatures (below −110 °F). The organic groups are hydrophobic, and thus, so is the polymer. Medical grade silicones are used widely in areas such as silicone rubber finger joints for those suffering from various forms of arthritis, mammary implants following radical mastectomy, hypodermic needles lubricated with silicone fluids to make insertion reasonably painless, and silicone rubber coatings encasing implants such as pacemakers and infusion pumps to name a few.

In this experiment, the chemistry of silicones will be investigated by preparing bouncing putty, a silicone polymer, via the hydrolysis of dichlorodimethylsilane. The silicone, containing residual hydroxyl groups will be cross-linked using boric acid. This trifunctional acid, $B(OH)_3$, which also contains hydroxyl groups, forms —Si—O—B— linkages resulting in a peculiar type of gum. The commercial bouncing putty found in novelty stores is compounded with softening agents, fillers, and coloring agents.

Prior Reading and Techniques

Section 5.D.6: Removal of Suspended Particles from Solution
Section 5.F: Concentration of Solutions

Related Experiments

Industrial Chemistry: Experiments 4, 12, and 34

EXPERIMENTAL SECTION

Safety Recommendations

Dichlorodimethylsilane (CAS No. 75-78-5): This compound is harmful if swallowed, inhaled, or absorbed through the skin. It is extremely destructive to the mucous membranes. The compound reacts violently with water. IHL-RAT LC50: 930 ppm/4H.

Boric acid (CAS No. 10043-35-3): While not generally considered dangerous, this compound is toxic if ingested. ORL-RAT LD50: 5.14 g/kg. Death has occurred from ingestion of 5–20 g in adults.

CHEMICAL DATA

Compound	FW	Amount (mL)	mmol	mp (°C)	bp (°C)	Density
Dichlorodimethylsilane	129.06	1	8.24	−16	70	1.064

Required Equipment
Magnetic stirring hot plate, 10-mL round-bottom flask, magnetic stirring bar, automatic delivery pipet, water condenser, Pasteur pipet, Pasteur filter pipet, litmus paper, 5-mL conical vial, sand bath.

Time Necessary for Experiment: 2.5 h.

EXPERIMENTAL PROCEDURE

NOTE: ***This reaction is carried out in the* HOOD.**

Place 2 mL of diethyl ether in a 10-mL round-bottom flask containing a magnetic stirring bar. Using an automatic delivery pipet, add 1 mL of dichlorodimethylsilane to the ether solvent and attach a water condenser to the flask.

NOTE: ***Dichlorodimethylsilane reacts rapidly with moisture to produce hydrogen chloride gas. Make the transfer swiftly.***

Arrange the assembly on a magnetic stirring hot plate, and with rapid stirring, **carefully** add 2 mL of water **dropwise,** from a Pasteur pipet, through the top of the condenser.

NOTE: ***Hydrogen chloride gas is evolved in this hydrolysis step. The addition must be made slowly at the beginning of the reaction or too vigorous an evolution of the HCl will occur. You will note that the ether is warmed to reflux temperature. This is the reason a cooled water condenser is used.***

Allow the resulting mixture to stir for an additional 10 min at room temperature after addition of the water is complete. Remove the reaction flask from the condenser. Using a Pasteur filter pipet, carefully remove the major portion of the lower aqueous layer. This is discarded. Reattach the flask to the reflux condenser and carefully add 1 mL of 10% sodium bicarbonate solution dropwise through the top of the condenser, with stirring.

NOTE: ***This step is to neutralize any residual acid remaining in the wet ether solution. Vigorous evolution of carbon dioxide gas is observed at this stage as the neutralization proceeds.***

After bubbling of the solution has stopped, remove the flask and using the Pasteur filter pipet from the previous step, separate the lower aqueous phase.

Test this phase with litmus paper. Repeat the neutralization step until the discarded aqueous layer is no longer acidic to litmus. Finally, wash the ether solution with 2 mL of water. Separate and discard the aqueous layer.

Transfer the wet ether layer in two portions to a Pasteur filter pipet column containing silica gel (bottom layer) and anhydrous sodium sulfate (~$\frac{3}{4}$ in. of each in the column). Collect the dried eluate in a tared 5-mL conical vial containing a boiling stone. Wash the column with an additional 3 mL of ether. When 2 mL of eluate has collected in the flask, concentrate the contents on a warm sand bath in the **HOOD** under a slow stream of nitrogen. Repeat this procedure for the other one half of the wet ether layer.

NOTE: *The concentration step is done in stages. If the flow of eluate is too slow the procedure may be speeded up by placing a pipet bulb on the top of the column and applying gentle pressure. This is a simple introduction to flash chromatography.*

Isolation of Product

Determine the weight of the clear residual silicone fluid. Add boric acid to the silicone fluid in an amount of 5% of the weight of the fluid, stirring constantly with a microspatula. Continue stirring for five minutes. Heat the resulting mixture to ~170–180 °C in a sand bath or on an aluminum block until a stiff silicone gum is obtained. This usually occurs within a 20-min time span. Allow the product to cool to room temperature.

Remove the product from the vial and roll the material into a ball. If the gum is somewhat brittle, continued kneading will produce the desired gumlike characteristics.

Perform the following tests on your product:

1. It should give a lively bounce on a hard surface.
2. Pulling sharply causes cleavage of the gum.
3. Pulling slowly results in stretching reminiscent of chewing gum.
4. Placed on a hard surface it will flow into a flat plate.
5. If test (4) is done on newspaper, careful removal of the flat gum will reveal the mirror image of the print.

QUESTIONS

1. By what process is pure silicon prepared in industry?
2. Make a comparison of the chemical and physical properties of SiO_2 and CO_2. Can you account for these differences?
3. What is the composition of carborundum, how is it made, and what major industrial use does it have?
4. Which would you expect to be the better Lewis base, C^{4-} or Si^{4-}? Explain.
5. Describe the molecular structure and hybridization of the Si atom in the following species. (a) $(CH_3)_2SiF_2$, (b) SiF_6^{2-}, and (c) SiF_4.
6. There is a wide variety of silicone polymers, both synthetic and naturally occurring. Carry out a literature search and locate representative examples of silicones that are used for each of the applications mentioned in the experimental introduction.

GENERAL REFERENCES

Eaborn, C., *Organosilicon Compounds,* Butterworths: London, 1960.

Meals, R. N.; Lewis, F. M., *Silicones,* Reinhold: New York, 1959.

Kirk–Othmer, *Encyclopedia of Chemical Technology,* 3rd ed., Wiley–Interscience: New York, Vol. 20, 1979, p. 922.

Rochow, E. G., *Chemistry of the Silicones,* 2nd ed., Wiley: New York, 1951.

Stark, F. O.; Falender, J. R.; Wright, A. P., "Silicones" in *Comprehensive Organometallic Chemistry,* G. Wilkinson, Ed., Pergamon Press: Oxford, 1982, Vol. 2, Chapter 9.3, p. 305.

Experiment 9 The Oxidation States of Tin

Part A: Preparation of Tin(IV) Iodide

Part B: Preparation of Tin(II) Iodide

INTRODUCTION

Like the transition metals, many of the main group elements also exhibit multiple oxidation states. An example of this occurs with the element tin, which has two common oxidation states: Sn(IV) (commonly named stannic) and Sn(II) (stannous). They are of approximately equal stability. Comparing the oxidation states of tin with those of other Group 14 (IVA) elements, we find that carbon, silicon, and germanium are nearly always found in the IV oxidation state. Lead, however, is most often found in the II oxidation state, with the IV state being fairly unstable [see Experiment 10 for a comparison of the stabilities of Sn(IV) and Pb(IV)]. A similar trend is found in the oxidation states of other main group families [Groups 13–16 (IIIA–VIA) elements], where the primary oxidation number decreases by two going down the group. It should be pointed out that in all cases here, we are not referring to ionic charges on the elements, as the compounds are appreciably covalent.

The trend can be rationalized in the following way, using the Group 14 (IVA) elements as examples. The electronic configuration of the family is ns^2np^2, with the electrons being arranged in orbitals as follows:

↿⇂	↿	↿	—
s	p_x	p_y	p_z

In this electronic state, the element can form two covalent bonds, using the two unpaired electrons in the p_x and p_y orbitals. The two electrons in the s orbital are not used in bonding and are sometimes termed an "inert pair." This bonding is seen for Sn(II) and Pb(II).

Alternatively, one of the s electrons can be promoted to the empty p_z orbital at the cost of an absorption of energy by the element, resulting in the following electronic arrangement.

↿	↿	↿	↿
s	p_x	p_y	p_z

Four covalent sp^3 hybridized bonds can now be formed (which through hybridization are equivalent) by the Group 14 (IVA) element. Bond formation occurs with the release of energy. The promotion of the s electron is thereby compensated for to some extent by the "return" of bond energy.

The likelihood of achieving the IV oxidation state is therefore related to two quantities.

1. The ease of electronic promotion: Easiest for the heavier family members, because of the greater distance of the electron from the nucleus.

and more importantly,

2. Bond strength: Strongest for the lighter family members, because of compact orbital size, short bond lengths, and good orbital overlap.

In the case of carbon, silicon, and germanium, the "cost" of electron promotion is more than compensated by the "return" of energy from the additional two bonds formed. These elements are thus found in the IV oxidation state. In the case of tin, the "cost" and "return" are of similar magnitude, hence, both states are of approximately equal stability. In the case of lead, the "cost" of electron promotion is not compensated for by the small energy return from the two additional bonds. Lead is therefore generally found in the II oxidation state, with Pb(IV) compounds being easily reduced.

Tin(IV) compounds may be prepared by direct reaction of metallic tin with mild oxidizing agents, such as iodine. The iodine is consequently reduced from the elemental state I(0) to the I(−I) state.

$$Sn + 2I_2 \rightarrow SnI_4$$

The direct reaction results in the slightly more stable IV oxidation state being formed.

Tin(II) iodide is easily obtained via the reaction of zinc iodide and tin(II) chloride solutions. The zinc iodide solution is, in turn, readily obtained by the direct oxidation of metallic zinc by iodine, in the presence of a small amount of water.

$$Zn(s) + I_2(s) \rightarrow Zn^{2+}(aq) + 2I^-(aq)$$

Tin(II) chloride solution is prepared by dissolving small, thin pieces of metallic tin in concentrated hydrochloric acid.

$$Sn\ (s) + 2HCl\ (aq) \rightarrow Sn^{2+}\ (aq) + 2Cl^-\ (aq) + H_2\ (g)$$

When the solutions of zinc iodide and tin(II) chloride are mixed together, an orange-red crystalline precipitate of tin(II) iodide is obtained. This reaction, where ions exchange partners, is called a **metathesis** or **double decomposition**,

$$Sn^{2+}\ (aq) + 2I^-\ (aq) \rightarrow SnI_2\ (s)$$

Prior Reading and Techniques

Section 2.F: Reflux and Distillation

Section 5.D.3 Isolation of Crystalline Product (Suction Filtration)

Related Experiments

Tin Chemistry: Experiments 6, 10, and 15

EXPERIMENTAL SECTION

Safety Recommendations

Tin (CAS No. 7440-31-5): Tin is not generally considered to be a dangerous material; however, the normal precautions (Section 1.A.3) should be taken.

Iodine (CAS No. 7553-56-2): Iodine is harmful if swallowed, inhaled, or absorbed through the skin. It is a lachrymating agent (makes you cry). ORL-RAT LD50: 14 g/kg. Ingestion of 2–3 g has been fatal.

Methylene chloride (CAS No. 75-09-2): This compound, also called dichloromethane, is harmful if swallowed, inhaled, or absorbed through the skin. Exposure may cause nausea, dizziness, and headache. It is a possible carcinogen. ORL-RAT LD50: 1600 mg/kg. Exposure to this compound should be minimized, as it is a narcotic at high concentrations.

Zinc (CAS No. 7440-66-6): Zinc is not generally considered to be a dangerous material; however, the usual safety precautions (Section 1.A.3) should be taken.

Part A: Preparation of Tin(IV) Iodide

CHEMICAL DATA

Compound	FW	Amount	mmol	mp (°C)	bp (°C)	Density
Sn	118.7	119 mg	1.00	232	2260	7.310[a]
I_2	253.8	475 mg	1.87	114	185	4.93

[a] Density is for white tin. The density of the other allotropic form, gray tin, is 5.75.

Required Equipment
Magnetic stirring hot plate, 10-mL round-bottom flask, boiling stone, reflux condenser, Pasteur pipet, ring stand, glass funnel, 10-mL Erlenmeyer flask, Hirsch funnel, clay tile, or filter paper.

Time Required for Experiment: 2 h.

EXPERIMENTAL PROCEDURE

Place 119 mg (1.00 mmol) of tin and 475 mg (1.87 mmol) of iodine into a 10-mL round-bottom flask containing a boiling stone and equipped with a reflux condenser (see Fig. 7.9). (A Keck clip may be used in place of a clamp to connect the round-bottom flask to the condenser.) Add 6.0 mL of methylene chloride, which acts as the solvent for this reaction, through the condenser using a Pasteur pipet.

Gently, heat the flask and contents using a hot water bath until a **mild** reflux is maintained. This can be detected through a moderate dripping rate from the

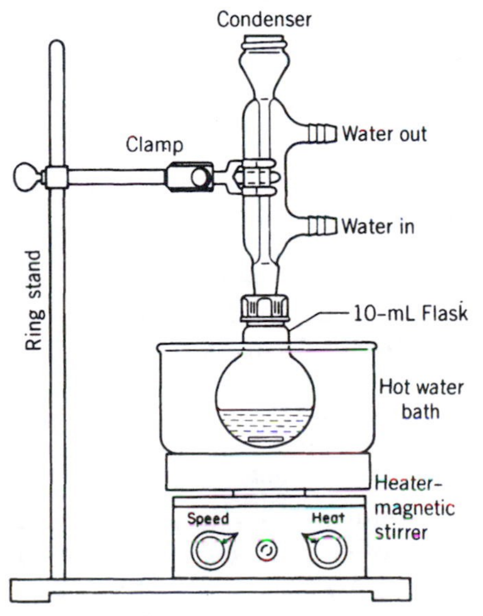

Figure 7.9. *Apparatus for Experiment 9.*

bottom of the condenser joint. Maintain the system at the reflux temperature until there is no visible violet color of iodine vapor in the condenser throat (~30–40 min).

Isolation of Product

Gravity filter the warm solution *rapidly* through a **loose** cotton or glass wool plug using a small glass funnel. Collect the filtrate (liquid) in a 10-mL Erlenmeyer flask. Any unreacted tin metal will remain in the funnel. Rinse the reaction flask with an additional 200 μL (automatic delivery pipet) of warm methylene chloride (**HOOD**), and also pass this solution through the filter, collecting the filtrate in the same flask.

Add a boiling stone to the filtrate and concentrate the solution on a sand bath (**HOOD!**) to approximately 2 mL. Cool the resulting solution in an ice–water bath, and collect the orange-red crystals of tin(IV) iodide by suction filtration using a Hirsch funnel. Wash the filter cake with two 0.5-mL portions of cold methylene chloride, and dry the crystals on a clay tile or on filter paper. Determine the melting point of the product, and calculate a percentage yield.

Characterization of Product

Dissolve a small amount of the product, SnI_4, in 5 mL of acetone. Divide this into two portions, A and B. To portion A, add a few drops of water. To portion B, add a few drops of saturated KI solution. Observe what happens in each case, and explain your observations in terms of the nature of the product.

Part B: Preparation of Tin(II) Iodide

CHEMICAL DATA

Name	FW	Amount	mmol	mp (°C)	Density
Zn	65.37	100 mg	1.53	420	7.140
Sn	118.7	80 mg	0.68	232	7.310[a]
I_2	253.8	100 mg	0.40	114	4.93

[a] Density is for white tin. The density of the other allotropic form, gray tin, is 5.75.

Required Equipment

Three 10-mL beakers, watch glass, magnetic stirring hot plate, magnetic stirring bar, ice–water bath, Pasteur filter pipet, Hirsch funnel, clay tile, or filter paper.

Time Required for Experiment: 3 h.

EXPERIMENTAL PROCEDURE

Cut a sample of tin foil weighing ~80 mg (0.68 mmol) into smaller pieces, and place them into a 10-mL beaker equipped with a magnetic stirring bar. In the **HOOD**, add 1.0 mL of concentrated HCl and 2–3 drops of 0.1*M* $CuSO_4$ solution.

> **NOTE: *Concentrated HCl is very corrosive. This reaction generates hydrogen gas, which forms explosive mixtures with air. Open flames must not be used in this experiment.***

Place the beaker on a magnetic stirring hot plate (**HOOD**), cover it with a watch glass, and slowly, with stirring, heat the mixture to just below boiling. Note the evolution of H_2 gas in the flask. While the reaction between the tin and HCl is underway, proceed to the next step of the synthesis.

In another 10-mL beaker, place 100 mg (1.53 mmol) of granular zinc, 1 mL of deionized (or distilled) water, and a magnetic stirring bar. Add 100 mg (0.40 mmol) of iodine crystals to the flask. Cool the flask in an ice–water bath. In the beginning, the reaction of iodine with zinc is slow. Stir the mixture slowly. After a few minutes, a brown color develops because of dissolved iodine. The reaction of iodine with zinc is an exothermic redox reaction and thus requires thorough cooling. The brown color of dissolved iodine will slowly disappear when the reaction is complete and the solution becomes light yellow. Some unreacted zinc remains at this point.

Transfer the ZnI_2 solution from the beaker to another 10-mL beaker using a Pasteur filter pipet. The filtered solution should be clear, having no suspension of metallic zinc. To ensure complete removal of ZnI_2, rinse the contents of the beaker with a few drops of deionized water, and transfer the washings using the same filter pipet to the beaker. (Do not use too much water to rinse the zinc residue; dilution will reduce the yield of product.) Unreacted zinc remaining in the beaker should be deposited in a waste container.

Check the tin–HCl mixture. If some metallic tin still remains unreacted, add a few more drops of concentrated HCl and warm the mixture on the magnetic stirring hot plate. (At the end of the reaction some black suspended particles will remain—this is normal.) Filter the tin solution using a clean Pasteur filter pipet and dropwise transfer it directly to the beaker containing the ZnI_2 solution. As soon as the $SnCl_2$ solution comes into contact with the ZnI_2 solution an orange-yellow precipitate of SnI_2 will form. Complete precipitation by cooling the solution in an ice bath. Collect the solid by suction filtration using a Hirsch funnel.

Recrystallization of the Product

Transfer the crude SnI_2 from the Hirsch funnel to the same beaker in which the precipitation was carried out, and add a magnetic stirring bar. Add 2–3 drops of concentrated HCl to the solid. Cool the beaker to room temperature and then continue to cool it in an ice bath to complete the recystallization. Collect the product by suction filtration on a Hirsch funnel and dry it on a clay tile or filter paper. Weigh the product and calculate the percentage yield. Obtain the melting point.

QUESTIONS

1. While SnI_4 is quite stable, $PbBr_4$ and PbI_4 do not exist as stable compounds. Explain.
2. Write the oxidation and reduction half-reactions for this experiment.
3. In the above characterization test, it was observed that SnI_4 is soluble in acetone. What does this tell you about the ionic nature of this compound? Is the charge on the tin actually 4^+ ? What was the effect of adding KI to the acetone solution?
4. In Part 9.A, the product is essentially pure SnI_4, with no SnI_2 being formed. Explain this experimental result.
5. Prepare a table comparing the physical properties (mp, bp, and density) of the chlorides of C, Si, Ge, Sn, and Pb. Are there any general trends in the table? Explain.
6. Consult the literature to determine the major commercial use of tin(II) fluoride. Why is it well suited to this purpose? What is the unusual structure of tin(II) fluoride? Compare it to the structures of tin(II) chloride, bromide, and iodide. The crystal structure of tin(II) iodide was reported by W. Moser and I. C. Trevena in 1969.

REFERENCE

1. Moeller, T.; Edwards, D. C. *Inorg. Syn.* **1953,** *4,* 119.

Experiment 10 Relative Stabilities of Tin(IV) and Lead(IV)

Part A. Preparation of Ammonium Hexachlorostannate(IV)

Part B. Preparation of Ammonium Hexachloroplumbate(IV)

INTRODUCTION

Although carbon compounds are usually restricted to being at most four coordinate, the heavier members of Group 14 (IVA) can achieve higher coordinations through the use of *d* orbitals. Six coordination is common for all heavier members of the group. In this experiment, six-coordinate complexes of tin and lead will be prepared. The tin(IV) complex, $(NH_4)_2SnCl_6$, is prepared through simple complexation of the covalent halide:

$$SnCl_4 + 2NH_4Cl \rightarrow (NH_4)_2SnCl_6$$

The analogous lead(IV) complex, $(NH_4)_2PbCl_6$, is not so easily prepared. Lead(IV) chloride is not a stable material, so the process must be carried out via a redox reaction of lead(II) chloride, with elemental chlorine used as a strong oxidizing agent:

$$PbCl_2 + 2HCl + Cl_2 \rightarrow H_2PbCl_6$$

The desired product is obtained via an ion exchange reaction with NH_4Cl:

$$H_2PbCl_6 + 2NH_4Cl \rightarrow (NH_4)_2PbCl_6 + H^+ + Cl^-$$

Information about the relative stability of the IV oxidation state can be found in the introduction to Experiment 9.

Prior Reading and Techniques

Section 1.B.4: Compressed Gas Cylinders and Lecture Bottles
Section 5.D.3: Isolation of Crystalline Products (Suction Filtration)

Related Experiments

Complexes of the Main Group Elements: Experiments 5 and 18
Tin Chemistry: Experiments 6, 9, and 15

EXPERIMENTAL SECTION

Part A: Preparation of Ammonium Hexachlorostannate(IV)[1]

Safety Recommendations

Tin(IV) chloride (CAS No. 7646-78-8): Tin(IV) chloride fumes in moist air, forming HCl. Loosely capped bottles often develop a coating of tin hydroxide. Tin(IV) chloride fumes are very corrosive, and must never be inhaled. Vapor pressure is 18.6 mm at 20 °C. IPR-MUS LD50: 101 mg/kg. The compound reacts violently with water and ethers, so use these materials *with caution*.

Ammonium chloride (CAS No. 12125-02-9): This compound is harmful if swallowed and causes eye irritation. ORL-RAT LD50: 1650 mg/kg. The usual safety precautions (see Section 1.A.3) should be taken.

CHEMICAL DATA

Compound	FW	Amount	mmol	mp (°C)	Density
$SnCl_4$	260.5	100 mg	0.38	−33	2.226
NH_4Cl	53.5	35 mg	0.65	340[a]	1.529

[a] Sublimes.

Required Equipment
Magnetic stirring hot plate, 10-mL beaker, magnetic stirring bar, Pasteur pipet, automatic delivery pipet, Hirsch funnel, clay tile, or filter paper.

Time Required for Experiment: 30 min.

EXPERIMENTAL PROCEDURE

Weigh 100 mg (0.38 mmol) of anhydrous $SnCl_4$ into a tared 10-mL beaker equipped with a magnetic stirring bar. **Cautiously (HOOD)**, dropwise (automatic delivery pipet) and with stirring, add 70-μL of water. Using a Pasteur pipet, add a solution of 35 mg (0.65 mmol) of NH_4Cl dissolved in 100 μL of water. The product, $(NH_4)_2SnCl_6$, will precipitate immediately.

Isolation of Product
Collect the crystals by suction filtration using a Hirsch funnel and wash them with two 0.5-mL portions each of cold ethanol and ether. Dry the product on a clay tile or on filter paper. Determine the melting point and calculate a percentage yield.

Part B: Preparation of Ammonium Hexachloroplumbate(IV)[1]

Safety Recommendations

Lead dichloride (CAS No. 7758-95-4): Lead dichloride, like **all** lead compounds, is a heavy metal poison. Contact of lead compounds with your skin should be avoided—only use this compound while wearing gloves. Avoid breathing the dust. Adverse effects of lead on human reproduction have been reported.

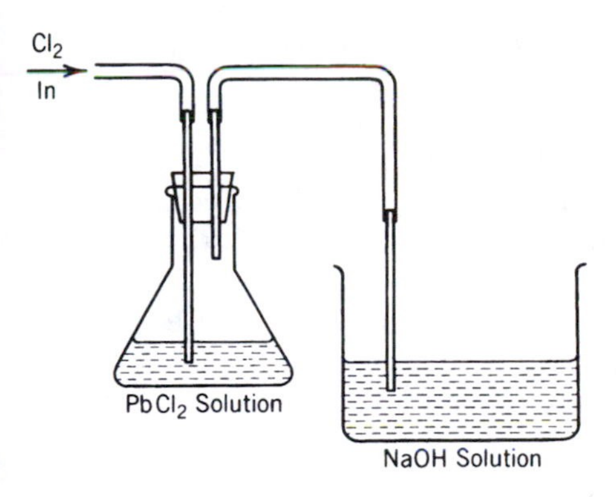

Figure 7.10. *NaOH trap for chlorination procedure.*

Chlorine (CAS No. 7782-50-5): Chlorine is a very hazardous, corrosive, and toxic gas. It must **only be used in a hood.** It should never be vented into the air, but rather, trapped in a NaOH solution, as shown in Figure 7.10. For safe handling of lecture bottles, see Section 1.B.4. The gas may be fatal if inhaled, and is extremely destructive to the tissue of the mucous membranes and upper respiratory tract, eyes, and skin. IHL-HMN LCLo: 500 ppm/5M.

Ammonium chloride (CAS No. 12125-02-9): This compound is harmful if swallowed and causes eye irritation. ORL-RAT LD50: 1650 mg/kg. The usual safety precautions (see Section 1.A.3) should be taken.

CHEMICAL DATA

Compound	FW	Amount	mmol	mp (°C)	bp (°C)	Density
$PbCl_2$	278.1	100 mg	0.36	501	950	5.850
Cl_2	70.9	Sufficient amount	Sufficient amount	−101	−34	
NH_4Cl	53.5	40 mg	0.75	340[a]		1.527

[a] Sublimes

Required Equipment

Two 10-mL Erlenmeyer flasks, magnetic stirring hot plate, magnetic stirring bar, ice–water bath, NaOH trap (see Fig. 7.10), chlorine lecture bottle, Tygon tubing, Hirsch funnel, clay tile or filter paper.

Time Required for Experiment: 30 min.

EXPERIMENTAL PROCEDURE

> **NOTE: *The following chlorination procedure should only be carried out in the HOOD.***

In a 10-mL Erlenmeyer flask equipped with a magnetic stirring bar, make up a suspension of 100 mg (0.36 mmol) of $PbCl_2$ in 1.5 mL of 12*M* HCl.

> **NOTE: *12M HCl is highly corrosive. Handle it with care.***

Cool the flask in an ice–water bath and pass a moderate stream of chlorine through the solution. The excess chlorine gas is led into a 6*M* NaOH trap, using the apparatus shown in Figure 7.10. Chlorine gas should be delivered from a lecture bottle (for safe handling of lecture bottles, see Section 1.B.4), through Tygon tubing, to a glass tube immersed in the stoppered $PbCl_2$ solution. A vent through a second glass tube, through Tygon tubing, to a beaker of 3*M* NaOH solution should be provided. The suspension will become a yellow solution as H_2PbCl_6 is formed and dissolves.

When all the $PbCl_2$ has dissolved, stop the chlorine flow and transfer the liquor (which contains the product), using a Pasteur pipet, into a 10-mL Erlenmeyer flask. Cool the solution in an ice bath. Add a solution of 40 mg (0.75 mmol) of NH_4Cl in 250 μL of water and let it stand for 10 min. Collect the precipitate using suction filtration with a Hirsch funnel. Wash the filter cake with two 0.5-mL portions of ice cold ethanol, followed by two 0.5-mL portions of ether. Dry the product on a clay tile or on filter paper. Determine the melting point and calculate a percentage yield.

Test for Comparative Stabilities of Sn(IV) and Pb(IV)

Number two small test tubes. Place a $\frac{1}{2}$ portion of the $(NH_4)_2SnCl_6$ you prepared in tube 1 and a similar portion of the $(NH_4)_2PbCl_6$ in tube 2. Add 100 μL of water to each tube and observe the results. Be sure to record any odors.

Heat the tubes in boiling water for 5 min, and then allow them to stand at room temperature for 30 min. Record your observations.

Carefully, heat a second set of similarly prepared test tubes over a microburner. Record your observations.

Recovery and Recycle of Lead(II) Chloride

All lead solids should be combined into a single 10-mL Erlenmeyer flask. Heating the ammonium hexachloroplumbate (**HOOD!**) will thermally decompose the material to form NH_3, HCl, and $PbCl_2$. The end of the reaction may be evidenced by the color change from yellow to white. Collect the $PbCl_2$, it may be used the next time the experiment is carried out.

QUESTIONS

1. The $(NH_4)_2SnCl_6$ can be prepared by simple complexation of the covalent solid, whereas the $(NH_4)_2PbCl_6$ requires a two-step redox reaction. Explain the difference.

2. Write balanced chemical reactions for each of the stability tests you performed.
3. From steric effects, one might expect the lead complex to be the more stable, as the central lead atom would have more room to accommodate the six chloride ligands than the smaller tin. It is the tin complex, however, which is more stable. Explain.
4. Unreacted chlorine reacts with a solution of NaOH in the chlorine trap. What are the products? What industrial process does this reaction illustrate?
5. How is chlorine gas commercially prepared?
6. In this experiment, we are very careful to recycle and recover all the lead used, because of the great environmental toxicity of this metal. Using the literature, locate and describe the physiological impact of lead poisoning. What are the major sources of environmental lead? (*Hint:* As with most toxicological questions, a good place to start is the *Merck Index.*)

REFERENCE

1. Experiment adapted from: Frey, J. E., *Basic Inorganic Synthesis Lab Manual CH 215*, Northern Michigan University: Marquette, MI.

Experiment 11 Preparation of Substituted 1,2,3,4-Thiatriazoles

Part A. Preparation of 5-Anilino-1,2,3,4-thiatriazole

Part B. Preparation of 5-Amino-1,2,3,4-thiatriazole

INTRODUCTION

The thiatriazole ring is an example of an inorganic heterocyclic ring system, containing nitrogen, sulfur, and carbon atoms. It was first prepared in 1896.[1] The system is thermally unstable; the compounds synthesized in this experiment undergo vigorous decomposition when heated. Extensive studies were carried out on the catalyzed and uncatalyzed decomposition of these materials.[2] This thermal instability led to the use of several of these substituted compounds as blowing agents to generate foamed materials.[3] They were also used as one of the ingredients in corrosion inhibiting formulations for copper and ferrous metals and in antihypertension compositions.[4] An additional use is in the formulation of materials for document reproduction.[5]

The heterocyclic ring is formed by the diazotization of the corresponding thiosemicarbazide.

$$R\text{—NH—C(=S)—NHNH}_2 + HCl + NaNO_2 \xrightarrow[<10^\circ C]{H_2O} \text{5-(RNH)-1,2,3,4-thiatriazole}$$

(Ring numbering: S1, N2, N3, N4, C5; C5 bears —N(H)—R.)

R = H and C_6H_5

Diazotization is a well-known oxidation reaction in organic chemistry.[6] It is used to prepared diazonium salts of alkane and arene primary amines, which are key intermediates in the preparation of a wide variety of materials.

Prior Reading and Techniques

Section 5.D.3: Isolation of Crystalline Products (Suction Filtration)

Section 5.D.4: The Craig Tube Method

Section 6.C: Infrared Spectroscopy

Related Experiments

Nitrogen Chemistry: Experiments 12 and 14

Organometallic Chemistry of the Main Group Elements: Experiments 6, 7, and 15

EXPERIMENTAL SECTION

Part A: Preparation of 5-Anilino-1,2,3,4-thiatriazole

Safety Recommendations

4-Phenyl-3-thiosemicarbazide (CAS No. 5351-69-9): This compound has not been extensively investigated, however, thiosemicarbazide is extremely toxic. It would be prudent to follow strict safety procedures and handle this material **only** with gloves. IPR-MUS LD50: 15 mg/kg.

Sodium nitrite (CAS No. 7632-00-0): Sodium nitrite is harmful if swallowed, inhaled, or absorbed through the skin. It has been shown to have effects on fertility and embryo or fetal development. It should be handled with care, wearing gloves. ORL-HMN LDLo: 71 mg/kg, ORL-RAT LD50: 85 mg/kg.

CHEMICAL DATA

Compound	FW	Amount	mmol	mp (°C)	Density
4-Phenyl-3-thiosemicarbazide	167.23	167 mg	1.0	138	
$NaNO_2$	69.00	69 mg	1.0	271	2.168

Required Equipment

Magnetic stirring hot plate, 10-mL round-bottom flask, magnetic stirring bar, 7.5-cm evaporating dish, thermometer, 10-mL Erlenmeyer flask, Pasteur pipets, Hirsch funnel, clay tile or filter paper.

Time Required for Experiment: 1 h.

EXPERIMENTAL PROCEDURE

Prepare a suspension of 167 mg (1 mmol) of 4-phenyl-3-thiosemicarbazide in 760 μL of 15% HCl in a 10-mL round-bottom flask containing a magnetic stirring bar. Place the flask in an an ice–salt bath, prepared in a 7.5-cm evaporating dish, set on a magnetic stirring hot plate. The bath temperature is monitored using a thermometer and the contents of the flask cooled to 5–10 °C with stirring.

Prepare a solution of sodium nitrite in a small test tube or 10-mL Erlenmeyer flask by dissolving 69 mg (1 mmol) of $NaNO_2$ in 150 μL of water (automatic delivery pipet). Add ~130 μL of this nitrite solution, dropwise, using a Pasteur pipet, to the stirred reaction mixture. The addition should be made over a time span of 8–12 min, such that the temperature of the mixture is not allowed to exceed 10 °C.

> **NOTE:** ***A rapid addition of the nitrite solution will result in the decomposition of the diazonium salt intermediate, resulting in a marked decrease in product formation.***

Isolation of Product

When the addition is complete, collect the white precipitate that has formed by suction filtration using a Hirsch funnel. Save the precipitate; it is the product.

Return the filtrate to the reaction flask using a Pasteur pipet and add the remaining sodium nitrite solution, dropwise. Addition is stopped when the reaction mixture becomes pale yellow (add additional nitrite and the solution becomes intense yellow). If further precipitation of product is observed, collect the product as before. Wash the combined filter cakes of the off-white 5-anilino-1,2,3,4-thiatriazole with three 0.5-mL portions of ice-cold distilled water, and then dry the product on a clay tile or on filter paper. The product may be further dried in a desiccator under vacuum over t.h.e. SiO_2 desiccant.

Purification of Product

The compound may be recrystallized from methanol using a Craig tube to yield beautiful, colorless needles. Dry the crystals on a clay tile or on filter paper.

Characterization of Product

Obtain the decomposition point for the product in the same manner as a melting point. The pure material decomposes vigorously at 142–143 °C. Obtain the IR spectrum of the product (KBr disk) and compare your data with the spectrum reported in the literature.[7]

To further characterize the product, the UV spectrum may be obtained in ethanol. The following data was reported[8]:

λ_{max} 241 nm ($\log \epsilon_{max}$ = 3.91)
λ_{max} 285 nm ($\log \epsilon_{max}$ = 3.91; appears as a shoulder)
λ_{max} 302 nm ($\log \epsilon_{max}$ = 4.03)

Part B: Preparation of 5-Amino-1,2,3,4-thiatriazole

Additional Safety Recommendations

Thiosemicarbazide (CAS No. 79-19-6): Thiosemicarbazide is **extremely toxic** and may be fatal if swallowed, inhaled, or absorbed through the skin. Handle **only** with gloves. ORL-RAT LD50: 9160 μg/kg, an exceptionally low value, although ORL-LD50s with other mammals are at least 10 times higher.

CHEMICAL DATA

Compound	FW	Amount	mmol	mp (°C)	Density
Thiosemicarbazide	91.14	91 mg	1.0	180	
$NaNO_2$	69.00	69 mg	1.0	271	2.168

Required Equipment: see Part 11.A.

Time Required for Experiment: 1 h.

EXPERIMENTAL PROCEDURE

Using the procedure outlined in Part 11.A, add dropwise (8–12 min) a solution of 69 mg (1 mmol) of sodium nitrite in 150 μL of water to a stirred and cooled (<10 °C) mixture of 91 mg (1 mmol) of thiosemicarbazide and 350 μL of 3*M* HCl.

Isolation of Product

Collect the fine white crystals of the product by suction filtration using a Hirsch funnel. Wash the filter cake with two 0.5-mL portions of ice-cold water and then dry on a clay tile or on filter paper. The material can be further dried over t.h.e. SiO_2 desiccant under vacuum. Purification by recrystallization is not recommended since decomposition easily occurs.

Characterization of Product

Obtain the decomposition point for the product in the same manner as a melting point. The pure material explodes at its melting point, 126 °C, so use a small sample. Obtain the IR spectrum of the product as a KBr pellet.

The UV spectrum has been reported[8] in ethanol:

$$\lambda_{max}\ 250\ \text{mm}\ (\log \epsilon_{max} = 3.51;\ \text{appears as a shoulder})$$

$$\lambda_{max}\ 267\ \text{mm}\ (\log \epsilon_{max} = 3.68)$$

QUESTIONS

1. Sulfur, in its common elemental state, is a polyatomic solid, while nitrogen is a diatomic gas. How does this fact reflect a key difference between first- and second-period elements?
2. Nitrogen gas and carbon monoxide are isoelectronic. However, their dissociation energies show a marked difference: CO = 1072 $kJ{\cdot}mol^{-1}$ and N_2 = 941 $kJ{\cdot}mol^{-1}$. Offer an explanation for this difference.
3. Why might you expect the thiatriazoles prepared in this experiment to decompose vigorously on heating?
4. Boron nitride, BN, is an extremely stable substance, thermally stable up to approximately 2000 °C. Can you explain this stability in terms of its structure?
5. Molecular nitrogen is an essentially inert species. However, it undergoes reaction at ambient temperatures with lithium. At elevated temperatures and pressures in the presence of selected catalysts, nitrogen reacts with hydrogen to form ammonia gas (Haber–Bosch process).
   - **a.** Write a balanced equation for the formation of lithium nitride and of ammonia based on the previous discussion.
   - **b.** Does the nitrogen act as an oxidation or reducing agent in each of these reactions?
   - **c.** Write the Lewis structure for the nitride and azide ions.
6. Although dinitrogen, N_2, and CO are isoelectronic, the first N_2 complex with a transition metal was not discovered until 1965 by A. D. Allan and C. V. Senoff. Explain why CO is a more stable ligand than N_2. Using the literature, detail how Allan and Senoff prepared the first metal–N_2 compound. What other metal–N_2 compounds were prepared since that time?

REFERENCES

1. Lieber, E.; Pillai, C. N.; Oftedahl, E.; Hites, R. D. *Inorg. Syn.* **1960,** *6,* 44 and references cited therein.
2. For example, see Holm, A.; Calsen, L.; Larsen, E. *J. Org. Chem.* **1978,** *43,* 4816.
3. Farbenfabriken Bayer Akt. Ges. *Br. Patent* 861,056, 1961; Schmidt, K. L.; Lober, F.; Muller, E.; Scheurlen, H. *Ger. Patent* 1,226,391, 1956.
4. Schoepke, H. G.; Swett, L. R., *US Patent* 3,265,576, 1966; also see *Br. Patent* 1,033,519, 1966.
5. Kendall, J. D.; Reynolds, K. *Br. Patent* 877,334, 1958.

6. Consult any basic or advanced organic text such as: Loudon, G. M., *Organic Chemistry,* 2nd ed., Addison–Wesley: Reading, MA, 1988; March, J., *Advanced Organic Chemistry,* 3rd ed., Wiley: New York, 1988.
7. Pouchert, C., Jr., Ed., *The Aldrich Library of IR Spectra,* 3rd ed., Aldrich Chemical Co: Milwaukee, WI, spectrum No. 1237H.; Lieber, E.; Rao, C. N. R.; Pillai, C. N.; Ramachandran, J.; Hites, R. D. *Can. J. Chem.* **1958,** *36,* 801.
8. Leiber, E.; Ramachandran, J; Rao, C. N. R.; Pillai, C. N. *Can. J. Chem.* **1959,** *37,* 563.

Experiment 12 Synthesis of Hexakis (4-nitrophenoxy)cyclotriphosphazene

Part A. Preparation of Potassium 4-Nitrophenoxide

Part B. Preparation of Hexakis(4-nitrophenoxy)cyclotriphosphazene

INTRODUCTION

Phosphazenes[1,2] are compounds that contain alternating nitrogen and phosphorus atoms, either in a chain or in a ring, with two substituents on each phosphorus. The structure of the cyclic trimer, the most stable member of the class, is shown in Figure 7.11. The compound shows more than a superficial resemblance to benzene, with the π-electron system being at least somewhat delocalized, and the ring being planar.

Hexachlorocyclotriphosphazene is the starting material for most reactions of this class of compounds, with the six chlorine ligands on the phosphorus atoms being fairly easy to replace with a large variety of other functional groups. One can, depending on the relative amounts of reagent used, substitute from one to all six of the chlorines. In intermediate cases, this leads to complex mixtures of products. If one were to react the ligand and the hexachlorocyclotriphosphazene in a 3:1 ratio, for example, while the predominant product would be trisubstituted, some of the tetrasubstituted, and disubstituted products would also be found (as well as the others in trace amounts). The situation is further complicated by the various isomeric possibilities of trisubstitution.

In this experiment, the cyclic trimer is hexasubstituted (to avoid product mixtures) using the salt of a phenol, potassium 4-nitrophenoxide.

$$(NPCl_2)_3 + 6KOC_6H_4NO_2 \rightarrow [NP(OC_6H_4NO_2)_2]_3$$

The product is shown in Figure 7.11.

One very unusual aspect of this reaction is that the product is obtained as a mixture of two crystalline modifications after recrystallization. The lower melting form (mp 249–250 °C) seems to be a metastable form of the higher melting

Figure 7.11. *Phosphazene and hexakis(4-nitrophenoxy)cyclotriphosphazene.*

form (mp 262–263 °C), and can be converted to the more stable form by crushing the crystals.[3]

Phosphazenes have a variety of industrial usages, one of the more interesting being intermediates in the preparation of fire-retardant polymers and resins.

Prior Reading and Techniques

Section 2.F: Reflux and Distillation

Section 5.B: Thermal Analysis

Section 5.D.3: Isolation of Crystalline Product (Suction Filtration)

Section 5.D.4: The Craig Tube Method

Section 5.I.5: Drying of the Wet Organic Layer

Related Experiments

Industrial Chemistry: Experiments 4, 8, and 34

Nitrogen Chemistry: Experiments 11 and 14

Phosphorus–Nitrogen Chemistry: Experiment 13

Thermal Analysis: Experiment 2

EXPERIMENTAL SECTION

Safety Recommendations

4-Nitrophenol (CAS No. 100-02-07): This compound is harmful if swallowed, inhaled, or absorbed through the skin. It is a possible mutagen. ORL-RAT LD50: 250 mg/kg.

Potassium hydroxide (CAS No. 1310-58-3): This compound is highly corrosive and very hygroscopic. Ingestion will produce violent pain in the throat. ORL-RAT LD50: 1.23 g/kg. If contacted with the skin, wash with large quantities of water.

Hexachlorocyclotriphosphazene (CAS No. 940-71-6): This compound is also known as phosphonitrilic chloride trimer. No toxicity data is available for this material, but it would be prudent to follow the usual precautions (Section 1.A.3).

Tetra-*n*-butylammonium bromide (CAS No. 1643-19-2): No toxicity data is available for this material, but it would be prudent to follow the usual precautions (Section 1.A.3).

CHEMICAL DATA

Compound	FW	Amount	mmol	mp (°C)	bp (°C)	Density
4-Nitrophenol	139.11	600 mg	4.31	113	279	
KOH	56.10	300 mg	5.35	380		

Part A. Preparation of Potassium 4-Nitrophenoxide

Required Equipment

Magnetic stirring hot plate, magnetic stirring bar, 10-mL Erlenmeyer flask, 10-mL beaker, calibrated Pasteur pipet, Hirsch funnel, clay tile.

Time Required for Experiment: 0.5 h.

EXPERIMENTAL PROCEDURE

Place 600 mg (4.31 mmol) of 4-nitrophenol in a 10-mL Erlenmeyer flask equipped with a magnetic stirring bar. Using a calibrated Pasteur pipet, add 2 mL of absolute ethanol, and stir to dissolve the solid.

In a 10-mL beaker, prepare a solution of 300 mg (5.35 mmol) of KOH dissolved in 1.5 mL of ethanol. Add this solution dropwise over 15 min, using a Pasteur pipet, to the stirred 4-nitrophenol solution. A yellow crystalline solid should separate over the course of the addition.

Isolation of Product

Cool the flask, with stirring, in an ice–water bath for 10 min. Collect the product crystals by suction filtration on a Hirsch funnel. Wash the yellow crystals with two 0.5-mL portions of chilled absolute ethanol. Dry the crystals on a clay plate. Use the dry product in Part 12.B. Wet crystals will result in a diminished yield of product.

> **NOTE:** ***A melting point should not be attempted for this salt, as it is known to violently decompose over a free flame.***

Part B. Preparation of Hexakis(4-nitrophenoxy)cyclotriphosphazene

CHEMICAL DATA

Compound	FW	Amount	mmol	mp (°C)	Density
Hexachlorocyclotriphosphazene	347.66	116 mg	0.33	113	1.98
Potassium 4-nitrophenoxide	177.20	385 mg	2.17	>300	

Required Equipment

Magnetic stirring hot plate, magnetic stirring bar, 10-mL round-bottom flask, $CaCl_2$ drying tube, calibrated Pasteur pipet, water condenser, Keck clip, sand bath, 25-mL beaker, Hirsch funnel, clay tile.

Time Required for Experiment: 3 h.

EXPERIMENTAL PROCEDURE[4]

> **NOTE:** ***While sodium 4-nitrophenoxide may be obtained commercially in approximately 90% purity, its substitution for the potassium salt is reported to give incomplete conversions to the hexakis product.***

Place 116 mg (0.33 mmol) of hexachlorocyclotriphosphazene and 385 mg (2.17 mmol) of potassium 4-nitrophenoxide in a 10-mL round-bottom flask equipped with a magnetic stirring bar. Add 20 mg of tetra-*n*-butylammonium bromide, which acts as a phase-transfer catalyst in this reaction, and 4 mL of dry THF. Attach a water condenser to the flask with a Keck clip, attach a $CaCl_2$ drying tube to the condenser, and clamp the apparatus in a sand bath set upon a magnetic stirring hot plate. With stirring, reflux the mixture for 1 h. The solution may initially become somewhat milky in appearance.

Isolation of Product

Cool the reaction mixture to room temperature and transfer the solution into 15 mL of ice–water contained in a 25-mL beaker. Use a Pasteur pipet to repeatedly wash the reaction flask with the ice–water from the beaker, transferring the washings into the beaker. The product, hexakis(4-nitrophenoxy)cyclotriphosphazene, is insoluble in water, but the main side product, KCl, will dissolve as will any unreacted potassium 4-nitrophenoxide. Filter the suspension using a Hirsch funnel, and wash the crystals with three 1-mL portions (Pasteur pipet) of water, followed by three 1-mL portions of methanol. Dry the product on a clay tile and then in an oven (85 °C) for 30 min.

Purification of Product

Recrystallization may be accomplished, if desired, by dissolving (Craig tube) the dry product in a minimum amount of hot DMF, and adding 10% of the DMF volume of 1-butanol to the hot solution. The polymorphic crystals will precipitate neatly as the solution cools to room temperature.

Characterization of Product

Obtain a DSC thermogram of the recrystallized product over the temperature range 25–300 °C.

> **NOTE:** ***Different results were obtained on different occasions, using identical preparations. The metastable crystalline form melts at 249 °C, the stable form at 262 °C. Crushing the metastable crystals converts them to the stable form.***

Obtain an IR spectrum of the product as a KBr pellet and compare the spectrum with that of 4-nitrophenol.

QUESTIONS

1. Phosphazine and borazine, $B_3N_3H_6$, have many similarities and some differences. Describe these.
2. There is some controversy over whether the π-electron system in phosphazines is truly delocalized. Present some arguments favoring both sides.
3. What is a phase-transfer catalyst? What types of materials are commonly used for this purpose? Describe the mechanism of operation for the catalyst.
4. Phosphazines are used in industry to prepare fire-resistant polymers. Based upon a review of the literature, what properties render a material fire resistant? What other compounds are used as flame retardants?

REFERENCES

1. Allcock, H. R. *Chem. Rev.* **1972,** *72,* 315.
2. Allcock, H. R., *Phosphorus–Nitrogen Compounds,* Academic Press: New York, 1972.
3. Bornstein, J.; Macaione, D. P.; Bergquist, P. R. *Inorg. Chem.* **1985,** *24,* 625.
4. The experimental method was adapted from that listed in Ref. 3. All changes (which speed up the reaction considerably) from the published work are through the courtesy of Dr. S. A. Leone, Merrimack College.

GENERAL REFERENCE

Padduck, N. L.; Searle, H. T., "The Phosphonitrilic Halides and Their Derivatives" in *Advances in Inorganic Chemistry and Radiochemistry,* H. J. Emeleus and A. G. Sharpe, Eds., Academic Press: New York, 1959, Vol. 1, p. 348.

Experiment 13 Synthesis of Ammonium Phosphoramidate

INTRODUCTION

Phosphorus acids containing bonds to atoms other than to oxygen are known with halogens, sulfur, and nitrogen. The phosphoramidates have one or more oxygen groups of the orthophosphate (PO_4^{3-}) ion replaced by NH_2 groups. The simplest members of this class are phosphoramidic acid, $H_2PO_3NH_2$, and phosphorodiamidic acid, $HPO_2(NH_2)_2$. These amino derivatives have been known for over a century.[1] Their investigation was somewhat hampered by the fairly complex, low yield synthesis first employed by Stokes,[2] involving the reaction of phosphoryl chloride with phenol to give a mixture of phosphorus esters, which subsequently react with ammonia and are then saponified to give the free phosphoramidate.

$$POCl_3 + 2C_6H_5OH \rightarrow (C_6H_5O)_2POCl \; [+ \text{ some } C_6H_5OPOCl_2, (C_6H_5)_3P]$$

$$(C_6H_5O)_2POCl + 2NH_3 \rightarrow (C_6H_5O)_2PONH_2$$

$$(C_6H_5O)_2PONH_2 + NaOH \rightarrow Na_2PO_3NH_2$$

The method employed in this experiment[3] is much simpler and proceeds in higher yield, involving the direct reaction of phosphoryl chloride with aqueous ammonia to form ammonium phosphoramidate:

$$POCl_3 + 5NH_3 + 2H_2O \rightarrow NH_4HPO_3NH_2 + 3NH_4Cl$$

The ammonium phosphoramidate is separated from the ammonium chloride byproduct by flooding the reaction mixture with acetone, in which the chloride is soluble, but the phosphoramidate is not. The ammonium salt is a more convenient product than the acid, as it is nonhygroscopic. This salt can be converted to phosphoramidic acid by reaction with perchloric acid and flooding with ethanol, in which the acid is completely insoluble.

The sodium salt $NaHPO_3NH_2$ has an interesting structure in the solid state, which indicates that the phosphoramidate anion exists as a zwitterion, $NH_3^+ \; PO_3^-$. The anion structure is quite similar to that of solid sulfamic acid, HSO_3NH_2, the sulfur equivalent to phosphoramidic acid.[4]

Prior Reading and Techniques

Section 5.D.3: Isolation of Crystalline Products (Suction Filtration)

Related Experiment

Phosphorus–Nitrogen Chemistry: Experiment 12

EXPERIMENTAL SECTION

Safety Recommendations

Phosphoryl Chloride (CAS No. 10025-87-3): This compound reacts violently with water. It is harmful if swallowed, inhaled, or absorbed through the skin. It is extremely destructive to the mucous membranes. As it has an irritating smell (vapor pressure = 28 mm at 20 °C), phosphoryl chloride should only be used in the **HOOD.** ORL-RAT LD50: 380 mg/kg.

Acetic acid (CAS No. 10908-8): Acetic acid is harmful if swallowed, inhaled, or absorbed through the skin. Concentrated acetic acid is very corrosive and has an unpleasant smell. It has been found to have effects on male fertility and to have behavioral effects on newborns. ORL-RAT LD50: 3530 mg/kg.

CHEMICAL DATA

Compound	FW	Amount	mmol	mp (°C)	bp (°C)	Density
$POCl_3$	153.33	183 (μL)	2.01	1	106	1.645
NH_3 (aq) (6*M*)[a]	35.05	3 (μL)	15.0[b]			

[a] Also commonly called NH_4OH.
[b] Millimoles (mmoles) of ammonia.

Required Equipment

Magnetic stirring hot plate, magnetic stirring bar, automatic delivery pipet, Pasteur pipet, 25-mL Erlenmeyer flask, 10-mL beaker, 10-mL graduated cylinder, Hirsch funnel, clay tile, ice–water bath.

Time Required for Experiment: 1.5 h.

EXPERIMENTAL PROCEDURE

Place 3 mL of 6*M* NH_3 (aq) (15.0 mmol of NH_3) (also commonly called NH_4OH) in a 25-mL Erlenmeyer flask containing a magnetic stirring bar. Set the flask in an ice–water bath on top of a magnetic stirring hot plate and cool the solution to 0 °C. Using a Pasteur pipet, add a previously measured (automatic delivery pipet) 183 μL (2.01 mmol) of phosphoryl chloride dropwise to the ammonia solution over a 5-min period.

NOTE: *The reaction will occur with considerable fuming and evolution of heat. Phosphoryl chloride has a disagreeable odor. Carry this step out in the HOOD.*

Vigorously stir the solution for 15 min. The milky solution will become clear over this time period.

Discontinue the stirring and add 10 mL of acetone to the solution. A two layer system will form, along with some precipitate of product.

NOTE: *The dividing line between the top layer (acetone) and the bottom layer (water and product) is difficult to see.*

Using a Pasteur pipet, transfer the bottom layer to a 10-mL beaker. Additional precipitate may form at this point. Add 80 μL of glacial acetic acid to bring the pH to about 6, and cool the beaker to 0 °C for 15 min.

Isolation of Product

Collect the white, crystalline precipitate in the acetone layer by suction filtration using a Hirsch funel. Dry the crystals on a clay tile. Next, collect the product that has precipitated in the beaker (from the aqueous phase) in the same manner and combine the two products. A second crop of precipitate may be obtained from the beaker's filtrate by diluting the filtrate with an equal volume of absolute ethanol. Determine the decomposition point of the product and calculate a percentage yield.

Characterization of Product

Obtain an IR spectrum of the product as a KBr pellet. Compare the position of the N—H stretching frequency of the ammonium group to that of the amide. How do the IR bands in this compound compare to similar bands in other phosphorus containing compounds? A useful reference is Corbridge, D. E. C.; Lowe, E. J. *J. Chem. Soc.* **1954,** 493.

QUESTIONS

1. The P—N bond length in the phosphoramidate anion is 1.78 Å, indicating no π-bond character. This is quite unlike the P—N bonds in phosphazenes, which are much shorter and possess substantial π character. Suggest a reason that the bond lengths are so different. (*Hint:* The P—O bond length in the phosphoramidate anion is 1.51 Å.)
2. How is phosphorodiamidic acid prepared?
3. When phosphazenes (see Experiment 12) are hydrolyzed, a series of cyclic phosphorimidates (also called metaphosphimic acids) are obtained. What is the structure of these compounds and how are they prepared?
4. Several organophosphoramidates play a role as energy storage compounds in vertebrates (phosphocreatine) and invertebrates (phosphoarginine). Search the literature on this subject. Write the formulas for these two bioinorganic materials and explain how they function to release energy to the muscles. An introductory reference is Szent-Gyorgyi, A. G. *Adv. In Enzymol.* **1955,** *16,* 346. In addition, O. Meyerhof did extensive work in this field in the early 1950s.

REFERENCES

1. Gladstone, J. H. *J. Chem. Soc.* **1850,** *2,* 131.
2. Stokes, H. N. *Am. Chem. J.* **1893,** *15,* 198.
3. Sheridan, R. C.; McCullough, J. F., Wakefield, Z. T. *Inorg. Syn.* **1972,** *13,* 23.
4. van Wazer, J. R., *Phosphorus and its Compounds,* Interscience: New York, 1958.

Experiment 14

Preparation of an Explosive: Nitrogen Triiodide Ammoniate

NOTE: THE PRODUCT IS AN EXTREMELY DANGEROUS EXPLOSIVE, WHEN DRY. DO NOT SCALE UP THE QUANTITIES USED IN THIS REACTION.

INTRODUCTION

Under normal circumstances it is quite difficult to prepare inorganic nitrogen compounds. This is due to the great stability of nitrogen gas (N_2) and the instability of bonds from nitrogen to oxidizing agents, such as oxygen, halogens, or other nitrogen atoms. Almost all nitrogen single bonds (with the main exception of N—H bonds) are extremely weak and very reactive. Most compounds containing an N—N single bond are prone to eliminate N_2, often with explosive force.

In order to explain this disparity in bond strengths it is necessary to consider single and multiple bonding in nitrogen compounds. Nitrogen, being a relatively small element with small orbitals, forms short bonds. When singly bonded, nitrogen forms four sp^3 hybrid orbitals. The normal coordination number for neutral nitrogen is three, with a lone pair of electrons also being present occupying the fourth sp^3 orbital. In hydrazine, N_2H_4, for example, a short N—N bond would be expected to be present, with a lone pair of electrons present on each nitrogen. The weakness of the N—N single bond is generally attributed to lone pair–lone pair repulsions. In hydrazine, these repulsions are reduced by the molecule's adoption of the gauche conformation. Similar bonds between nitrogen and other elements with lone pairs of electrons (O, F, Cl, Br, I) will be weak for similar reasons. The weak bonds are quite reactive, and the thermodynamically favorable reaction to form nitrogen gas further adds to the instability of most singly bonded nitrogen compounds. For this reason, such compounds find use as explosives and as rocket fuels.

Nitrogen triiodide ammoniate, $NI_3 \cdot NH_3$, is an example of such a nitrogen explosive. Like many such nitrogen compounds, it is fairly stable in solution, but becomes explosive when dry. Detonation can be accomplished by merely touching the dry surface with a feather (or even by a door slamming some distance away).

Related Experiments

Nitrogen Chemistry: Experiments 11 and 12.

EXPERIMENTAL SECTION

Safety Recommendations

Ammonium hydroxide (CAS No. 1336-21-6): This compound is harmful if swallowed, inhaled, or absorbed through the skin. More properly named as hydrated ammonia, it has the pungent, stinging smell of ammonia gas and should only be used in the **HOOD.** ORL-RAT LD50: 350 mg/kg, IHL-HMN LCLo: 5000 ppm.

Iodine (CAS No. 7553-56-2): Iodine is harmful if swallowed, inhaled, or absorbed through the skin. It is a lachrymating agent (makes you cry). ORL-RAT LD50: 14 g/kg. Ingestion of 2–3 g has been fatal.

CHEMICAL DATA

Compound	FW	Amount	mmol	mp (°C)	bp (°C)	Density
$NH_3(aq)$[a]	35.05	1 mL	25.68			0.9
I_2	253.81	100 mg	0.394	113.5	184.3	4.93

[a] Also commonly called NH_4OH.

Required Equipment

Magnetic stirring hot plate, magnetic stir bar, Pasteur pipet, 10-mL beaker, filter paper, yardstick.

Time Required for Experiment: 1.5 h.

EXPERIMENTAL PROCEDURE

NOTE: ***Carry out this reaction only in a*** **HOOD. THE PRODUCT IS AN EXTREMELY DANGEROUS EXPLOSIVE, ESPECIALLY IN LARGER QUANTITIES. EXERCISE EXTREME CARE.**

Place 100 mg (0.394 mmol) of iodine in a 10-mL beaker equipped with a magnetic stirring bar. Set the beaker on a magnetic stirring hot plate in the hood. With a calibrated Pasteur pipet, add 1 mL of concentrated ammonia (**Caution:** *Caustic!*). Stir the suspension for 5 min.

Isolation of Product

Spread the moist solid out over several thicknesses of filter paper (paper towels may be used equally well for this purpose), and allow the material to dry for 1 h in the **HOOD**.

> **NOTE: *When dry, the product is a high explosive, and likely to detonate under any and all conditions.* STAND WELL BACK!**

The red-brown solid that forms is nitrogen triiodide monoamine. **While standing at some distance** from the dry product, tap it gently with the end of a yardstick. It will detonate immediately. No characterization should be carried out for this product.

QUESTIONS

1. Nitrogen trifluoride is a much weaker base than nitrogen trichloride. It is also a more stable compound. Explain.
2. While N—N and O—O bonds are quite weak, P—P and S—S single bonds are generally stronger. Explain.
3. Unlike N—N single bonds, the N—N triple bond in nitrogen gas is quite stable. Explain why.
4. Calculate the bond order of each of the following nitrogen compounds: N_2, N_2^+, NO, NO^+, and NO^-.
5. Elemental nitrogen and elemental phosphorus are quite different in their structures. Describe the structures and explain why they are different.
6. Nitrogen compounds are used extensively as explosives. Searching the literature, detail what classes of nitrogen compounds are used for this purpose. Why are they so explosive?

REFERENCE

1. Chen, P. S., *Entertaining and Educational Chemical Demonstrations,* Chemical Elements Publishing Co.: Camarillo, CA, 1974.

GENERAL REFERENCES

Jander, J., "Recent Chemical and Structural Investigation of Nitrogen Triiodide, Tribromide, Trichloride and Related Compounds" in *Advances in Inorganic Chemistry and Radiochemistry,* H. J. Emeleus and A. G. Sharpe, Eds., Academic Press, New York, 1976, Vol. 19, p. 1.

Jolly, W. L., *The Inorganic Chemistry of Nitrogen,* Benjamin: New York, 1964.

Experiment 15 Synthesis of Trichlorodiphenylantimony(V) Hydrate

INTRODUCTION

A large number of organometallic compounds of the heavier Group 15 (VA) elements can be made employing organic-transfer reagents, of which the Grignard and organolithium reagents are the most familiar examples. Another excellent class of organic-transfer reagents are the tetraorganotin(IV) compounds. Several of these compounds are commercially available. The organotin compound will readily transfer two organic groups to a metal chloride, or under more forcing conditions, transfer all four organic groups.

In this experiment, tetraphenyltin is used to prepare trichlorodiphenylantimony(V), by reacting tetraphenyltin and antimony pentachloride in a 1 : 1 ratio. This experiment is therefore an example of the first type of organotin reaction:

$$Sn(C_6H_5)_4 + SbCl_5 \rightarrow SbCl_3(C_6H_5)_2 + SnCl_2(C_6H_5)_2$$

Compounds such as trichlorodiphenylantimony(V) have no nitrogen analogs, as nitrogen has no low energy *d* orbitals and cannot achieve five coordination. Furthermore, nitrogen does not form stable bonds to halogens (see Experiment 14), because of extremely strong lone pair–lone pair repulsions, weakening the relatively short N—X bonds. The heavier elements form longer bonds, lessening lone pair repulsions when present. Furthermore, the low energy *d* orbitals allow lone pair electron density to be diffused, lessening repulsions further. Antimony therefore forms a large number of compounds that are more stable than their nitrogen analogs.

Prior Reading and Techniques

Section 2.F: Reflux and Distillation

Section 5.D.3: Isolation of Crystalline Products (Suction Filtration)

Section 6.C: Infrared Spectroscopy

Related Experiments

Organometallic Chemistry of the Main Group Elements: Experiments 6, 7, and 11

Tin Chemistry: Experiments 6, 9, and 10

EXPERIMENTAL SECTION

Safety Recommendations

Tetraphenyltin (CAS No. 595-90-4): This compound's toxicity data is not known. It would be prudent to handle it with care, as many tin compounds are toxic. Do not contact with the skin.

Antimony pentachloride (CAS No. 7647-18-9): This compound is commercially available as a 1*M* solution in methylene chloride (Aldrich Chemical) and may conveniently be used in that form. The compound is harmful if swallowed, inhaled, or absorbed through the skin. ORL-RAT LD50: 1115 mg/kg.

Methylene chloride (CAS No. 75-09-2): The compound, also called dichloromethane, is harmful if swallowed, inhaled, or absorbed through the skin. Exposure may cause nausea, dizziness, and headache. It is a possible carcinogen. ORL-RAT LD50: 1600 mg/kg. Exposure to this compound should be minimized, as it is a narcotic at high concentrations.

CHEMICAL DATA

Compound	FW	Amount	mmol	mp (°C)	Density
$SbCl_5$, 1*M* in CH_2Cl_2	299.02	250 μL	0.25		1.442
$(C_6H_5)_4Sn$	427.11	107 mg	0.25	224	

Required Equipment
Magnetic stirring hot plate, 10-mL round-bottom flask, magnetic stirring bar, 1-mL syringe, water condenser, Keck clip, Pasteur pipet, $CaCl_2$ drying tube, Hirsch funnel, clay tile.

Time Required for Experiment: 3 h.

EXPERIMENTAL PROCEDURE

Place 107 mg (0.25 mmol) of tetraphenyltin into a 10-mL round-bottom flask equipped with a magnetic stirring bar. Add 2 mL of hexane with a Pasteur pipet. Attach a water condenser equipped with a $CaCl_2$ drying tube using a Keck clip. Using a sand bath and with stirring, heat the solution to a gentle reflux. When reflux temperature is reached, briefly remove the drying tube. Add 250 μL of 1*M* (0.25 mmol) antimony pentachloride solution in dichloromethane **drop-wise** through the condenser using a syringe. Replace the drying tube.

NOTE: ***Antimony pentachloride is extremely sensitive to moisture. Be sure that the syringe is scrupulously dry. Flush the pipet with hexane prior to use. Clean the syringe and needle immediately after use.***

Heat the resulting solution at reflux, with stirring, for an additional 2 h. The reaction mixture should turn gray-black and a fine precipitate will develop over this time period.

Isolation of Product
Allow the product mixture to cool to room temperature. Isolate the fine, gray metallic crystals of $SbCl_3(C_6H_5)_2{\cdot}H_2O$ by filtration under suction using a Hirsch funnel.

NOTE: ***The product has a tendency to adhere to the sides of the round-bottom flask and is sometimes difficult to remove. It is soluble in acetone and can be washed from the sides with a small amount of this solvent. The acetone is then evaporated.***

Dry the product on a clay tile. The product is sufficiently pure for further characterization. Obtain a melting point and calculate the percentage yield.

Purification of Product
If desired, the trichlorodiphenylantimony(V) hydrate product may be recrystallized from a minimum amount of hot 5*M* HCl using a Craig tube.

Characterization of Product
Obtain the IR spectrum of the product as a KBr pellet and compare it with the spectra of $(C_6H_5)_4Sn$ and $(C_6H_5)_3Sb$.

QUESTIONS

1. Antimony has two common oxidation states (III and V), with representative compounds having vastly different properties. Compare and contrast, as an example, SbF_3 and SbF_5.

2. Antimony metal has a-very small coefficient of expansion and thereby finds use in type metal. Why?
3. The mid-IR spectra of $(C_6H_5)_3Sb$ and $(C_6H_5)_4Pb$ are nearly identical. Suggest why this might be.
4. The monohydrate produced in this experiment has octahedral symmetry. Heating the product under vacuum produces the anhydrous compound, which is dimeric. Suggest a structure for the dimer.
5. Using the literature, prepare a report detailing the various uses of antimony metal in industry. How is the metal prepared from its ores?
6. Antimony(V) halides are powerful Lewis acids. Locate in the literature specific examples of their use. A useful starting point is Yakobsen, G. G.; Furin, G. G. *Synthesis* **1980,** 345.

REFERENCE

1. Haiduc, I.; Silverstru, C. *Inorg. Syn.* **1985,** *23,* 194.

GENERAL REFERENCE

Doak, G. O.; Freedman, L. D., *Organometallic Compounds of Arsenic, Antimony and Bismuth,* Wiley: New York, 1970.

Experiment 16 Preparation of Sodium Tetrathionate

Part A: Determination of Reaction Quantities

Part B: Quantitative Preparation of Sodium Tetrathionate

INTRODUCTION

Sulfur forms a large variety of oxo-anions, the best known of which are the sulfate anion, SO_4^{2-}, and the sulfite anion, SO_3^{2-}. The polythionates are a second class of sulfur–oxygen anions having a general formula $S_nO_6^{2-}$, where n ranges to greater than 20. These anions, containing more than one sulfur, are normally named according to the number of sulfur atoms present. Thus, the anion $S_4O_6^{2-}$ is named the tetrathionate anion. Polythionates are stable only as salts—the free acids cannot be isolated.

In general, polythionates are obtained by the reaction of thiosulfate, $S_2O_3^{2-}$, solutions with sulfur dioxide in the presence of As_2O_3. Oxidizing agents like H_2O_2 and I_2 also react with thiosulfate solutions to form polythionate salts.

In the thiosulfate anion, sulfur is in the II oxidation state. This anion can easily be oxidized to the tetrathionate anion, $S_4O_6^{2-}$, where the sulfur atoms are in a mean oxidation state of 2.5. In this reaction, iodine is used as the oxidizing agent:

$$2Na_2S_2O_3 + I_2 \rightarrow Na_2S_4O_6 + 2NaI$$

The reaction, which is generally carried out in aqueous medium, is a quantitative one. This is the basis of the quantitative application (iodometric method of titration) of this reaction in analytical chemistry.

Prior Reading and Techniques

Section 5.B: Thermal Analysis

Section 5.D.3: Isolation of Crystalline Product (Suction Filtration)

Section 6.C: Infrared Spectroscopy

EXPERIMENTAL SECTION

Safety Recommendations

Sodium thiosulfate pentahydrate (CAS No. 10102-17-7): This compound is not normally considered dangerous. IPR-MUS LD50: 5600 mg/kg. The normal precautions should be observed (Section 1.A.3).

Manganese dioxide (CAS No. 1313-13-9): This compound may be harmful by inhalation, ingestion, or skin absorption. SCU-MUS LD50: 422 mg/kg.

Potassium iodide (CAS No. 7681-11-0): The compound is harmful if swallowed, inhaled, or absorbed through the skin. No toxicity data is available. It has been shown to have deleterious effects on newborns and on pregnancy.

Iodine (CAS No. 7553-56-2): Iodine is harmful if swallowed, inhaled, or absorbed through the skin. It is a lachrymating agent (makes you cry). ORL-RAT LD50: 14 g/kg. Ingestion of 2–3 g has been fatal.

CHEMICAL DATA

Compound	FW	Amount	mmol	mp (°C)	bp (°C)	Density
I_2	253.81	100 mg	0.39	113.5	184.3	4.93
$Na_2S_2O_3$	248.18	To be determined	To be determined			1.729

Required Equipment

Magnetic stirring hot plate, 20-mL volumetric flask, two 10-mL beakers, Pasteur pipet, 25-mL buret, Hirsch funnel.

Time Required for Experiment: 3 h.

Part A: Determination of Reaction Quantities

EXPERIMENTAL PROCEDURE

Dissolve an accurately weighed sample of 100 mg (0.39 mmol) of iodine in 20 mL of water containing excess KI (1 g) in a 25-mL beaker. The solution is a deep brown in color.

Accurately weigh 250–300 mg of sodium thiosulfate pentahydrate, $Na_2S_2O_3 \cdot 5H_2O$, into a 20-mL volumetric flask. Fill the flask to the mark with water.

> **NOTE:** ***If a volumetric flask is not available, a graduated pipet may be used to transfer 20 mL of water to a 25-mL beaker for this purpose.***

Prepare a 1.0% solution of soluble starch according to the following procedure: Place 1.0 mL of water in a 10-mL beaker and bring it to a boil on a sand bath. Weigh out 10 mg of soluble starch, suspend it in a drop or two of water, stir it, and finally add it to the hot water with a Pasteur pipet. Heat the mixture for a minute or two to obtain an almost clear solution, adding water, if necessary, to compensate for the loss due to evaporation.

Quickly titrate the iodine solution prepared earlier with the thiosulfate solution, using either a buret or a graduated pipet. Add the thiosulfate solution dropwise to the iodine solution until the deep red-brown color of the iodine solution becomes light yellow, but not colorless.

> **NOTE:** ***If the solution becomes colorless, add a few crystals of iodine that were weighed previously (not more than 10–12 mg) to the solution to regenerate the iodine color.***

When the titrated solution assumes a light yellow color, add 1.0 mL of the starch solution to it. The solution will become purple or blue in color because of the formation of a blue-violet complex between I_2 and starch. Continue the titration by adding thiosulfate solution dropwise to an iodine–starch solution until the solution becomes colorless. The mass of sodium thiosulfate in milligrams required for reacting completely with the amount of iodine taken may now be calculated. A sample calculation follows.

Suppose we took 275 mg of of sodium thiosulfate in 20 mL of water. Assume that 14.8 mL of this solution is required to titrate 104 mg of iodine in 20 mL of solution.

$$\text{mass of thiosulfate per milliliter} = 275 \text{ mg}/20 \text{ mL} = 13.75 \text{ mg}\cdot\text{mL}^{-1}$$

$$\text{mass of thiosulfate in 14.8 mL} = 14.8 \text{ mL} \times 13.75 \text{ mg}\cdot\text{mL}^{-1} = 203.5 \text{ mg}$$

This is the mass of sodium thiosulfate that is required to completely reduce 104 mg of iodine.

In Part 16.B this calculation is used to determine the amount of reagent used.

Part B: Quantitative Preparation of Sodium Tetrathionate

EXPERIMENTAL PROCEDURE

The experiment works best when a slight excess of iodine is present. In a 10-mL beaker, weigh out 105% of the amount of iodine used in Part 16.A, and dissolve it in 2 mL of 95% ethanol. Using an agate mortar and pestle, grind a sample of sodium thiosulfate to a fine powder. Weigh out the same amount of powdered sodium thiosulfate as was used in Part 16.A. Add the powdered thiosulfate in several parts to the iodine solution, vigorously agitating the mixture using a glass rod or spatula before the addition of the next portion. Since thiosulfate is insoluble in alcohol, thorough mixing of the reactants is necessary to ensure completion of the reaction. The mixture may be warmed (but do not boil, iodine may sublime off) to increase the rate of reaction.

At the end of the reaction, the slight excess of iodine should be left unreacted as indicated by a faint yellow color of the solution. If the mixture turns completely colorless, add one or two crystals of iodine to the solution to regenerate a persistent faint yellow color of iodine. This will ensure that no thiosulfate is left unreacted.

Isolation of Product

Collect the microcrystals of sodium tetrathionate by suction filtration using a Hirsch funnel. Wash the beaker and the product with several 500-μL portions of ethanol, transferring the product as quantitatively as possible to the funnel. Wash the product with 500 μL of ether and dry the product on the filter under suction. Determine the percentage yield of the product.

Characterization of Product

Obtain the IR spectrum of the product as a KBr pellet and compare the spectrum to that of sodium thiosulfate. Obtain a TGA thermogram of the product and of sodium thiosulfate, and also determine the level of hydration of both materials.

Determination of the Presence of Sodium Iodide

Sodium iodide is a byproduct of this reaction. Although it is fairly soluble in alcohol, it is important to make sure that the product is not contaminated with traces of NaI. Take a small amount of the product in a test tube. Add 1 or 2 drops of concentrated H_2SO_4 directly to the product, followed by a small amount of solid MnO_2. Warm the mixture on a flame. If no violet fumes of I_2 are observed, there is no iodide in the mixture.

> **NOTE:** ***If desired, this procedure may be used to obtain elemental iodine from sodium or potassium iodide, by simply collecting the subliming iodine on the bottom of a watch glass containing ice.***

QUESTIONS

1. What effect does the increase in oxidation state of the sulfur from 2 (thiosulfate) to 2.5 (tetrathionate) have on the IR frequency of the S—O stretch?
2. The polythionate free acids are not stable. What do they decompose into?
3. The polythionates can be viewed as being derivatives of the sulfanes. What is a sulfane and how is one prepared?
4. Dithionic acid, $H_2S_2O_6$, appears to be a simple acid analog to the polythionates; however, the acid and its salts do not show similar chemical behavior. From a search of the literature, discuss the similarities and differences of dithionic acid and the polythionates.

REFERENCE

1. Janickis, J. *Acc. Chem. Res.* **1969,** *2,* 316.

Experiment 17 Thione Complexes of Cobalt(II) Nitrate Hexahydrate

Part A: Synthesis of $Co(mimt)_4(NO_3)_2 \cdot H_2O$

Part B: Synthesis of $Co(mimt)_2(NO_3)_2$

INTRODUCTION

Sulfur and nitrogen heterocyclic compounds, such as 2-mercapto-1-methylimidazole (also called 1,3-dihydro-1-methyl-2*H*-imidazole-2-thione or **mimt**), are examples of species containing ligands that can bond in more than one manner. Such ligands are termed **ambidentate.** The structure of this ligand is shown below.

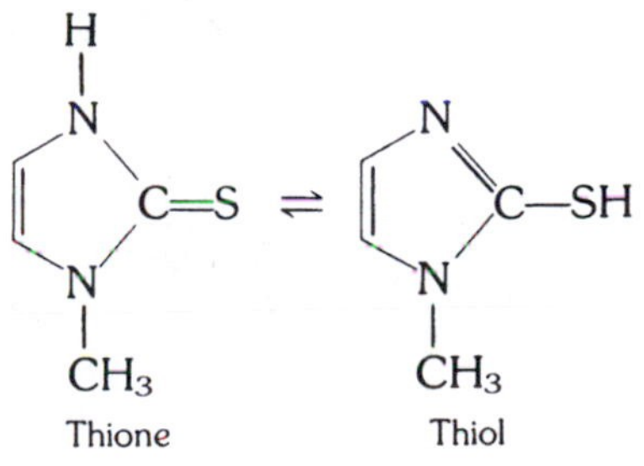

H–N / C=S / N–CH₃ ⇌ N / C—SH / N–CH₃

Thione Thiol

As seen above, the mimt ligand exists in two forms, similar to the keto–enol tautomerism observed in organic chemistry. Most commonly, the mimt ligand will bond to metals by donation of electrons at the sulfur atom, although several cases of N-bonding are known.[1] The nature of bonding can easily be seen in the IR spectral region, where the C═S bond of the parent at 745 cm^{-1} is shifted to lower frequency through donation of electrons to metals (see Experiment 20 for more examples of this type of frequency shift). New bands corresponding to the M—S bond appear at very low frequency.

Cobalt(II) is commonly found in both octahedral and tetrahedral coordination because of the similar crystal field stabilization energies for the d^7 species. In this experiment, tetrahedral mimt complexes of Co(II) will be prepared. The complex obtained depends on the quantity of mimt available for reaction and on the solvent used. Structures of the two complexes are found in Ref. 2.

Prior Reading and Techniques

Section 2.F: Reflux and Distillation

Section 5.A: Microscale Determination of Magnetic Susceptibility

Section 5.B: Thermal Analysis

Section 5.D.3: Isolation of Crystalline Products (Suction Filtration)

Section 6.C: Infrared Spectroscopy

Related Experiments

Cobalt Complexes: Experiments 7, 26, 27, 30, 35, and 47

EXPERIMENTAL SECTION

Safety Recommendations

Cobalt(II) nitrate hexahydrate (CAS No. 10026-22-9): This compound is harmful if inhaled or swallowed. ORL-RAT LD50: 691 mg/kg.

2-Mercapto-1-methylimidazole (mimt) (CAS No. 60-56-0): This compound is harmful if inhaled, swallowed, or absorbed through the skin. It was shown to have effects on embryo or fetal development and to cause tumors when present in large amounts. ORL-RAT LD50: 2250 mg/kg.

Ethyl acetate (CAS No. 141-78-6): This compound is not generally considered dangerous, although the usual precautions should be taken (Section 1.A.3). ORL-RAT LD50: 5620 mg/kg.

Triethyl orthoformate (CAS No. 122-51-0): This compound is flammable and moisture sensitive. It may be harmful if inhaled, ingested, or absorbed through the skin. SKN-RBT LD50: 20 g/kg.

CHEMICAL DATA

Compound	FW	Amount	mmol	mp (°C)	Density
$Co(NO_3)_2 \cdot 6H_2O$	291.03	73 (mg)	0.25	55	1.88
mimt (Reaction A)	114.17	114 (mg)	1.00	144	
mimt (Reaction B)	114.17	55 (mg)	0.48	144	

Required Equipment

Magnetic stirring hot plate, two 10-mL Erlenmeyer flasks, 10-mL round-bottom flask, Keck clip, magnetic stirring bar, sand bath, water condenser, $CaCl_2$ drying tube, ice–water bath, Hirsch funnel, Pasteur pipet, clay tile, or filter paper.

Part A: Synthesis of $Co(mimt)_4(NO_3)_2 \cdot H_2O$

Time Required for Experiment: 1.5 h.

EXPERIMENTAL PROCEDURE[3]

In a 10-mL Erlenmeyer flask, prepare the solvent for this reaction by mixing 3.5 mL of absolute ethanol with 190 μL of triethylorthoformate.

In a 10-mL round-bottom flask equipped with a magnetic stirring bar, dissolve 73 mg (0.25 mmol) of hydrated cobalt(II) nitrate in 2.5 mL of the solvent prepared above. Add 114 mg (1 mmol) of **mimt** dissolved in 1.25 mL of the same solvent to the solution, using a Pasteur pipet. Attach a water condenser equipped with a $CaCl_2$ drying tube to the round-bottom flask using a Keck clip, and clamp the apparatus in a sand bath set atop a magnetic stirring hot plate.

Heat the resulting mixture at reflux, with stirring, for 30 min. Transfer the hot liquid (Pasteur pipet) to a 10-mL Erlenmeyer flask containing a boiling stone. Reduce the volume by 10% by heating on the sand bath (**HOOD**). Allow the solution to cool to room temperature and then cool it further in an ice–water bath. Collect the resulting emerald green crystals by suction filtration using a Hirsch funnel, and wash them with two 1-mL portions of cold, absolute ethanol. Initially, dry the product on a clay tile or on filter paper. Further drying may be carried out under vacuum (16 mm) at 50 °C for 30 min. Calculate the percentage yield.

Characterization of Product

Take the melting point of the product. Make a KBr pellet of the material and obtain the IR spectrum. Compare the IR spectrum with that of **mimt.** If available, a far-IR spectrum should be obtained in order to observe the metal–ligand bands (300–325 cm^{-1}).

This complex exhibits an interesting thermal decomposition pattern. If available, obtain the TGA thermogram between ambient temperature and 700 °C. Reference 3 details the steps that occur in the thermal decomposition.

Determine the magnetic moment of this material. How many unpaired electrons are present? Does this correspond to the predicted number?

Part B: Synthesis of $Co(mimt)_2(NO_3)_2$

Time Required for Experiment: 1.5 h.

EXPERIMENTAL PROCEDURE

In a 10-mL Erlenmeyer flask, prepare the solvent for this reaction by mixing 3.5 mL of ethyl acetate with 190 μL of triethyl orthoformate.

In a 10-mL round-bottom flask equipped with a stirring bar, dissolve 73 mg (0.25 mmol) of hydrated cobalt(II) nitrate in 2.5 mL of the solvent prepared above. Add 55 mg (0.5 mmol) of **mimt** dissolved in 1.25 mL of solvent to the solution using a Pasteur pipet. Attach a water condenser to the round-bottom flask using a Keck clip, and clamp the apparatus in a sand bath set atop a magnetic stirring hot plate.

Heat the resulting mixture at reflux, with stirring, for 30 min. Transfer the hot liquid to a 10-mL Erlenmeyer flask containing a boiling stone (Pasteur pipet) and reduce the volume by 10% by heating in a sand bath (**HOOD**). Allow the solution to cool to room temperature, and then cool it further in an ice–water bath. Collect the resulting dark blue crystals by suction filtration using a Hirsch funnel and wash them with two 1-mL portions of cold, absolute ethanol. Dry the product on a clay tile or on filter paper.

Characterization of Product

Take a melting point of the product. Make a KBr pellet of the crystals, and obtain the IR spectrum. How does it compare to that of **mimt** itself and of the product from Part 17.A?

This complex exhibits an interesting thermal decomposition pattern. If available, obtain the TGA thermogram between ambient temperature and 700 °C. Reference 3 details the steps that occur in the thermal decomposition.

Determine the magnetic moment of this material. How many unpaired electrons are present? Does this correspond to the predicted number?

QUESTIONS

1. Compare the crystal field stabilization energies for d^1 through d^{10} complexes in octahedral and tetrahedral configurations (recall that $\Delta_t = 4/9\ \Delta_o$). For what electronic configuration is the difference smallest?
2. In what direction does the C=S band (745 cm^{-1}) shift in the IR spectrum of the products prepared in this experiment? What does this tell you about how **mimt** bonds to cobalt in these compounds?
3. Based on the information given in the literature, compare and contrast the thione–thiole and the keto–enol tautomerism found in organic chemistry. When is each form favored?

REFERENCES

1. For example, Dehand, J.; Jordonov, J. *Inorg. Chim. Acta* **1976,** *17,* 37.
2. Raper, E. S.; Nowell, I. W. *Inorg. Chim. Acta* **1980,** *43,* 165.
3. Raper, E. S.; Creighton, J. R. *Inorg. Syn.* **1985,** *23,* 171.

Experiment 18

Positive Oxidation States of Iodine: Preparation of Dipyridineiodine(I) Nitrate

INTRODUCTION

The valence shell electron configuration for the halogens is ns^2–np^5. They are therefore one electron short of possessing a complete valence shell and are quite stable as the anions, X^-. It is possible, however, to remove electrons from all of the halogens except fluorine, and form compounds with the halogen atoms in positive oxidation states. The element with the greatest capacity to be oxidized is the bottom element in any family. For the halgoens, the bottom elements are astatine (very rare, and very radioactive) and iodine. Generally, this oxidation takes place in combination with elements that are more electronegative than the halogen. In the periodate ion, IO_4^-, for example, the oxidation state of the iodine is VII. The iodine in periodate ion has been oxidized by the more electronegative oxygen.

In this experiment, an iodine complex salt is synthesized with the iodine in a positive oxidation state. In the presence of silver nitrate, $AgNO_3$, and pyridine, C_5H_5N(py), iodine reacts, forming the iodine(I) cation and iodide ion. Silver iodide is insoluble in the experimental solvent mixture and precipitates from solution.

$$I_2 + Ag^+ = AgI\ (s) + I^+$$

The iodine(I) cation is stabilized by complex formation with pyridine and, in ether, precipitates as the nitrate.

$$I(py)_2^+ + NO_3^- = I(py)_2NO_3$$

Prior Reading and Techniques

Section 5.D.3: Isolation of Crystalline Products (Suction Filtration)

Section 6.C: Infrared Spectroscopy

Related Experiments

Iodine Chemistry: Experiment 19

Complexes of the Main Group Elements: Experiments 5 and 10

EXPERIMENTAL SECTION

Safety Recommendations

Iodine (CAS No. 7553-56-2): Iodine is harmful if swallowed, inhaled, or absorbed through the skin. It is a lachrymating agent (makes you cry). ORL-RAT LD50: 14 g/kg. Ingestion of 2–3 g has been fatal.

Pyridine (CAS No. 110-86-1): Pyridine is harmful if swallowed, inhaled, or absorbed through the skin. It has a noxious smell, and is a general anesthetic. **Dispense it only in the HOOD. Wash all utensils in contact with the pyridine *in the HOOD* with acetone.** ORL-RAT LD50: 891 mg/kg.

Silver nitrate (CAS No. 7761-88-8): Like most silver compounds, silver nitrate is a heavy metal poison. It may be fatal if ingested. ORL-MUS LD50: 50 mg/kg.

CHEMICAL DATA

Compound	FW	Amount	mmol	mp (°C)	Density
I_2	253.8	250 mg	1.0	113.5	4.93
C_5H_5N	79.1	500 μL	6.2	−42	0.98
$AgNO_3$	169.9	170 mg	1.0	212	4.35

Required Equipment

Two 10-mL Erlenmeyer flasks, automatic delivery pipet, magnetic stirring bar, Pasteur pipet, Hirsch funnel, ice–water bath, clay tile or filter paper, five small test tubes.

Time Required for Experiment: 3 h.

EXPERIMENTAL PROCEDURE

> **NOTE: *Do all solution preparations and reaction steps in the HOOD, including the filtrations.***

In a 10-mL Erlenmeyer flask containing a stirring bar, dissolve 170 mg (1.0 mmol) of $AgNO_3$ in 500 μL of pyridine (automatic delivery pipet). In a separate 10-mL Erlenmeyer flask, dissolve 250 mg (1 mmol) of iodine in 5 mL of chloroform.

> **NOTE: *Chloroform has narcotic vapors. Avoid breathing the fumes. It is also highly flammable. Chloroform is listed as a carcinogen by the EPA.***

Add the chloroform solution slowly, with stirring, to the pyridine solution using a Pasteur pipet. A yellow precipitate of AgI will form. Remove the silver iodide precipitate from the solution by suction filtration using a Hirsch funnel. It is the filtrate *solution* that contains the product. Save the precipitate as well, as it will be tested to confirm that it is silver iodide.

Isolation of Product

Add 5 mL of diethyl ether to the filtrate and stopper the filter flask. Shake vigorously (or mix on a Vortex mixer) and allow the solution to stand. The dipyridineiodine(I) nitrate product will crystallize very slowly from solution. Allow at least 30 min for complete crystallization (during the last 10 min, the Erlenmeyer flask should be placed in an ice–water bath). Decant the mother liquor, retaining the yellow crystalline product. Wash the crystals with two additional 500-μL portions of ether and decant the washings. Warm the Erlenmeyer flask containing the product on a hot plate **at the lowest setting** to vaporize any residual ether. Weigh the crystals, and calculate a percentage yield. Determine the melting point.

Characterization of Product

Test for Silver Halides In a series of three small test tubes, place ~10 mg of one each of the following: (a) silver iodide precipitate, (b) silver chloride, and (c) silver bromide. To each test tube, add five drops of 7.5*M* NH_3. Which silver halide(s) begins to dissolve? Now add 1 mL of 15*M* NH_3 to each tube. Which salts dissolve now? Devise a test to distinguish between the silver halides.

Determination of Iodine Add a small portion of your dipyridineiodine(I) nitrate product to two small test tubes (~10 mg each). To one tube, add 1 mL of 6*M* HCl; to the other, add 1 mL of dilute NaOH. What do you observe? To each tube, add 1 mL of a saturated aqueous KI solution. What can you conclude about the stability of iodine cations in acidic and basic media?

Infrared Analysis Prepare a KBr pellet of the product. Compare the IR spectrum of the product with the **published** IR spectrum of pyridine. (**Do not run the IR spectrum of pyridine yourself**.)

QUESTIONS

1. The reaction between the iodine(I) cation and the pyridine is best understood as a Lewis acid–Lewis base reaction, which forms an adduct. Write this reaction.
2. Iodine shows the least tendency of the halogens to form the iodide anion, I^-. Often, it forms the triiodide ion, I_3^-. Explain this fact.
3. Using the IR spectra of your product and of pure pyridine, how can one show that the pyridine is acting as an electron donor to the iodine?
4. Based upon the amount of I_2 used, the **maximum** theoretical yield of product is only 50%. Explain this fact.
5. Several compounds are known with iodine polycations (e.g., I_3^+). Based on the literature, describe these compounds, and draw their Lewis dot structures.

REFERENCES

1. Kauffman, G. G.; Stevens, K. L. *Inorg. Syn.* **1963,** *7*, 176.
2. Zingaro, R. A.; Witmer, W. B. *Inorg. Syn.* **1963,** *7*, 169.

GENERAL REFERENCE

Downs, A. J.; Adams, C. J., "I⁺ Cations" in *Comprehensive Inorganic Chemistry,* J. C. Bailar, et al., Eds., Pergamon: Oxford, 1973, Vol. 2, Chapter 26, Section 4.A.6, p. 1345.

Experiment 19 Synthesis of Interhalogens: Iodine Trichloride

INTRODUCTION

Interhalogens are compounds that have one halogen atom bonded to another. The general formula for interhalogen compounds is XX'_y, where X is the more easily oxidized halogen, and X′ is the more oxidizing halogen. In order to completely fill all valence orbitals, there must be an even total number of halogens (in order that there be an even number of valence electrons). This restricts y to being an odd number.

Interhalogen compounds are generally rather unstable, with physical properties intermediate between those of the two halogens present in the compound. The most readily accessible interhalogens are those of iodine, as iodine is the most easily oxidized of the halogens.

In most cases, interhalogens are prepared by direct reaction of the two halogens. This poses handling problems in the laboratory, as fluorine and chlorine are corrosive gases, and bromine is a corrosive liquid. The product in this experiment, ICl_3, may be prepared using $KClO_3$ as the chlorine source, rather than the harder to handle chlorine gas; the chlorate ion oxidizes elemental iodine (reaction not balanced):

$$ClO_3^- + I_2 \rightarrow ICl_3$$

Prior Reading and Techniques

Section 5.D.3: Isolation of Crystalline Products (Suction Filtration)

Section 5.D.4: The Craig Tube Method

Related Experiment

Iodine Chemistry: Experiment 18

EXPERIMENTAL SECTION

Safety Recommendations

Iodine (CAS No. 7553-56-2): Iodine is harmful if swallowed, inhaled, or absorbed through the skin. It is a lachrymating agent (makes you cry). ORL-RAT LD50: 14 g/kg. Ingestion of 2–3 g has been fatal.

Potassium chlorate (CAS No. 3811-04-9): Potassium chlorate is harmful if swallowed, inhaled, or absorbed through the skin. ORL-RAT LD50: 1870 mg/kg. Potassium chlorate forms explosive salts with many metals, ammonia, and several other materials. Handle with care!

CHEMICAL DATA

Compound	FW	Amount	mmol	mp (°C)	bp (°C)	Density
I_2	253.8	500 mg	1.97	114	184	4.93
$KClO_3$	122.6	250 mg	2.00	356	400[a]	2.33

[a] Decomposes.

Required Equipment

Magnetic stirring hot plate, magnetic stirring bar, automatic delivery pipet, 10-mL Erlenmeyer flask, thermometer, water bath, fritted glass filter, Craig tube, clay tile.

Time Required for Experiment: 1.5 h.

EXPERIMENTAL PROCEDURE[1]

> **NOTE: *ICl_3 has a penetrating, pungent odor.* Work only in the HOOD. *The compound is very corrosive to skin, and leaves painful, brown patches. Be sure to wear gloves when working with this compound, and wash carefully afterwards.***

Spread a layer of 250 mg (2.0 mmol) of finely powdered $KClO_3$ over the bottom of a 10-mL Erlenmeyer flask equipped with a magnetic stirring bar. Add a layer of 500 mg (1.97 mmol) of powdered iodine over the first layer, and then add 250 μL of water (automatic delivery pipet) to the flask. Set the flask atop a magnetic stirring hot plate and commence stirring.

Insert a thermometer into the flask; the temperature of the reaction must be maintained below 40 °C by cooling in a water bath, if necessary. Slowly, add 1 mL of concentrated HCl dropwise using a Pasteur pipet over a 30-min period. The purple iodine should disappear and an orange solution form. Near the end of the stirring period, yellow crystals of product appear.

Isolation of Product

Cool the solution using an ice–water bath. Collect the crude ICl_3 product under suction using a fritted glass filter. The impure solid product is recrystallized using a Craig tube by dissolving the material in a minimum amount of hot ethanol, and cooling slowly to room temperature and then in ice. The crystals are dried on a clay tile. The product is air stable for short periods of time; decomposition at room temperature will occur after ~1 h.

QUESTIONS

1. Most interhalogens are quite unstable. Why? Why is ICl_3 reasonably stable?
2. Iodine forms the largest variety of interhalogen compounds. Why?
3. Given the single bond energies X—F in the table below, explain the trend.

X—F Compound	Bond Strength ($kJ \cdot mol^{-1}$)
ClF_3	175
BrF_3	200
IF_3	270

4. For the series IF, IF_3, IF_5, and IF_7, indicate the oxidation number of iodine in each species, and also the geometry it would be expected to have.
5. Balance the redox reaction used in this experiment (see discussion).
6. Some interhalogens were proposed for use as alternate solvent systems. From the literature, determine which ones. Why are they well suited to this task?

REFERENCE

1. Bauer, G., *Handbook of Preparative Inorganic Chemistry*, Academic Press: New York, 1963.

GENERAL REFERENCES

Wiebenga, E. H.; Havinga, E. E.; Boswijk, K. H., "Structures of Interhalogen Compounds and Polyhalides" *Advances in Inorganic Chemistry and Radiochemistry,* H. J. Emeleus and A. G. Sharpe, Eds., Academic Press: New York, 1961, Vol. 3, p. 133.

Downs, A. J.; Adams, C. J. "Interhalogens" in *Comprehensive Inorganic Chemistry,* J. C. Bailar, et al., Eds., Pergamon: Oxford, 1973, Vol. 2, Chapter 26, Section C, p. 1476.

Chapter 8
Chemistry of the Transition Metals

Experiment 20 Metal Complexes of Dimethyl Sulfoxide

Part A: Preparation of $CuCl_2 \cdot 2DMSO$

Part B: Preparation of $PdCl_2 \cdot 2DMSO$

Part C: Preparation of $RuCl_2 \cdot 4DMSO$

INTRODUCTION

The infrared (IR) spectrum is a valuable tool for determining the nature of bonding in a particular compound. As an example of the use of IR spectroscopy in determining the nature of bonding in a compound, compare the IR spectra of acetone, CH_3COCH_3, and acetyl chloride, CH_3COCl, shown in Figure 8.1. Both compounds have a C═O double bond and would be expected to have a major IR absorbance at about 1700 cm^{-1}. For acetyl chloride, however, the carbon atom would bear a partial positive charge because of electron donation to the electronegative chlorine, and would be a poorer electron source. The second resonance form (below) is therefore relatively unimportant for acetyl chloride, but would be more important for acetone.

$$R{-}\overset{\overset{\displaystyle :\!O\!:}{\|}}{C}{-}\ddot{\underset{\cdot\cdot}{Cl}}: \longleftrightarrow R{-}\underset{+}{\overset{\overset{\displaystyle :\!\ddot{O}\!:^{-}}{|}}{C}}{-}\ddot{\underset{\cdot\cdot}{Cl}}: \qquad (20.1)$$

Thus, the CO bond order is higher in the case of acetyl chloride than it is in the case of acetone, and the IR absorbance comes at higher frequency ($\sim$1800 cm^{-1}).

In this experiment, IR spectroscopy is used to investigate a series of DMSO complexes (DMSO, CH_3SOCH_3). Dimethyl sulfoxide is structurally similar to acetone, with a sulfur replacing the carbonyl carbon. The normal absorption of the S═O bond occurs at 1050 cm^{-1}. This is lower than the C═O frequency, since the SO bond has a larger reduced mass than the CO bond resulting in the frequency shift.

Metals can bond to DMSO either through its oxygen or its sulfur. If the bonding is to the sulfur, the metal donates electrons from its π orbitals (the t_{2g}) into an empty π orbital on the DMSO ligand, thereby increasing the S—O bond order. Thus, if the metal is bonded to the DMSO at the sulfur, the frequency of the S═O absorption increases. If the bonding is to the oxygen of the DMSO, the metal forms a bond with one of the lone pairs on the oxygen, and thereby withdraws electron density from the oxygen. This favors the second resonance form in Eq. 20.1, since the oxygen will "seek" to gain electrons to compensate for the electrons donated to the metal. The net effect is that the S═O bond order declines and the S═O absorption appears at lower frequency.

Three different metal complexes of DMSO are synthesized. The metals used are copper (as anhydrous $CuCl_2$), palladium (as $PdCl_2$), and ruthenium (as $RuCl_3$). In each case, the metal forms an adduct with DMSO.

$$CuCl_2 + 2(CH_3)_2S{=}O \rightarrow CuCl_2 \cdot 2(CH_3)_2S{=}O$$

$$PdCl_2 + 2(CH_3)_2S{=}O \rightarrow PdCl_2 \cdot 2(CH_3)_2S{=}O$$

$$RuCl_3 + 4(CH_3)_2S{=}O \rightarrow RuCl_2 \cdot 4(CH_3)_2S{=}O$$

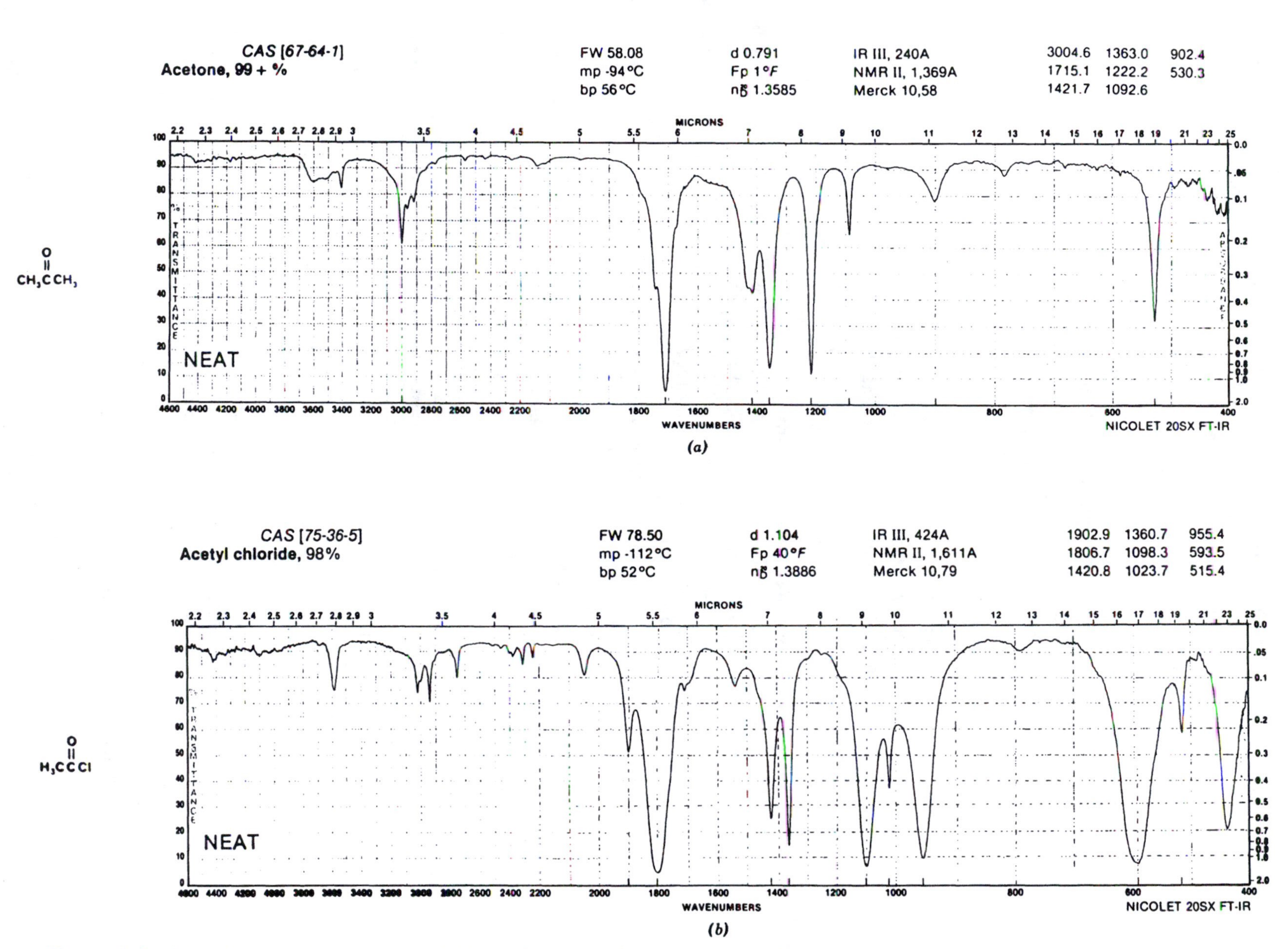

Figure 8.1. *Infrared spectra of (a) acetone and (b) acetyl chloride. (Courtesy of Aldrich Chemical Co., Milwaukee, WI.)*

The ruthenium reaction is somewhat unusual, as it is also a redox reacton. The ruthenium is reduced from Ru(III) to Ru(II) and some of the excess DMSO in solution is oxidized to sulfone.

Prior Reading and Techniques

Section 2.F: Reflux and Distillation

Section 5.D.3: Isolation of Crystalline Products (Suction Filtration)

Section 6.C: Infrared Spectroscopy

Related Experiments

Copper Chemistry: Experiments 24, 47, and 49

Palladium Chemistry: Experiments 39, 41, and 46

Ruthenium Chemistry: Experiment 44

EXPERIMENTAL SECTION

Safety Recommendations

Copper(II) chloride (CAS No. 7447-39-4): This compound is harmful if swallowed or inhaled. IPR-MUS LD50: 7400 μg/kg.

Palladium(II) chloride (CAS No. 7647-10-1): This compound may be fatal if swallowed, inhaled, or absorbed through the skin. It may be carcinogenic. ORL-RAT LD50: 2704 mg/kg.

Ruthenium(III) chloride trihydrate (CAS No. 14898-67-0): This compound is harmful if swallowed, inhaled, or absorbed through the skin. IPR-RAT LD50: 360 mg/kg.

Dimethyl sulfoxide (CAS No. 67-68-5): This compound is harmful if swallowed, inhaled, or absorbed through the skin. ORL-RAT LD 50: 14,500 mg/kg.

CHEMICAL DATA

Compound	FW	Amount	mmol	bp (°C)	mp (°C)	Density
$CuCl_2$	134.5	150 (mg)	1.11		620	3.386
$PdCl_2$	177.3	135 (mg)	0.75			4.00
$RuCl_3 \cdot 3H_2O$	261.4	100 (mg)	0.38			
$(CH_3)_2SO$	78.1	Various	Various	189	18.4	1.101

NOTE: ***It is convenient in this experiment to do Parts 20.A–C simultaneously.***

Part A: Preparation of $CuCl_2 \cdot 2DMSO$

Required Equipment

Magnetic stirring hot plate, 10-mL Erlenmeyer flask, calibrated Pasteur pipet, automatic delivery pipet, magnetic stirring bar, Hirsch funnel, clay tile or filter paper.

Time Required for Experiment: 0.5 h.

EXPERIMENTAL PROCEDURE[1]

Place 150 mg (1.11 mmol) of copper(II) chloride in a 10-mL Erlenmeyer flask equipped with a magnetic stirring bar. Add 1 mL of absolute ethanol (calibrated Pasteur pipet) and stir until the copper(II) chloride dissolves. Slowly, add 250 μL of DMSO (automatic delivery pipet). The immediate exothermic reaction yields a light green precipitate. Stir the mixture for several minutes.

Isolation of Product

Collect the product by suction filtration using a Hirsch funnel. Wash the product crystals with two 500-μL portions of cold ethanol. Dry the material on a clay tile or on filter paper. Calculate a percentage yield and determine the melting point.

Part B: Preparation of $PdCl_2 \cdot 2DMSO$

Required Equipment

Magnetic stirring hot plate, 10-mL Erlenmeyer flask, 10-mL graduated cylinder, automatic delivery pipet, magnetic stirring bar, ice–water bath, Hirsch funnel, clay tile or filter paper.

Time Required for Experiment: 3 h.

EXPERIMENTAL PROCEDURE[1]

Place 1.25 mL of DMSO (automatic delivery pipet—it may be necessary to add this in two portions) in a 10-mL Erlenmeyer flask equipped with a magnetic stirring bar. Slowly add, with stirring, finely powdered palladium(II) chloride (135 mg, 0.75 mmol). The solution turns dark brown, and after about 2.5 h of stirring, yields an orange crystalline product. Any unreacted $PdCl_2$ forms a dark, heavy, brown mass that will stay at the bottom of the flask.

Isolation of Product

Decant the orange suspension of product complex into a Hirsch funnel, taking care not to transfer the unreacted $PdCl_2$. Suction filter the product and wash it with two 500-μL portions of ether. Dry the material on a clay tile or on filter paper. Calculate the percentage yield and determine the decomposition point.

Part C: Preparation of $RuCl_2 \cdot 4DMSO$

Required Equipment

Magnetic stirring hot plate, magnetic stirring bar, 10-mL round-bottom flask, water condenser, Keck clip, sand bath, calibrated Pasteur pipet, Pasteur filter pipet, 10-mL beaker, ice bath.

Time Required for Experiment: 2 h.

EXPERIMENTAL PROCEDURE

Place 100 mg (0.383 mmol) of $RuCl_3 \cdot xH_2O$ in a 10-mL round-bottom flask equipped with a magnetic stirring bar. Attach a water condenser with a Keck clip. Place the apparatus in a sand bath, set atop a magnetic stirring hot plate. Add 1 mL of DMSO through the condenser using a calibrated Pasteur pipet.

Heat the mixture to reflux for 5 min. The red solution quickly turns orange-yellow. Cool the solution, and transfer it, using a Pasteur filter pipet, to a 10-mL beaker. Reduce the volume of solution to about 0.5 mL by passing a gentle stream of N_2 gas over the warmed liquid. (Yellow crystals may separate out at this point.)

Isolation of Product

Add 2 ml of dry acetone dropwise, to form two layers. Cool the mixture in an ice bath. On standing for 10–15 min, yellow crystals of product will form. Collect the product by suction filtration using a Hirsch funnel. Wash the product with one 500-μL portion of acetone, followed by the same portion of ether. Calculate the percentage yield and determine the melting point.

Characterization of Products

Acquire the IR spectrum of each product and determine the position of the S═O band. Assign all major bands. You will find it helpful to compare the spectrum to DMSO itself. Determine whether the DMSO is coordinated at the sulfur or oxygen in each case.

Deuterium Analogs

$RuCl_2 \cdot 4DMSO\text{-}d_6$ may be prepared in a similar fashion to $RuCl_2 \cdot 4DMSO$, using DMSO-d_6 as the solvent. What changes are observed in the IR spectrum? How can deuterium substitution be an aid to assignment of band frequencies?

QUESTIONS

1. Which element (sulfur or oxygen) would you expect platinum, mercury, iron, and zinc halides to coordinate with in DMSO?
2. Dimethyl sulfoxide is an aprotic dipolar solvent that readily dissolves many inorganic salts. Water is a protic dipolar solvent that dissolves inorganic salts. Compare the solubility characteristics of these two compounds in terms of the dissolution process.
3. Why would you expect some metals to complex at the sulfur and some at the oxygen? Look up hard–soft acid–base rules in the literature and determine which metals fall into each category.

REFERENCE

1. Boschmann, E.; Wollaston, G. *J. Chem. Educ.* **1982,** *59,* 57.

GENERAL REFERENCES

Ebsworth, E. A. V.; Rankin, D. W. H.; Cradock, S., *Structural Methods in Inorganic Chemistry,* Blackwell: Oxford, 1987.

Reynolds, W. R., "Dimethyl Sulfoxide in Inorganic Chemistry" in *Progress in Inorganic Chemistry,* S. J. Lippard, Ed., Interscience: New York, 1970, Vol. 12, p. 1.

Experiment 21 Preparation of *Trans*-dichlorotetrapyridinerhodium(III) Chloride

INTRODUCTION

Rhodium(III) forms an extensive variety of complexes. Nearly all complexes of rhodium(III) are octahedral. Rhodium(III) is very stable and kinetically inert because of its extremely favorable d^6 configuration, and is always low spin and diamagnetic.

The starting material in this synthesis, $RhCl_3 \cdot nH_2O$, is a hydrate of variable composition, n usually falling between 3 and 4. In order to calculate a proper percentage yield, it is necessary to have an accurate assay of the starting material. In this experiment, it will be *assumed* that the starting material is a trihydrate.

Complexes of $RhCl_3$ are readily formed by either the direct reaction of the trichloride hydrate with Lewis bases (pyridine, CO, phosphines, etc.) or by the oxidative addition of Rh(I) complexes. In this experiment, the direct reaction with pyridine forms a cationic complex.

$$RhCl_3 \cdot 3H_2O + 4py = [RhCl_2(py)_4]^+Cl^-$$

The complex forms as the trans- geometrical isomer (see Experiment 26 for a preparation of both cis and trans isomers in the cobalt family).

The product complex (as well as other complexes of formula *trans*-$[RhL_4X_2]Y$ [L = pyridine type ligand, X = Cl, Br, Y = Cl, Br, NO_3^-, ClO_4^-]) has been shown to have high levels of antibacterial activity against Gram-positive organisms and *Escherichia coli.*[1] A similar complex, $RhCl_3(py)_2DMSO$, showed considerable activity against leukemia in mice.[2]

Prior Reading and Techniques

Section 5.D.3: Isolation of Crystalline Products (Suction Filtration)

Section 5.D.4: The Craig Tube Method

Section 6.C: Infrared Spectroscopy

Related Experiments

Rhodium Chemistry: Experiments 24A, 34, and 42

EXPERIMENTAL SECTION

Safety Recommendations

Rhodium(III) chloride hydrate (CAS No. 20765-98-4): This compound is harmful if swallowed, inhaled, or absorbed through the skin. ORL-RAT LD50: 1302 mg/kg. It is a possible mutagen, although this has not been definitively established.

Pyridine (CAS No. 110-86-1): The compound has a noxious odor and should only be used in the **HOOD**. It is harmful if swallowed, inhaled, or absorbed through the skin. ORL-RAT LD50: 891 mg/kg.

Sodium hypophosphite hydrate (CAS No. none): The compound may be harmful by inhalation, ingestion, or skin absorption. IPR-MUS LD50: 1584 mg/kg.

CHEMICAL DATA

Compound	FW	Amount	mmol	mp (°C)	bp (°C)	Density
$RhCl_3 \cdot 3H_2O$	263.26[a]	50 mg	0.19			
Pyridine	79.10	180 mg	2.28	−42	115	0.978

[a] Formula weight for the trihydrate.

Required Equipment

Magnetic stirring hot plate, 10-mL beaker, sand bath, magnetic stirring bar, automatic delivery pipet, ice–water bath, Hirsch funnel, clay tile or filter paper.

EXPERIMENTAL PROCEDURE[3]

NOTE: *Perform all steps involving pyridine in the* HOOD. *Wash all materials coming into contact with pyridine in the* HOOD.

Place 50 mg (0.19 mmol) of rhodium(III) chloride trihydrate in a 10-mL beaker equipped with a magnetic stirring bar. Add 1 mL of water (Pasteur pipet) and heat the mixture gently, with stirring, on a sand bath until the solid dissolves. Add an excess of pyridine (180 mg, 2.28 mmol), using an automatic delivery pipet. A pink-red precipitate forms initially and dissolves over a few moments to give an orange solution of trichlorotripyridinerhodium(III).

Add a crystal of solid sodium hypophosphite hydrate to the warm orange solution, followed by 1 mL of water (Pasteur pipet). Bring the solution to a boil. After a few seconds, the solution will suddenly turn bright yellow. Allow the solution to cool to room temperature and then continue to cool in an ice–water bath for 10 min. Yellow crystals of the product compound precipitate at this point.

Isolation of Product

Collect the crude product by suction filtration using a Hirsch funnel. The product may be recrystallized by dissolving it (Craig tube) in a minimum of hot water, and cooling it in ice for 30 min. Wash the material with a 0.5-mL portion of ice–water and dry it on a clay tile or filter paper.

Characterization of Product

Obtain an IR spectrum (KBr pellet) of the product and compare the spectrum to that of pyridine.

QUESTIONS

1. It is stated in the experimental introduction that cobalt(III) complexes may be high (only in rare cases) or low spin, but rhodium(III) complexes are always low spin. Explain. Why is high spin so unusual for cobalt(III)?
2. Complexes that have d^6 electron arrangements are generally kinetically inert. Why?
3. Upon treatment of the product with $AgNO_3$, how many moles of AgCl would one theoretically expect to precipitate?
4. Rhodium is a member of the "platinum metals" group. Search the literature to determine the major commercial use for rhodium and the other platinum metals.

REFERENCES

1. Bromfield, R. J.; Dainty, R. H.; Gillard, R. D.; Heaton, B. T. *Nature* **1969,** *223,* 735.
2. Colamarino, P.; Orioli, P. *J. Chem. Soc. Dalton Trans.* **1976,** 845.
3. Gillard, R. D.; Wilkinson, G. *Inorg. Syn.* **1967,** *10,* 64.

Experiment 22 Synthesis of Metal Acetylacetonates

Part A: Preparation of Tris(2,4-pentanedionato)chromium(III)

Part B: Preparation of Tris(2,4-pentanedionato)manganese(III)

INTRODUCTION

Coordination compounds (or complexes) consist of a central atom surrounded by various other atoms, ions, or small molecules (called ligands). There is only a tenuous distinction at best between coordination complexes and molecular compounds. The most common dividing line is that complexes have more ligands than the central atom oxidation number. Silicon tetrafluoride, SiF_4, would not be a coordination compound, as there are four ligands on the Si(IV). But $[SiF_6]^{2-}$ would be considered a coordination compound as there are six ligands on the Si(IV). In this experiment, the coordination compounds tris(2,4-pentanedionato)chromium(III) and tris(2,4-pentanedionato)manganese(III) are synthesized.

In the presence of base, 2,4-pentanedione, acacH, readily loses a proton to form the acetylacetonate anion, acac, as shown.

acacH $\xrightarrow{-H^+}$ acac

Hydrogen atoms on α-carbon atoms that are adjacent to carbonyl, C=O, groups are relatively acidic. The three different representations of the acetyl acetonate anion are called resonance forms (they differ only in the location of the electrons).

In this experiment, the basic solution needed to remove the proton from the acac is provided by generating ammonia, NH_3, via the hydrolysis of urea:

$$(NH_2)_2C{=}O + H_2O = 2NH_3 + CO_2$$

In water, ammonia acts as a base:

$$NH_3 + H_2O = NH_4^+(aq) + OH^-(aq)$$

Acetyl acetonate is an example of a bidentate (*bi*-two, *dent*-teeth) ligand, since it can bond to a metal via both oxygen atoms. Ligands of this type are also often called chelating (*chelos*-claw) ligands. Three acac ligands are therefore needed to complete the octahedral coordination about the central metal ion, giving formula $[M(acac)_3]^{n+}$. The structure of the chromium(III) complex is shown in Figure 8.2. Since the outer part of the complex consists of organic groups, most metal acetylacetonates are hydrophobic, and insoluble in water.

In Part 22.B, tris(2,4-pentanedionato)manganese(III) is synthesized. Manganese(III) is a normally unimportant and unstable oxidation state, but can be stabilized in aqueous solution by use of complexing anions such as acetate, acetylacetonate, or oxalate. The so-called tris(2,4-pentanedionato)manganese(III) is easily prepared through an oxidation–reduction reaction of Mn(II) and Mn(VII). The compound is a moderately strong oxidizing agent, seeing some use in organic reactions of phenols. Its structure is quite different (see Question 6) from the chromium analog prepared in Part 22.A.

Experiment 23 utilizes the tris(2,4-pentanedionato)chromium(III) prepared in Part 22.A. In this experiment the complex is brominated and the bromination is monitored via GC.

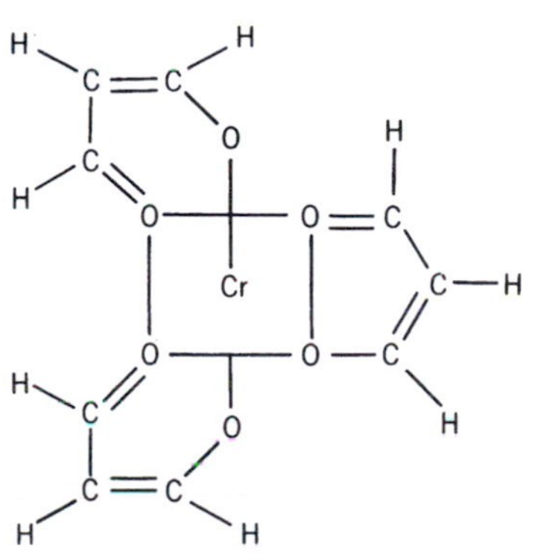

Figure 8.2. *Structure of Cr(acac)$_3$.*

Prior Reading and Techniques

Section 5.A: Microscale Determination of Magnetic Susceptibility

Section 5.D.3: Isolation of Crystalline Products (Suction Filtration)

Section 5.D.4: The Craig Tube Method
Section 6.C: Infrared Spectroscopy

Related Experiments

Chromium Chemistry: Experiments 23, 29, and 33
Manganese Chemistry: Experiment 43

EXPERIMENTAL SECTION

Part A: Preparation of Tris(2,4-pentanedionato)chromium(III)

SAFETY RECOMMENDATIONS

Chromium(III) chloride hexahydrate (CAS No. 10060-12-5): Chromium compounds are considered mildly toxic. The reagent $CrCl_3 \cdot 6H_2O$ has MLD_{iv} in mice of 801 mg/kg. Chromium(III) compounds, in general, have little toxicity. Certain of them, however, have been listed as carcinogens by the EPA.

2,4-Pentanedione (CAS No. 123-54-6): Also known as acetylacetone. The compound is a mild irritant to the skin and mucous membranes. It is a flammable liquid. ORL-RAT LD50: 590 mg/kg.

Urea (57-13-6): Urea is not generally considered dangerous and is classified as a diuretic. ORL-RAT LD50: 8471 mg/kg. The usual safety precautions (see Section 1.A.3) should be taken.

CHEMICAL DATA

Compound	FW	Amount	mmol	bp (°C)	mp (°C)	Density
$CrCl_3 \cdot 6H_2O$	266.4	130 mg	0.49		83	1.760
Urea	60.06	500 mg	8.3		133–135	1.335
2,4-Pentanedione	100.12	400 μL	3.84	140	−23	0.975

Required Equipment
Magnetic stirring hot plate, 10-mL Erlenmeyer flask, microwatch glass, magnetic stirring bar, 10-mL graduated cylinder, automatic dispensing pipet, 150-mL beaker, Hirsch funnel, clay tile, or filter paper.

Time Required for Experiment: 2.5 h.

EXPERIMENTAL PROCEDURE[1]

In a 10-mL Erlenmeyer flask fitted with a microwatch glass cover and containing a magnetic stirring bar, place 2.0 mL of distilled water (graduated cylinder) and 130 mg (0.49 mmol) of chromium(III) chloride hexahydrate. When the chromium complex has dissolved, add 500 mg (8.3 mmol) of urea and 400 μL (3.84 mmol) of acetylacetone (automatic dispensing pipet). A large excess of acacH is used, as it helps the reaction go to completion.

NOTE: *The acacH should be dispensed in the* HOOD.

Clamp the Erlenmeyer flask in a 150-mL beaker of boiling water set on a magnetic stirring hot plate. Heat the mixture, with stirring, for ~1 h. As the urea releases ammonia and the solution becomes basic, deep maroon crystals will begin to form. These form as a crust at the surface of the reaction mixture.

Isolation of Product

Cool the reaction flask to room temperature. Collect the crystalline product by suction filtration using a Hirsch funnel. Wash the crystals with three 200-μL portions of distilled water. Dry the product on a clay tile or on filter paper and determine the percentage yield. Take the melting point of the material.

Characterization of Product

Obtain the IR spectrum of the product and of pure 2,4-pentanedione as either Nujol mulls or KBr pellets. Determine the magnetic moment of this complex and compare it to the manganese complex prepared in Part 22.B.

Part B: Preparation of Tris(2,4-pentanedionato)manganese(III)

Safety Recommendations

Manganese(II) chloride tetrahydrate (CAS No. 13446-34-9): This compound is harmful if swallowed, inhaled, or absorbed through the skin. ORL-RAT LD50: 1484 mg/kg.

2,4-Pentanedione (CAS No. 123-54-6): This compound is a mild irritant to the skin and mucous membranes. It is a flammable liquid. ORL-RAT LD50: 590 mg/kg.

Potassium permanganate (CAS No. 7722-64-7): This compound is a powerful oxidizing agent and should be handled with care. It is harmful if swallowed, inhaled, or absorbed through the skin. It is extremely destructive to the mucous membranes and skin. ORL-RAT LD50: 1090 mg/kg, ORL-HMN LDLo: 143 mg/kg.

Sodium acetate trihydrate (CAS No. 6131-90-4): This compound is not generally considered dangerous. The normal precautions should be observed. ORL-RAT LD50: 3530 mg/kg.

CHEMICAL DATA

Compound	FW	Amount	mmol	bp (°C)	mp (°C)	Density
$MnCl_2 \cdot 4H_2O$	197.91	100 mg	0.50		58	2.010
$KMnO_4$	158.04	20 mg	0.13			2.703
$NaC_2H_3O_2 \cdot 3H_2O$	136.08	5.20 mg	3.8		58	1.45
2,4-Pentanedione	100.12	400 μL	3.8	140.4	−23	0.975

Required Equipment

Magnetic stirring hot plate, 10-mL Erlenmeyer flask, magnetic stirring bar, 10-mL graduated cylinder, automatic delivery pipet, 10-mL beaker, Pasteur pipet, Hirsch funnel, clay tile, or filter paper.

Time Required for Experiment: 2 h.

EXPERIMENTAL PROCEDURE[2]

Add 100 mg of manganese(II) chloride tetrahydrate (0.5 mmol) and 260 mg (1.9 mmol) of sodium acetate trihydrate to a 10-mL Erlenmeyer flask equipped with a magnetic stirring bar. Place the flask atop a magnetic stirring hot plate, add 4 mL of water (graduated cylinder) and stir the mixture to dissolve all the solids. When dissolution is complete, using an automatic delivery pipet, add 400 μL (3.84 mmol) of acetylacetone.

NOTE: ***The acacH should be dispensed in the* HOOD.**

In a 10-mL beaker, prepare a solution of 20 mg (0.127 mmol) of potassium permanganate in 1 mL of water. Add this solution slowly, dropwise (Pasteur pipet), with stirring, to the reaction mixture. After stirring for 5 min, add a second portion of 260 mg (1.92 mmol) of sodium acetate trihydrate dissolved in 1 mL of water (Pasteur pipet) dropwise.

> **NOTE: *Be sure that all the permanganate was added. After addition is complete, remove 1 mL of the reaction solution using the same pipet, rinse the permanganate beaker, and return it to the reaction flask.***

Heat the mixture to near boiling on a magnetic stirring hot plate for 10 min. Allow the mixture to cool to room temperature.

Isolation of Product

Collect the dark brown precipitated solid by suction filtration using a Hirsch funnel, and wash it with a 1-mL portion of water. After drying on a clay tile or filter paper, the product is suitable for characterization without further purification. Calculate the percentage yield and determine the melting point.

If desired, recrystallization can be accomplished by dissolving the solid in a minimum of warm toluene (**HOOD!**) and filtering, if necessary, using a Pasteur filter pipet. The solid is reprecipitated by cooling the toluene solution to room temperature and adding 1.5 mL of petroleum ether.

Characterization of Product

Obtain the IR spectrum of the product and of pure 2,4-pentanedione as either Nujol mulls or KBr pellets. Determine the magnetic moment of this complex and compare it to the chromium complex prepared in Part 22.A.

QUESTIONS

1. Write and balance the half-reactions for any redox reactions in this experiment.
2. Chromium has several common oxidation states other than III. What are they? What color are solutions of these species? Suggest an easy way of determining whether an oxidation or reduction of a chromium containing solution has taken place.
3. In acetone, the alkyl hydrogen atoms are quite difficult to remove in the presence of base. In acetylacetone, however, a proton is readily lost, forming the acac anion. Why is there a difference between these two similar compounds?
4. Manganese(II) (d^5) is nearly colorless, whereas Mn(VII) (d^0) is dark violet. Explain.
5. Explain why Mn(II) and Mn(VII) are used in a roughly 4:1 ratio in this experiment.
6. The structures of $Cr(acac)_3$ and $Mn(acac)_3$ are quite different. What is the true structure of the so-called "manganic acetate?"
7. Would either the $Cr(acac)_3$ or $Mn(acac)_3$ species exhibit the Jahn–Teller effect? Explain.
8. Interaction of the $Mn(acac)_3$, prepared in Part 22.B, with bipyridine (bipy) in acetonitrile, yields $Mn_4O_2(O_2CR)_7bipy_2$, which is of tremendous theoretical interest in the area of photosynthesis. From the literature, describe these complexes and their biochemically significant properties. The pioneering work in this area was published by G. Christou and co-workers, in 1987.

REFERENCES

1. Fernelius, W. C.; Blanch, J. E. *Inorg. Syn.* **1957,** *5,* 130.
2. Charles, R. G. *Inorg. Syn.* **1963,** *7,* 183.

GENERAL REFERENCE

Fackler, J. P., Jr., "Metal β-Ketoenolate Complexes" in *Progress in Inorganic Chemistry,* F. A. Cotton, Ed., Interscience: New York, 1966, Vol. 7, p. 471.

Experiment 23 Gas Chromatographic Analysis of Brominated Tris(2,4-pentanedionato)chromium(III)

INTRODUCTION

Metal bonded acetylacetonato, acac, ligands readily undergo electrophilic substitution. Substitution generally occurs at the hydrogen on the central or γ-carbon atom of the chelated ligand. Depending on the amounts of the electrophilic reagents, mono-, di-, and trisubstituted chelates can be prepared. All of these derivatives, including the parent compound, may exist in solution simultaneously at certain points along the reaction coordinate. These species can be separated by chromatographic methods.

Gas chromatography (GC) lends itself well to the separation of the reaction products because of the ready volatility of the components. In this experiment, the bromination of $Cr(acac)_3$, prepared in Experiment 22.A, is carried out and studied by GC analysis, as a demonstration of the utility of GC in inorganic chemistry. Brominated $Cr(acac)_3$ species can be successfully eluted and separated at temperatures below 170 °C from columns packed with a silicone liquid phase on Chromosorb W or a DC-750 column. Tetracosane is used as an internal reference.

Prior Reading and Techniques

Section 5.G.3: Gas Chromatography

Related Experiments

$Cr(acac)_3$ Preparation and Use: Experiments 22 and 29

Chromatographic Analysis: Experiments 28, 34, and 40

EXPERIMENTAL SECTION

Safety Recommendations

Tris(2,4-pentanedionato)chromium(III) (CAS No. 21679-31-2): This compound is harmful if swallowed or inhaled. ORL-RAT LD50: 3360 mg/kg.

Tetracosane (CAS No. 646-31-1): No toxicity data is available for this compound. The usual precautions (Section 1.A.3) should be observed.

Carbon tetrachloride (CAS No. 56-23-5): This compound is classified as a carcinogen. Avoid contact with the skin. Avoid breathing the fumes. Use only in the **HOOD**. ORL-RAT LD50: 2350 mg/kg.

***N*-Bromosuccinimide (NBS)** (CAS No. 128-08-5): This compound is harmful if swallowed, inhaled, or absorbed through the skin. No toxicity data is available. The usual precautions (Section 1.A.3) should be observed.

CHEMICAL DATA

Compound	FW	Amount	mmol	mp (°C)	bp (°C)	Density
$Cr(acac)_3$	349.33	49 (mg)	0.14	210	340	
Tetracosane	338.66	50 (mg)	0.14	49	391	
NBS[a]	177.99	150 (mg)	0.28	180		

[a] *N*-Bromosuccinimide = NBS.

Required Equipment

Two 25-mL Erlenmeyer flasks, injection syringe.

Time Required for Experiment: 3 h.

EXPERIMENTAL PROCEDURE

> **NOTE:** ***Any GC equipped with on-column injection may be used in this experiment. A glass column[1] (1 m × 3 mm), packed with 1% OV101 on Chromosorb W or DC-750 and conditioned at 200 °C should be used. The mobile phase consists of N_2 gas, maintained at 45 mL·min^{-1}. The injector temperature should be set at 160 °C and the oven temperature set at 200–220 °C.***

Prepare the first mixture by dissolving 49 mg (0.14 mmol) of $Cr(acac)_3$ and 50 mg of tetracosane in 20 mL of CCl_4 in a 25-mL Erlenmeyer flask. In a second 25-mL Erlenmeyer flask, dissolve 150 mg (0.28 mmol) of NBS in 20 mL of CCl_4. Both these mixtures should be placed in a room temperature bath to stabilize their temperatures. Inject 1 μL of the first mixture into the GC column. Adjust the chromatograph so that the retention time for $Cr(acac)_3$ has a value of ~1 min, and so that the retention time for the tetracosane is 10–13 min. The chart speed should be 10 mm·min^{-1}.

When the temperature of both the flasks has stabilized, mix the solutions together. Increase the sensitivity of the detector by a factor of 2 to compensate for the dilution of $Cr(acac)_3$. Immediately, inject 1 μL of the mixture onto the column. Record the time of injection. The chromatogram should be obtained until the peak due to tetracosane appears. Repeat the measurements as frequently as possible for about 60 min, using a fresh sample from the reaction flask each time. After 60 min of reaction, allow 5 min for chromatographic acquisition **after** the tetracosane peak appears. This is done in order to detect the peak resulting from the trisubstituted product, which elutes later than tetracosane. Continue injections until a peak for $Cr(Bracac)_3$ is observed. A chromatogram obtained at 160 °C will eventually contain peaks due to $Cr(acac)_3$, $Cr(Bracac)(acac)_2$, $Cr(Bracac)_2(acac)$, tetracosane, and $Cr(Bracac)_3$, in that order. Calculate the relative amounts of each component.

FURTHER WORK

The experiment can be modified to determine the rate curves, as described in Reference 1. If GC MS is available, the mass spectra of the products can also be determined.

QUESTIONS

1. Why is tetracosane used in the chromatographic analysis?
2. List several alternative substitution reactions that might be carried out on the acac ligand.
3. Account for the elution order of the brominated products.

4. Suggest a suitable reaction mechanism for the bromination of the acac ligand.
5. Give the structure and name of the principal byproduct generated in the reaction of NBS and $Cr(acac)_3$.
6. From the current year's literature, find two papers describing the use of GC for the separation of inorganic compounds.

REFERENCE

1. Cardwell, T. J.; Lorman, T. H. *J. Chem. Educ.* **1986,** *63,* 90.

Experiment 24 Determination of Magnetic Moments in Metal–Metal Bonded Complexes

Part A: Synthesis of Rhodium(II) Acetate Ethanolate

Part B: Synthesis of Copper(II) Acetate Monohydrate

INTRODUCTION

The idea that metal atoms could individually bond to other metal atoms was one that arose comparatively late in the development of inorganic chemistry.[1] It was not until 1913 that a compound was discovered to have a metal–metal bond, and not until 1963 that compounds with metal–metal bonds were thought to be anything more than oddities.

The first compound with metal–metal bonds was found to have the formula $Ta_6Cl_{14}·7H_2O$ by H. S. Harned. Earlier, the compound was thought to be $TaCl_2·2H_2O$. Clearly, Harned's work indicated that there was some interaction between the tantalum atoms in this compound. The structure was shown some 40 years later to consist of an octahedron of mutually bonded tantalum atoms, with each edge of the octahedron being bridged by a chlorine. The formula would be best represented by $[Ta_6Cl_{12}]Cl_2$.

With the discovery of rhenium cluster compounds in 1963, the area of metal–metal bonding exploded with interest, with hundreds of such compounds currently known. Compounds are currently known containing not only M—M single bonds, but also double, triple, and even quadruple bonds. This experiment (and Experiment 25) allows us to synthesize and characterize a series of compounds containing metal–metal bonds.

Rhodium(II) and iridium(II) do not form simple complexes similar to those of the well-known cobalt species, $[Co(NH_3)_6]^{2+}$ or $[CoCl_4]^{2-}$. The most common complexes of Rh and Ir formed in the II oxidation state are bridged species that contain a metal–metal bond. The most familiar of these is tetrakis(acetato)-dirhodium(II), whose structure is shown in Figure 8.3.

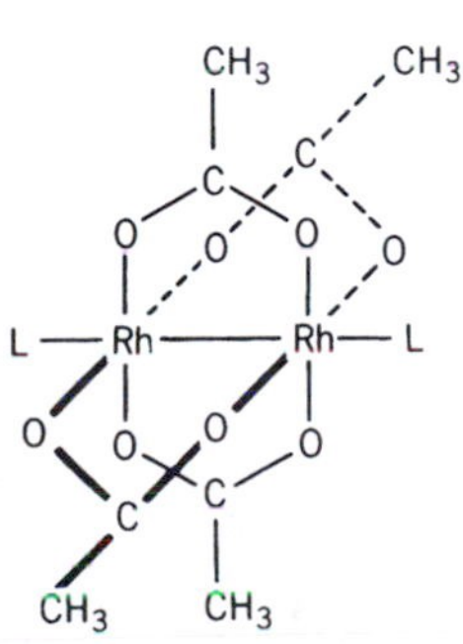

Figure 8.3. *Structure of rhodium(II) acetate alcoholate.*

In this complex, the unpaired electrons of rhodium(II) (d^7) are strongly coupled, and the complex is diamagnetic. (For a discussion of magnetic susceptibility, see Section 5.A.) The metal–metal bond is quite short, only 2.386 Å, leading to the conclusion that a rhodium–rhodium bond is present in this compound. Several other metals form similar acetates with metal–metal bonds.

This structure is similar to that exhibited by copper(II) acetate dihydrate, which *may* contain a copper–copper bond. In the copper complex, there is only a weak coupling of the unpaired electrons on the Cu(II) ions (d^9). Thus, while the ground state is diamagnetic, there is a low energy excited state that is paramagnetic. This excited state is appreciably populated at room temperature and the complex therefore appears to be paramagnetic. As the temperature increases, the magnetic moment increases as well. (The alternative explanation to this behavior, not involving Cu—Cu bonding, is described as an antiferromagnetic coupling of the unpaired spins.) Structural determination shows that the two copper atoms are separated by a distance of 2.64 Å. This compares to a 2.56-Å interatomic distance in metallic copper, making the assumption of a Cu—Cu bond more controversial.

The rhodium(II) acetate ethanolate is prepared via the direct reaction of rhodium(III) chloride hydrate and sodium acetate trihydrate in the presence of ethanol. It is the ethanol that is oxidized. The copper(II) acetate hydrate is synthesized via a less direct route, involving the intermediate formation of a tetraamine complex, conversion of the tetraamine complex to a precipitated hydroxide, and subsequent reaction with acetic acid.

$$CuSO_4 \cdot 5H_2O + 4NH_3\ (aq) \rightarrow [Cu(NH_3)_4]^{2+} + SO_4^{2-}$$

$$[Cu(NH_3)_4]^{2+} + NaOH \rightarrow Cu(OH)_2(s)$$

$$Cu(OH)_2 + CH_3CO_2H \rightarrow [Cu(CH_3CO_2)_2 \cdot H_2O]_2$$

The solvent ligands, L (see Fig. 8.3), can be easily removed by heating in vacuum to yield the nonadducted complexes. In the case of the rhodium(II) complexes, if the ligand is an oxygen-bonded species (such as ethanol in the synthesis below), the complexes are blue-green in color. If the ligand is a π acid, such as triphenylphosphine, the complex is red.

Rhodium(II) acetate ethanolate, as well as other tetrakis(μ-carboxylato)-dirhodium(II) species, show some antitumor behavior by inhibiting DNA synthesis. The base adenine's nitrogen N7 hydrogen bonds to the carboxylate oxygen of the rhodium(II) species. The rhodium species then bridges between the DNA strands.[2]

Prior Reading and Techniques

Section 2.F: Reflux and Distillation

Section 5.A: Microscale Determination of Magnetic Susceptibility

Section 5.C.2: Purging with an Inert Gas

Section 5.D.3: Isolation of Crystalline Products (Suction Filtration)

Section 5.D.4: The Craig Tube Method

Section 6.C: Infrared Spectroscopy

Related Experiments

Copper Chemistry: Experiments 20, 47, and 49

Metal–Metal Bonding: Experiments 25 and 45

Rhodium Chemistry: Experiments 21, 34, and 42

EXPERIMENTAL SECTION

Part A: Synthesis of Rhodium(II) Acetate Ethanolate

Safety Recommendations

Rhodium(III) chloride hydrate (CAS No. 20765-98-4): This compound is harmful if swallowed, inhaled, or absorbed through the skin. ORL-RAT LD50:

1302 mg/kg. It is a possible mutagen, although this has not been definitively established.

Sodium acetate trihydrate (CAS No. 6131-90-4): This compound is not generally considered dangerous. The normal precautions should be observed. ORL-RAT LD50: 3530 mg/kg.

CHEMICAL DATA

Compound	FW	Amount	mmol	mp (°C)	Density
$RhCl_3 \cdot xH_2O$	263.26[a]	50 mg	0.19	100[b]	
$NaC_2H_3O_2 \cdot 3H_2O$	136.08	100 mg	0.73	58	1.45

[a] Based on calculations for the trihydrate.
[b] Decomposes.

Required Equipment

Magnetic stirring hot plate, 10-mL side arm round-bottom flask, magnetic stirring bar, Keck clip, automatic delivery pipet, source of nitrogen, sand bath, water condenser.

Time Required for Experiment: 2.5 h, plus overnight crystallization time

EXPERIMENTAL PROCEDURE[3]

Thoroughly flush a 10-mL side arm round-bottom flask equipped with a stirring bar with N_2 gas. Dissolve 50 mg (~0.19 mmol) of rhodium(III) chloride hydrate and 100 mg (0.73 mmol) of sodium acetate trihydrate in 1 mL of glacial acetic acid and 1 mL of absolute ethanol (automatic delivery pipet). Transfer this to the side arm flask using a Pasteur pipet and attach a water condenser with a Keck clip. Place the apparatus in a sand bath atop a magnetic stirring hot plate. Stir the mixture at room temperature, maintaining a positive pressure of nitrogen, until a red color develops.

Reflux the solution gently, under nitrogen, for 1 h. The red solution should become green, and a blue-green solid precipitates during this time.

Isolation of Product

Collect the blue-green solid product, $[Rh(OCOCH_3)_2]_2 \cdot 2C_2H_5OH$, by suction filtration using a Hirsch funnel.

Purification of Product

The rhodium(II) acetate ethanolate may be recrystallized by dissolving the product in the minimum amount of boiling methanol (~6 mL), and filtering if necessary. Concentrate the solution (**HOOD**) to a volume of approximately 4 mL using a gentle stream of nitrogen. Store the solution in a refrigerator overnight. Collect the first crop of crystals by suction filtration using a Hirsch funnel. Further concentration and cooling yields a second crop of crystals. The nonsolvated complex, $[Rh(OCOCH_3)_2]_2$, can be generated by heating at 45 °C in a vacuum for 20 h.

Characterization of Product

Acquire the IR spectrum of the product as a KBr pellet. Determine the magnetic moment of the product (see Section 5.A). Either the ethanolate or the nonsolvated complex may be used. Does it correspond to that of a diamagnetic complex?

Part B: Synthesis of Copper(II) Acetate Monohydrate

Safety Recommendations

Copper(II) sulfate pentahydrate (CAS No. 20919-8): This compound is not normally considered dangerous, but the usual precautions should be taken. ORL-RAT LD50: 300 mg/kg, ORL-HMN LDLo: 1088 mg/kg.

Sodium hydroxide (CAS No. 1310-73-2): This compound is harmful if swallowed, inhaled, or absorbed through the skin. It is extremely caustic, especially on wet surfaces, forming a strongly alkaline solution. Solid sodium hydroxide is hygroscopic. IPR-MUS LD50: 40 mg/kg.

Acetic acid (CAS No. 10908-8): Acetic acid is harmful if swallowed, inhaled, or absorbed through the skin. Concentrated acetic acid is very corrosive and has an unpleasant smell. It has been found to have effects on male fertility and to have behavioral effects on newborns. ORL-RAT LD50: 3530 mg/kg.

CHEMICAL DATA

Compound	FW	Amount	mmol	bp (°C)	mp (°C)	Density
$CuSO_4 \cdot 5H_2O$	249.6	250 mg	1.0		110[a]	2.284
$NH_3(aq)$[b], 50%	35.05	Sufficient amount				0.900
NaOH	40.0	80 mg	2.0		318	2.130
CH_3CO_2H	60.1	Minimum amount		116	16	1.049

[a] Loses 4 equivalents of H_2O.
[b] Also known as NH_4OH.

Required Equipment

Magnetic stirring hot plate, 10-mL beaker, magnetic stirring bar, Pasteur pipet, Hirsch funnel, clay tile, or filter paper.

Time Required for Experiment: 1.5 h.

EXPERIMENTAL PROCEDURE[4]

In a 10-mL beaker equipped with a magnetic stirring bar, dissolve 160 mg (1.0 mmol) of copper(II) sulfate or 250 mg (1.0 mmol) of copper(II) sulfate pentahydrate in 5.0 mL of water. Stir the mixture, and warm it to 40–50 °C on a sand bath to aid the dissolution.

Using a Pasteur pipet, add 50% NH_3 (aq) to the warm, stirred, light blue solution, until the intense blue color of the copper ammonium complex is evident. During this addition, a precipitate of copper hydroxide may form initially, but it will dissolve on further addition of the NH_3 (aq) solution.

Add 80 mg (2.0 mmol) of sodium hydroxide flakes to the deep blue solution, and stir the mixture for 15–20 min at 55–65 °C. A light blue solid of copper(II) hydroxide precipitates during this time. Allow the mixture to cool to room temperature and collect the precipitate by vacuum filtration using a Hirsch funnel. Wash the blue solid with three 2-mL portions of warm water.

Transfer the solid $Cu(OH)_2$ to a 10-mL beaker, and dissolve it in the minimum amount of 10% acetic acid. Warming on the sand bath with stirring aids the dissolution process. Concentrate the solution nearly to dryness (**HOOD**) by warming it on a sand bath under a slow stream of nitrogen. Collect the beautiful deep blue crystals that form by filtration using a Hirsch funnel. Dry the product on a clay plate or on filter paper.

Characterization of Product

Acquire the IR spectrum of the product as a KBr pellet. Determine the magnetic moment of the product (see Section 5.A). Does it correspond to that of a diamagnetic or paramagnetic complex at room temperature? If paramagnetic, how many unpaired electrons seem to be present?

QUESTIONS

1. What physical indications, other than magnetic moment, might lead one to conclude that a metal–metal bond is present?
2. Even when the magnetic susceptibility seems to indicate that a metal–metal bond is present, the low magnetic susceptibility might be due to other reasons. Discuss this point.
3. One of the largest classes of metal–metal bonded compounds are the metal carbonyl clusters. Discuss the bonding in two such members of this class.
4. Metal clusters are under active investigation as "mimics" to bulk metals in catalysis. Perform a literature search and discuss several examples of metal clusters that were studied in this manner.

REFERENCES

1. A good introduction to this interesting area may be found in the following references.
   a. Cotton, F. A., "Multiple Bonds and Metal Clusters" in *Reactivity of Metal–Metal Bonds,* M. H. Chisholm, Ed., ACS Symposium Series No. 155, American Chemical Society: Washington, DC, 1981.
   b. Cotton, F. A., Walton, R. A., *Multiple Bonds Between Metal Atoms,* Krieger: Malabar, FL, 1988.
2. Hughes, R. G., Bear, J. L.; Kimball, A. P. *Am. Assoc. Cancer Res.* **1972,** *13,* 120.
3. Rempel, G. A.; Legzdins, P.; Smith, H.; Wilkinson, G. *Inorg. Syn.* **1973,** *13,* 90.
4. Catterick, J.; Thornton, P. *Adv. Inorg. Chem. Radiochem.* **1977,** *20,* 291. Kato, M.; Jonassen, H. B.; Fannin, J. C. *Chem. Rev.* **1969,** *64,* 99.

GENERAL REFERENCES

Baird, M. C., "Metal–Metal Bonds in Transition Metal Complexes" in *Progress in Inorganic Chemistry,* F. A. Cotton, Ed., Interscience: New York, 1968, Vol. 9, p. 1.

Felthouse, T. R., "The Chemistry, Structure and Metal–Metal Bonding in Compounds of Rhodium(II)" in *Progress in Inorganic Chemistry,* S. J. Lippard, Ed., Interscience: New York, 1982, Vol. 29, p. 74.

Experiment 25 Multiply Bonded Series: Preparation of Tetrabutylammonium Octachlorodirhenate(III)

INTRODUCTION

The existence of a quadruple bond in inorganic systems was first recognized in 1964 in the case of $[Re_2Cl_8]^{2-}$. The complex was actually discovered in early 1954 at the Kurnikoff Institute in the Soviet Union, but mistakenly characterized as a Re(II) compound, K_2ReCl_4. The formula and structure were correctly explained 10 years later by F. Albert Cotton as being a species containing a Re—Re quadruple bond.[1] The structure of the complex is shown in Figure 8.4.

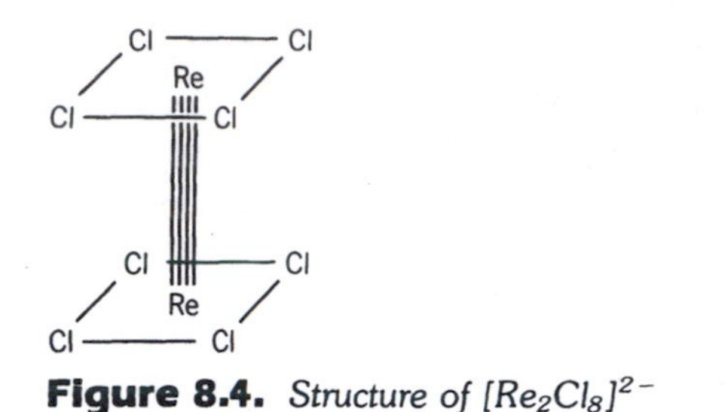

Figure 8.4. *Structure of* $[Re_2Cl_8]^{2-}$

The bonding can most easily be explained by considering the space orientations of the *d* orbitals. Each rhenium is slightly displaced above (or below) the center of a square planar array of four chloride ions. The metal $d_{x^2-y^2}$ orbital has the appropriate symmetry to bond to the chlorides. The remaining metal *d* orbitals are the d_{z^2} (a σ orbital), the d_{xz} and the d_{yz} orbitals (which are of π symmetry), and the d_{xy} orbitals. The two d_{xy} orbitals are parallel to each other, and overlap to form a type of bond not seen in organic chemistry, called a δ (delta) bond. The four orbitals (one σ, two π, and one δ) are filled when eight electrons are available, therefore, four electrons per metal ion. Rhenium(III) (d^4) has four electrons available. The quadruple bond is quite strong, short, and stable. The bond persists through a great variety of reactions.

In order for overlap between the two d_{xy} orbitals to be maximized, the $ReCl_4$ two square planes must be eclipsed relative to each other. Despite the fact that interatomic repulsions between the chlorides are maximized in this orientation, the ability to quadruple bond is the overriding factor. (Cases are known, however, where the two planes are not strictly eclipsed, with deviations of up to 20°.)

Prior Reading and Techniques

Section 2.F: Reflux and Distillation

Section 5.A: Microscale Determination of Magnetic Susceptibility

Section 5.C.2: Purging with an Inert Gas

Section 5.D.3: Isolation of Crystalline Products (Suction Filtration)

Section 5.F.2: Evaporation Techniques

Section 6.C: Infrared Spectroscopy

Related Experiments

Metal–Metal Bonding: Experiments 24 and 45

EXPERIMENTAL SECTION

Safety Recommendations

Tetrabutylammonium perrhenate(VII) (CAS No. 16385-59-4): No toxicity data is available for this compound, however, rhenium compounds are known to be heavy metal poisons, and should be handled with care.

Benzoyl chloride (CAS No. 98-88-4): This material is harmful if inhaled, swallowed, or absorbed through the skin, and is a possible carcinogen. Since the compound has a high vapor pressure (1 mm at 32 °C), it is a lachrymator and has a disagreeable odor. It should only be used in the **HOOD**. No LD50 data is available.

Tetra-*n*-butylammonium bromide (CAS No. 1643-19-2): No toxicity data is available for this material, but it would be prudent to follow the usual precautions (Section 1.A.3).

CHEMICAL DATA

Compound	FW	Amount	mmol	mp (°C)	bp (°C)	Density
Tetra-*n*-butylammonium perrhenate (VII)	492.67	100 mg	0.2			
Benzoyl chloride	140.57	1 mL	0.86	−1	198	1.211
Tetra-*n*-butylammonium bromide	322.38	170 mg	0.53	103		

Required Equipment
Magnetic stirring hot plate, 10-mL side arm round-bottom flask, nitrogen source, magnetic stirring bar, water condenser, $CaCl_2$ drying tube, mercury bubbler, automatic delivery pipet, graduated cylinder, 10-mL beaker, Hirsch funnel, clay tile or filter paper.

Time Required for Experiment: 3.5 h.

EXPERIMENTAL PROCEDURE[2]

Place 100 mg (0.2 mmol) of tetra-*n*-butylammonium perrhenate(VII) (available from Ref. 3) in a 10-mL side arm flask equipped with a magnetic stirring bar and attached through the side arm to a source of nitrogen. Attach a water condenser to the flask, a $CaCl_2$ drying tube to the condenser, and the drying tube to a mercury bubbler, as shown in Figure 8.5. The mercury in the bubbler should be covered with a layer of mineral oil.

> **NOTE: *Do not use a mineral oil bubbler. Mercury is needed to increase the pressure and therefore the boiling point of the benzoyl chloride reagent. With a mineral oil bubbler, yields are drastically reduced.***

Purge the reaction vessel with nitrogen for 15 min. A positive pressure of nitrogen should be maintained throughout the reaction.

Momentarily remove the drying tube and quickly add (automatic delivery pipet) 1 mL (0.86 mmol) of benzoyl chloride down the condenser. Gently reflux the resulting mixture for 90 min. The boiling point of the benzoyl chloride should

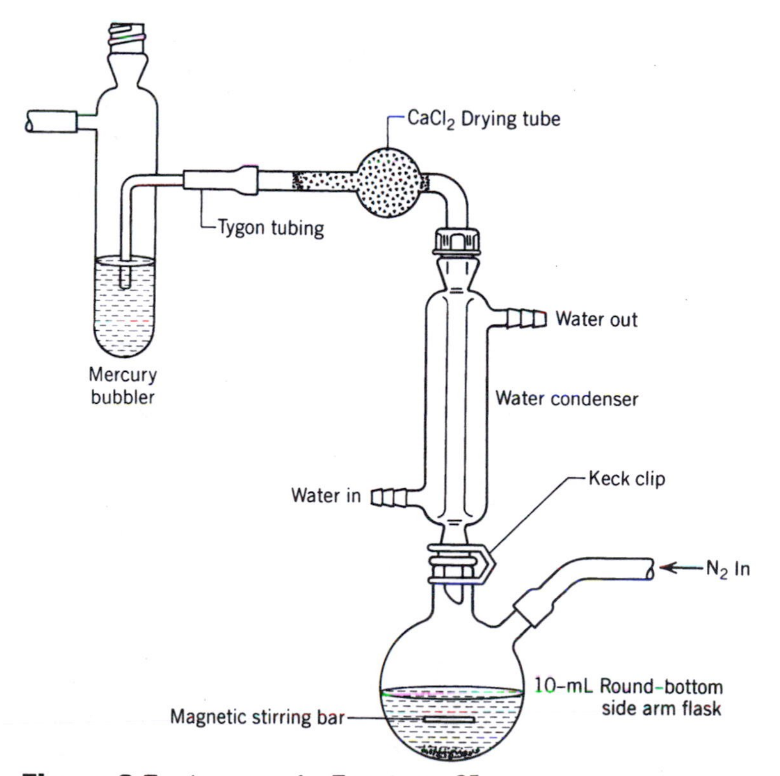

Figure 8.5. *Apparatus for Experiment 25.*

be ~209 °C. The solution will turn from yellow, through various intermediate colors, to dark green over this time period.

Allow the solution to cool to room temperature. In a 10-mL beaker, prepare a solution of 170 mg (0.53 mmol) of tetra-*n*-butylammonium bromide dissolved in 2.5 mL of ethanol (graduated cylinder) that was saturated by bubbling HCl gas through it for 1 min **[HOOD]**.

> **NOTE: *If no lecture bottle of HCl is available, this step may be conveniently accomplished as follows: Place 1 mL of concentrated HCl in a 10-mL side arm test tube, which was capped. Attach a piece of Tygon tubing to the side arm and attach a Pasteur pipet to the other end of the tubing. Place the tip of the pipet in the ethanol—tetra-n-butylammonium bromide solution. Remove the cap momentarily and add a few drops of concentrated H_2SO_4 to the HCl. Immediately replace the cap. A vigorous reaction producing HCl gas occurs and HCl gas will bubble through the solution.***

Immediately, add the HCl saturated solution to the reaction mixture, using a Pasteur pipet. Reflux the resulting mixture for 1 h, still under N_2. The color should change to a dark brown-black over this time period.

Isolation of Product

Evaporate the mixture to one half of its original volume by gently warming it under a stream of N_2. Collect the resulting blue-green crystals of tetra-*n*-butylammonium octachlorodirhenate(III) (which are air stable) by suction filtration on a Hirsch funnel.

> **NOTE: *The solution is often so dark and viscous that the crystals cannot be seen. It is quite easy to be fooled into thinking that no crystallization has occurred.***

Rinse the reaction flask with three 0.5-mL portions of ethanol and use the washings to rinse the product. Follow this with a 0.5-mL portion of ether. Dry the crystals on a clay tile or filter paper and determine the percentage yield.

Characterization of Product

Obtain the IR spectrum of the product and compare it to published sources.[4] Determine the magnetic moment of the compound (see Section 5.A).

QUESTIONS

1. Quadruple bonds are fairly common in transition metal complexes, but are never seen in organic chemistry. Why?
2. Rhenium was the last nonradioactive element to be discovered. Why did it take so long?
3. The $[Re_2X_8]^{2-}$ ion can be reduced to form both $[Re_2X_8]^{3-}$ and $[Re_2X_8]^{4-}$ species. What structure would you expect these ions to have? What would you expect the Re—Re bond order to be?
4. Several transition metals other than rhenium form complexes containing quadruple bonds. Cite two examples from the literature, and discuss the structure and bonding in each.

REFERENCES

1. Cotton, F. A. *Inorg. Chem.* **1965**, *4*, 334. Cotton, F. A. *Inorg. Chem.* **1967**, *6*, 924.
2. Barder, T. J.; Walton, R. A. *Inorg. Syn.* **1985**, *23*, 116.
3. Aldrich Chemical Co., No. 25,022-8, $21.30/g.
4. *The Aldrich Library of FT-IR Spectra*, Vol. II, Pouchert, C. J. Ed., Aldrich Chemical Co.: Milwaukee, WI, 1985, Spectrum No. 1293A.

GENERAL REFERENCES

Baird, M. C., "Metal–Metal Bonds in Transition Metal Complexes" in *Progress in Inorganic Chemistry*, F. A. Cotton, Ed., Interscience: New York, 1968, Vol. 9, p. 1.

Kepert, D. L.; Vrieze, I. C., "Compounds of the Transition Elements Involving Metal–Metal Bonds" in *Comprehensive Inorganic Chemistry*, J. C. Bailar et al., Eds., Pergamon: Oxford, 1973, Vol. 4, Chapter 47, p. 197.

Experiment 26 Geometric Isomerism

Part A: Synthesis of *trans*-Dichloro*bis*(ethylenediamine)cobalt(III) Chloride

Part B: Synthesis of *cis*-Dichloro*bis*(ethylenediamine)cobalt(III) Chloride

INTRODUCTION

The modern era of inorganic chemistry can be said to have begun at the turn of the 20th century with Alfred Werner's pioneering work on metal complex structure and coordination.[1] Werner proved (in 1911) that compounds containing six ligands connected to a central metal atom were indeed octahedral by an elegant resolution of the complex $[Co(en)_2(NH_3)X]^{2+}$. This type of geometry had been theorized earlier (1875) by van't Hoff, who suggested that appropriately substituted octahedral molecules should exhibit geometric isomerism.

Compounds having the same formulas but different structures are isomeric. With geometrical isomers, it is the arrangement of ligands on the central atom that differs. In an octahedral compound of formula MA_4B_2 (M = metal, A and B = ligands), where a central metal is surrounded by four of one type of ligand and two of another, there are two ways to arrange the groups, as shown in Figure 8.6*a*. In the cis isomer, the two B groups are adjacent to each other, while in the trans isomer, the two B groups are opposite each other.

Geometrical isomers are totally different compounds, having different physical properties, and often having different colors. In most syntheses, both isomers are obtained. Separation can be a problem, but because of the (usually) different solubilities and reactivities of the isomers, separation is possible. The first geometrical isomers were also identified by Werner, who in 1893 determined the structure of the inorganic geometric isomer pair *cis*- and *trans*-$[Pt(NH_3)_2Cl_2]$.

Geometrical isomers frequently contain bidentate ligands, which occupy two coordination sites. Ethylenediamine (en = $H_2NCH_2CH_2NH_2$) is such a bidentate ligand. The two geometrical isomers of the compound to be synthesized in this experiment, dichlorobis(ethylenediamine)cobalt(III) chloride, $[Co(en)_2Cl_2]Cl$, are shown in Figure 8.6*b*. This pair was also first investigated by Werner. Experiment 27 deals with the separation of optical isomers of this complex. Werner received the Nobel Prize in 1913 for his pioneering work in structural inorganic chemistry.

Numerous complexes of cobalt(III) are known and nearly all have octahedral structures. In solution, these ions undergo ligand substitution reactions rather

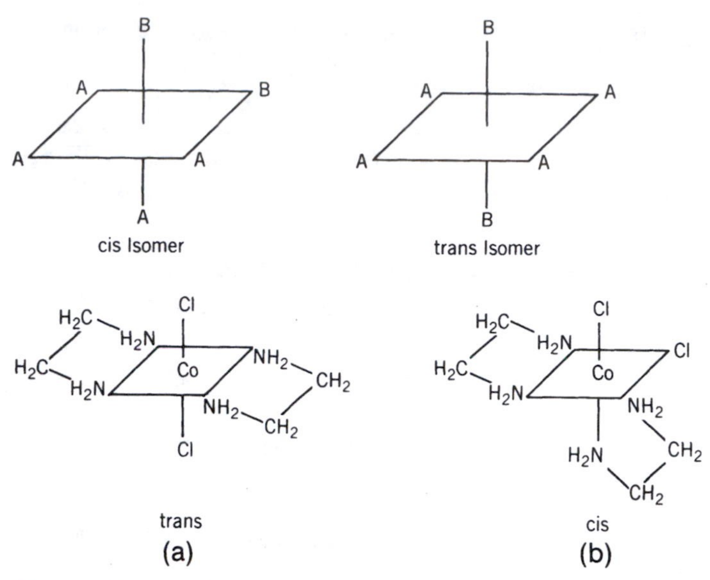

Figure 8.6. a. *The* cis *and* trans *isomers of* MA_4B_2. b. *Geometric isomers of* $[Co(en)_2Cl_2]^+$.

slowly compared to the complexes of many other transition metals. Because of this relative stability, they are of particular interest, as they may be easily studied. Indeed, much of our knowledge of and the theory concerning octahedral complexes in general was derived from studies of cobalt(III) species.

Cobalt(III) exhibits a particular tendency to coordinate with ligands containing nitrogen. A majority of these complexes have ammonia, amines, or nitrogen bonded NCS^- groups. Several of these compounds have cis and trans isomers and one of them, dichlorobis(ethylenediamine)cobalt(III) chloride, is particularly appropriate for demonstrating geometric isomerism in transition metal complexes (Parts 26.A and B). It is of further interest to realize that the cis-isomer of this geometric pair exists as an enantiomorphic (optically active) pair of isomers. The racemic mixture is obtained when the cis-isomer is prepared (Part 26.B), but the mixture can be resolved and one of the enantiomers separated (Experiment 27).

The *trans*-dichlorobis(ethylenediamine)cobalt(III) chloride, shown in Figure 8.6b, is prepared[2] by the air oxidation of an aqueous solution of cobalt(II) chloride hexahydrate and ethylenediamine, followed by the addition of concentrated hydrochloric acid. The synthesis uses a Co^{2+} species rather than a Co^{3+} salt, because the cobaltic ion reacts with water and is therefore unstable in the presence of moisture.

$$4Co^{3+}\ (aq) + 2H_2O = 4Co^{2+}\ (aq) + 4H^+\ (aq) + O_2\ (g)$$

Once Co^{3+} has coordinated with ethylenediamine and chloride ligands, it shows little or no tendency to oxidize water.

Prior Reading and Techniques

Section 5.D.3: Isolation of Crystalline Products (Suction Filtration)

Section 5.F.2: Evaporation Techniques

Section 6.C: Infrared Spectroscopy

Related Experiments

Cobalt Chemistry: Experiments 7B, 17, 27, 30, 35, and 47B
Isomerism: Experiments 27, 37, 46, and 49

EXPERIMENTAL SECTION

Part A: Synthesis of *trans*-Dichlorobis(ethylenediamine)cobalt(III) Chloride

Safety Recommendations

Cobalt(II) chloride hexahydrate (CAS No. 7791-13-1): This compound is harmful if swallowed, inhaled, or absorbed through the skin. ORL-RAT LD50: 766 mg/kg.

Ethylenediamine (CAS No. 107-15-3): This compound is harmful if swallowed, inhaled, or absorbed through the skin. ORL-RAT LD50: 500 mg/kg. It has an irritating ammonia odor (vapor pressure is 10 mm at 20 °C), so it should only be used in the **HOOD**.

CHEMICAL DATA

Compound	FW	Amount	mmol	bp (°C)	mp (°C)	Density
$CoCl_2 \cdot 6H_2O$	237.85	300 mg	1.26		87	1.920
$NH_2CH_2CH_2NH_2$	60.10	1 mL/10% solution	1.50	118	8.5	0.899

Required Equipment
Side arm test tube, Pasteur pipets, magnetic stirring hot plate, water aspirator, ice bath, Hirsch funnel, watch glass.

Time Required for Experiment: 1.5 h.

EXPERIMENTAL PROCEDURE[2,3]

NOTE: ***If Experiment 27 is to be performed, double all quantities in Part 26.A. Alternatively, combine the product from two students to have enough trans product for Experiment 27.***

In a side arm test tube equipped with an air inlet (Fig. 8.7), place 300 mg (1.26 mmol) of $CoCl_2 \cdot 6H_2O$, 2 mL of water (graduated cylinder), and 1.0 mL of 10% ethylenediamine (automatic delivery pipet, **HOOD**!).

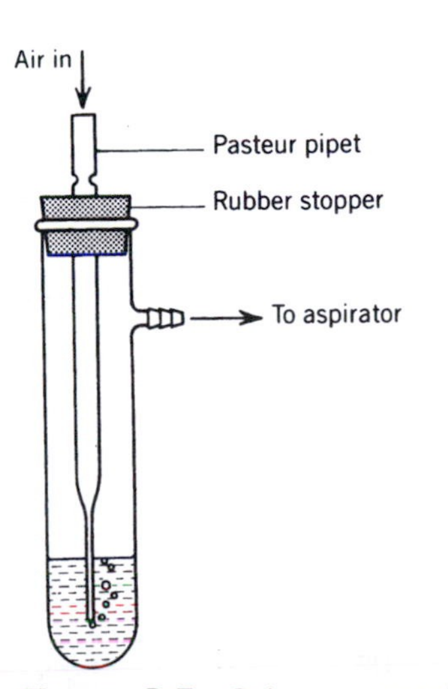

Figure 8.7. *Side arm test tube apparatus.*

Clamp the reaction tube in a hot water bath (90–95 °C) and connect the side arm to a water aspirator through a water trap. Turn on the aspirator so as to draw air through the solution at a slow but steady rate. The reacting system, which is purple in color, is maintained under these conditions for a period of 1.0 h. From time to time, add additional water to the reaction flask (down the air inlet tube) to maintain the water volume. After the 1.0-h heating time, disconnect the aspirator and remove the reaction tube from the water bath.

NOTE: ***Do not turn off the aspirator until the tube is disconnected. This will prevent any water from being sucked back into the reaction tube due to the change in pressure. This is a good practice to follow even though the water trap is used.***

Allow the tube to cool to approximately 50–60 °C. Using a Pasture pipet, slowly add 600 μL of concentrated HCl down the inlet tube. Swirl the reaction assembly by hand for several minutes and place it back into the hot water bath. Reconnect the aspirator and adjust it so that a steady stream of air is once again pulled through the solution. This procedure is continued until the volume of the solution is decreased to the point that crystals of the product are evident in the tube.

Disconnect the tube from the aspirator, remove it from the water bath and place it in an ice bath to cool. Scrape the resulting solid from the tube and collect it by suction filtration using a Hirsch funnel. Wash the crystals with two 2-mL portions of cold methanol, which is added to the reaction flask to assist in the removal of additional crystalline product. This is followed by washing with two 2-mL portions of cold diethyl ether. The beautiful green crystals that form are actually the hydrochloride salt of the desired product. To obtain the *trans*-dichlorobis(ethylenediamine)cobalt(III) chloride, place the crystals on a small watch glass, and heat them in an oven at 110 °C for 1.5 h.

Characterization of Product

Record the IR spectrum (KBr pellet) of the compound for comparison with the cis isomer prepared in Part 26.B of this experiment. The Visible spectrum may also be obtained and compared.

Part B: Synthesis of *cis*-Dichlorobis(ethylenediamine)cobalt(III) Chloride

Safety Recommendations: See Part 26.A

Required Equipment

Hot water bath, 5-cm watch glass, 10-mL beaker, magnetic stirring hot plate, magnetic stirring bar.

Time Required for Experiment: 45 min.

EXPERIMENT PROCEDURE

NOTE: *If Experiment 27 is to be performed, increase the amounts used in Part 26.B by a factor of 7–8.*

Place 10 mg (0.06 mmol) of the green *trans*-dichlorobis(ethylenediamine)cobalt(III) chloride on a 5-cm watch glass. Dissolve this solid material in 300 μL of water (automatic delivery pipet) and allow the solution to stand for about 10 min at room temperature. Place the watch glass on a hot water bath prepared from a 10-mL beaker containing a magnetic stirring bar filled with boiling water. The beaker is previously positioned in a sand bath on a magnetic stirring hot plate. Heat the green solution and concentrate it to dryness. A deep violet, glassy material is formed on the watch glass.

Isolation of Product

Cool the product and carefully scrape it from the glass surface (**Caution**—*the glassy product tends to scatter*). Weigh the material and calculate the percentage yield. A quantitative yield is usually obtained.

Characterization of Product

Obtain the IR spectrum of the material (KBr pellet) and compare it to that of the trans isomer prepared in Part 26.A. The Visible spectrum may also be obtained and compared.

QUESTIONS

1. How many isomers exist of the complex having formula MA_3B_3? Draw and name them.
2. It was stated that Co^{3+} complexes are very stable and the ligands in such complexes are not labile. Define labile and explain the relative inertness of Co^{3+} complexes.
3. Define the symmetry elements of the trans and cis isomers prepared in this experiment and assign the isomers to symmetry point groups.
4. In place of oxygen as the oxidizing agent, hydrogen peroxide may be used in this reaction. Balance the following oxidation–reduction reaction using this reagent.

$$Co^{2+} + H^+ + H_2O_2 \rightarrow Co^{3+} + H_2O$$

   Which species is the reducing agent in this reaction?
5. In the absence of the ethylenediamine ligands, the hexaaquocobalt(III) ion reacts rapidly with water according to the following scheme.

$$[Co(H_2O)_6]^{3+} + H_2O \rightarrow [Co(H_2O)_6]^{2+} + O_2 + H^+$$

   Balance the equation and determine which species is the reducing agent.
6. Today, chemists take for granted the octahedral configuration of most transition metal complexes. It should be known, however, that this was no easy matter to prove. One of the great chemical literature debates took place between S. M. Jørgensen and Alfred Werner about the true structure of cobalt amine complexes. Describe both men's arguments and detail how Werner's views eventually won out. There is also an interesting analogy between the lives of Werner and August Kekulé (of benzene fame). See the following for details: Kauffman, G. B. *J. Chem. Educ.* **1976,** *53*, 445. George Kauffman has written extensively about chemical history.

REFERENCES

1. For an interesting first hand account of this area, see Bailer, J. C., Jr., "Research in the Sterochemistry of Cobalt Complexes" in *Stereochemistry of Optically Active Transition Metal Compounds,* ACS Symposium Series No. 119, Bodie E. Douglas and Saito Yoshihiko, Eds., American Chemical Society: Washington, DC, 1980.
2. Bailer, J. C., Jr., *Inorg. Syn.* **1946,** *2,* 222.
3. Baldwin, M. E. *J. Chem. Soc.* **1960,** 4369.

GENERAL REFERENCES

Harrowfield, J. MacB.; Wild, S. B., "Isomerism in Coordination Chemistry" in *Comprehensive Coordination Chemistry,* G. Wilkinson, Ed., Pergamon: Oxford, 1987, Vol. 1, Chapter 5, p. 179.

Pratt, J. N.; Thorp, R. G., "Cis and Trans Effects in Cobalt(III) Complexes" in *Advances in Inorganic Chemistry and Radiochemistry,* H. J. Emeleus and A. G. Sharpe, Eds., Academic Press: New York, 1969, Vol. 12, p. 375.

Experiment 27

Optical Isomers: Separation of an Optical Isomer of *cis*-Dichlorobis(ethylenediamine) cobalt(III) Chloride

INTRODUCTION

The cis isomer of $[Co(en)_2Cl_2]Cl$ (see Experiment 26 for the synthesis of this isomer) actually consists of two isomers that are mirror images of each other.

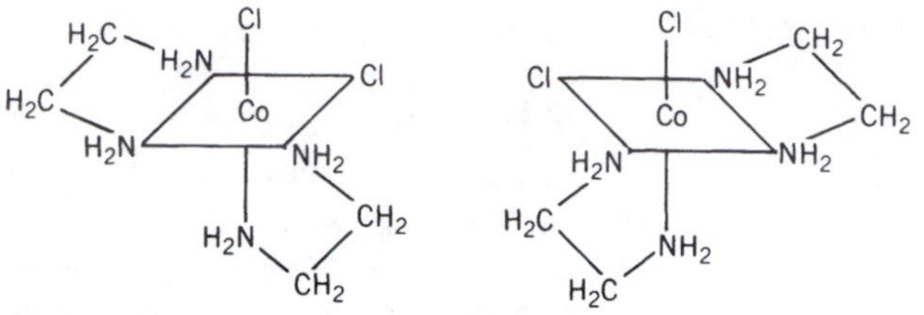

Figure 8.8. *Optical isomers of* cis-$[Co(en)_2Cl_2]^+$.

Such isomers are termed optical isomers. The two forms are shown in Figure 8.8.

These two forms, much like your right and left hands, cannot be superimposed upon each other. Optical isomers generally have physical properties that are quite similar, and are generally very difficult to separate (resolve). The most common way to resolve optical isomers is by treating the mixture (called a racemic mixture) with another compound that is optically active, with which one of the isomers will form an insoluble salt. In this experiment, the two optical isomers are treated with potassium antimonyl-*d*-tartrate, which selectively precipitates one of the two optical isomers, the *d* isomer. Werner was also active in this area, having reported (in 1912) on several such compounds, including conversions of *levo* (left handed) into *dextro* (right handed) complexes.

By addition of potassium antimonyl-*d*-tartrate, the *d* isomer of *cis*-dichlorobis(ethylenediamine)cobalt(III) chloride can be selectively precipitated from the racemic mixture of the two optically active isomers.

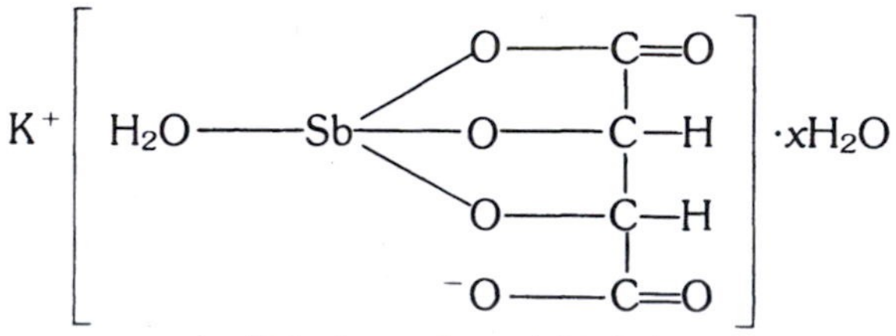

Potassium antimonyl-d-tartrate

Prior Reading and Techniques

Section 5.D.3: Isolation of the Crystalline Products (Suction Filtration)

Related Experiments

Cobalt Chemistry: Experiments 7B, 17, 26, 30, 35, and 47B

Isomerism: Experiments 26, 37, 46, and 49

EXPERIMENTAL SECTION

Safety Requirements

Potassium antimonyl-*d*-tartrate hydrate (CAS No. none): This compound is harmful if swallowed, inhaled, or absorbed through the skin. ORL-HMN LDLo: 2 mg/kg. ORL-RAT LD50: 115 mg/kg.

CHEMICAL DATA

Compound	FW	Amount	mmol	mp (°C)	Density
cis-Dichlorobis(ethylenediamine)cobalt(III) chloride	285.45	145 mg	0.50		
Potassium antimonyl-*d*-tartrate hydrate	333.93	167 mg	0.50		2.607

Required Equipment

Magnetic stirring hot plate, 10-mL Erlenmeyer flask, magnetic stirring bar, water bath, Pasteur filter pipet, 10-mL graduated cylinder, Hirsch funnel, clay tile or filter paper.

Time Required for Experiment: 2 h.

EXPERIMENTAL PROCEDURE[1,2]

In a 10-mL Erlenmeyer flask containing a magnetic stirring bar, dissolve 167 mg (0.5 mmol) of potassium antimony-*d*-tartrate hydrate in 2 mL of water (graduated cylinder). Clamp the flask in a water bath set on a magnetic stirring hot plate.

In a 10-mL beaker, dissolve 145 mg (0.5 mmol) of *cis*-dichlorobis(ethylenediamine)cobalt(III) chloride in 4 mL of water.

Transfer the solution in the beaker, by use of a Pasteur filter pipet, to the solution in the Erlenmeyer flask. Heat the solution in the water bath to 70–80 °C, with stirring, for 45 min. A pale violet precipitate will form over this time period. Cool the flask to room temperature.

Isolation of Product

Collect the crystalline product by suction filtration using a Hirsch funnel. Wash the filter cake with two 0.5-mL portions of chilled water. Dry the product on a clay tile.

Characterization of Product

Determine the specific rotation of the product using a polarimeter. The compound may be dissolved in very hot water, and the specific rotation must be measured immediately, as the product will precipitate upon cooling.

QUESTIONS

1. In order for an organic molecule to be optically active, what configuration must be present?
2. What symmetry elements make a compound optically inactive?
3. What is circular dichroism and how does it relate to optical activity?
4. Once an optically active compound is isolated, how does one know which configuration it has? From the literature, discuss the Cotton Effect in terms of assignment of absolute configuration.

REFERENCES

1. Bailer, J. C., Jr., *Inorg. Syn.* **1946,** *2,* 222.
2. Baldwin, M. E. *J. Chem. Soc.* **1960,** 4369.

Experiment 28 Ion Exchange Separation of the Oxidation States of Vanadium

INTRODUCTION

Vanadium, due to the beautiful colors of its various oxidation states, is named for the Scandinavian goddess Vanadis, the goddess of beauty. First discovered in 1801 by del Rio, it was originally thought that this discovery was in error and that del Rio had simply obtained impure chromium. The element was rediscovered by Sefstrom in 1830. Most vanadium is used in the making of specialty steels, as the alloy ferrovanadium. The metal is also found in several living systems, most notably in the ascidian family (sea squirts and tunicates). In this experiment, the various oxidation states of vanadium are investigated using ion exchange chromatography.

Vanadium has four common oxidation states: V, IV, III, and II. Starting with vanadium(V) in the form of ammonium metavanadate, NH_4VO_3, a series of reductions is carried out in this experiment. In the first step, using hydrochloric acid, the vanadium is converted from VO_3^- to VO_2^+, which in turn is then partially reduced to form the VO^{2+} ion. Ion exchange chromatography separates the two ions. The VO^{2+} is subsequently reduced further to V(III) and V(II), which are separated in a like manner.

Prior Reading and Techniques

Section 5.G.4: Liquid Chromatography

Section 6.B: Visible Spectroscopy

Related Experiments

Chromatography: Experiments 23, 34F and 40D

EXPERIMENTAL SECTION

Safety Recommendations

Ammonium metavanadate (CAS No. 7803-55-6): This compound is harmful if swallowed, breathed, or absorbed through the skin. The ORL-RAT LD50: 160 mg/kg. Avoid breathing the dust.

Mercury(II) chloride (CAS No. 7487-94-7): This compound is toxic, as are all mercury compounds. One or 2 g is frequently fatal. Do not breathe the dust. Do not handle the zinc–mercury amalgam with bare hands. ORL-RAT LD50: 1 mg/kg.

Zinc (CAS No. 7440-66-6): Zinc is not generally considered to be a dangerous material, however, the usual safety precautions (Section 1.A.3) should be taken.

CHEMICAL DATA

Compound	FW	Amount	mmol	mp (°C)	Density
NH_4VO_3	117.0	200 mg	1.71	200[a]	2.326

[a] Decomposes.

Required Equipment

Dowex AG50W-X2 cation exchange resin, 25-mL buret, glass wool, three 100-mL flasks, 150-mm test tube, microburner, 25-mL Erlenmeyer flask, 12 small test tubes, Pasteur pipet.

Time Required for Experiment: 3.0 h.

EXPERIMENTAL PROCEDURE[1,2]

Instructor Preparation of Zinc–Mercury Amalgam

Dissolve 200 mg of $HgCl_2$ in 30 mL of water and 200 μL of concentrated nitric acid. Add 25 g of zinc (30 mesh) to this solution. Shake the solution briefly and decant the supernatant liquid. Wash the resulting zinc–mercury amalgam twice with water. Store it under water until used. This procedure makes enough amalgam for 10 student preparations.

Preparation of the Chromatographic Column and Solutions

Using a glass rod, push a small wad of glass wool down to the stopcock of a 25-mL buret. With the stopcock of the buret closed, fill the buret with an aqueous slurry of AG50W-X2 cation exchange resin (100–200 mesh, H^+ form). It may be necessary to soak the resin in HCl and rinse it thoroughly with water before use. Allow the resin to settle. Open the stopcock and continue adding the slurry of resin until the amount of the settled resin is ~5–6 in. high. Make sure the column is tightly packed. Close the stopcock as necessary to keep the level of liquid above the top of the resin.

> **NOTE:** ***During the course of this experiment, do not allow the level of liquid to fall below the top of the settled resin, or it will start to channel.***

When the resin level is sufficiently high, wash the resin with distiled water until the eluate is clear.

In separate flasks, make up 100 mL each of 3, 1, and 0.4*M* HCl, starting with dilute HCl (6*M*).

First Separation

In the hood, add 2.0 mL of concentrated HCl to 200 mg (1.71 mmol) of NH_4VO_3 in a 150-mm test tube. Heat the mixture to boiling over a microburner for 2–3 min, add 10 mL of distilled water, and mix well. To prevent bumping, do not heat the test tube at the bottom. The solution should change from its original yellow color to a bright green. (What reaction has occurred?)

Pour the bright green solution onto the cation exchange column. Allow the level of liquid to fall to about 1 in. above the top of the resin, and then add *0.4M HCl* as necessary to keep the buret full. Two bands will develop on the column. The upper band is readily detected by its green color (it is actually blue, but the orange color of the resin makes it appear to be green). The lower band is yellow and is somewhat difficult to see because of the orange resin. When the eluate from the column becomes yellow, begin collecting fractions of 5–6 mL each in clean test tubes. When all the yellow species has eluted, or when the solution begins to turn green or blue, change the solution on the top of the column to *1.0M HCl*. Continue collecting fractions of 5–6 mL until all of the green band has eluted. Combine the blue fractions (you should have ~25–30 mL in all). Combine the darkest yellow fractions, but do not combine in any fractions that are yellow-green or blue-green. Make sure that the column is still wet.

Second Separation

Place two thirds of the blue solution (~15–20 mL) in a 25-mL Erlenmeyer flask and add 2 g of zinc–mercury amalgam. Cork the flask and shake vigorously. A green, then violet color should eventually form. Stop the shaking at the first appearance of violet color (do not allow it to become all violet), so that a mixture of the two oxidation states is present. This solution is then decanted onto the chromatographic column used in the first separation. Allow the level of the liquid to fall to about 1 in. above the top of the resin, and then add *1.0M HCl* as necessary to keep the buret full. Distinct green and violet bands will form. Collect the violet form in 5-mL fractions as it elutes. After the violet band has eluted from the column, use *3.0M HCl* to elute the green species. Collect this in the same way. Combine the individual fractions in the same manner as in the first separation.

To regenerate the resin, pass 20–30 mL of water through the column. Remove the resin from the buret and collect the resin through gravity filtration. Return it to the original bottle. Do not dry the resin!

Characterization of Products

Place 1 mL each of the four solutions in a series of small test tubes. Add dropwise (Pasteur pipet) 0.001*M* $KMnO_4$. Which solutions decolorize the permanganate? Why? Acquire the Visible spectrum of each of the four ions, noting the absorbance maxima for each oxidation state.

Alternate Method

A portion of the bright green solution obtained from NH_4VO_3 and HCl may be directly reduced to V(II) with the zinc–mercury amalgam. Exposure of the solution to air will oxidize V(II) to V(III). Samples of each may thereby be obtained.

Recycling

Any leftover zinc–mercury amalgam should be collected, washed with dilute HCl, and stored under water for future use.

QUESTIONS

1. Identify the ions formed in each reaction step, along with their colors.
2. Balance the following redox reaction in acidic solution.

$$MnO_4^- + VO^{2+} \rightarrow MnO_2 + VO_2^+$$

   Write balanced equtions for the other redox reactions in this experiment.
3. What is an amalgam? Give an everyday example of the use of an amalgam.
4. Based on literature electrochemical potentials, make a plot of energy (y axis) versus oxidation number (x axis) for the various oxidation states of vanadium. Based upon your diagram, which of these states (if any) are unstable with respect to disproportionation? What line shape would this correspond to in the diagram?

REFERENCES

1. Cornelius, R. *J. Chem. Educ.* **1980,** *57,* 316.
2. Hentz, F. C., Jr.; Long, G. G. *J. Chem. Educ.* **1978,** *55,* 55.

GENERAL REFERENCE

Clark, R. J. H., "Vanadium," in *Comprehensive Inorganic Chemistry,* J. C. Bailar, Ed., Pergamon: Oxford, 1973, Vol. 3, Chapter 34, p. 491.

Experiment 29 Determination of Δ_o in Cr(III) Complexes

INTRODUCTION

The *d* orbitals of a metal ion in an octahedral ligand or crystal field (i.e., surrounded by an octahedral array of ligands) are split into a higher energy (e_g) and lower energy (t_{2g}) set, as shown in the following energy level diagram.

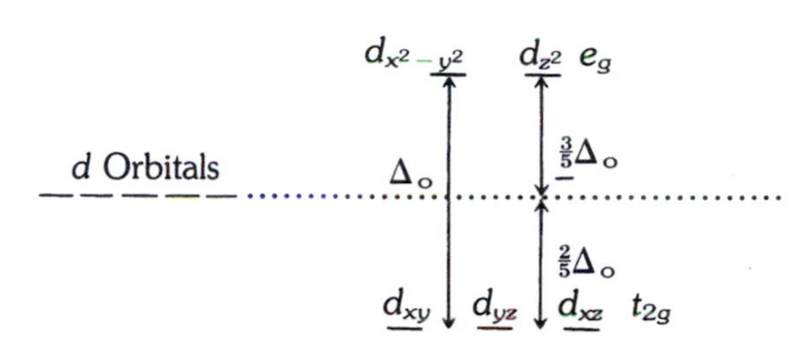

The energy difference between the upper and lower energy levels is designated as Δ_o (pronounced del-oh), the octahedral crystal field splitting. In the older literature, this energy difference is also known as 10 *Dq*.

The degree of splitting of the *d* orbitals (the magnitude of Δ_o) in octahedral complexes depends on several factors, including the charge on the metal, the size of the metal, and the nature of the ligand. The situation is simplified considerably by considering a series of compounds of the same metal in a given oxidation state. The only major variable, in this case, is the ligands bonded to the metal. From the study of the spectra of such complexes it is possible to arrange the various ligands in a sequence according to their ability to cause *d*-orbital splitting. This series is known as the **spectrochemical series** and is given below:

$$\text{halides} < OH^- < C_2O_4^{2-} < H_2O < NCS^- < py < NH_3 < en < phen < NO_2^- < CN^-, CO$$

It should be emphasized that the spectrochemical series is only an empirical correlation and is determined from experimental data. The magnitude of Δ_o increases by a factor of about 2 as one moves from halide to CN^- in the spectrochemical series. Cyanide anion has the strongest ligand field of all common ligands.

The interpretation of the electronic spectrum of a d^1 coordination compound is straightforward. The spectrum shows only one *d*–*d* band, corresponding to a transition from the t_{2g} level to the e_g level. In complexes with electronic configurations d^n where $1 < n < 9$, the spectral interpretation becomes more difficult, because of additional interactions between the *d* electrons. More than one absorption band occurs in such systems.

In the case of Cr^{3+} (d^3) compounds, the spectral characteristics are reasonably easy to interpret. It is easily seen from the energy level diagram (Fig. 8.9) that the lowest energy state (the ground state) is labeled $^4A_2$ (pronounced quartet-A-two), corresponding to an electronic configuration of three electrons in the t_{2g} level. There are several excited states as well. Keeping in mind the Visible selection rules, the major bands seen in the Visible spectrum will correspond to transitions to other quartet states. These states in order of ascending energy, are labeled $^4T_2$ (two electrons in the t_{2g}, one in the e_g), $^4T_1$, and another $^4T_1$ (one electron in the t_{2g}, two in the e_g). **The energy separation between the two lowest energy levels, $^4A_2$ and $^4T_2$, is Δ_o.** This band will be the one with the longest wavelength in the spectrum. Thus, although Cr(III) complexes show more complicated spectra than that shown by d^1 systems, the value of Δ_o may be directly calculated from the longest wavelength band.

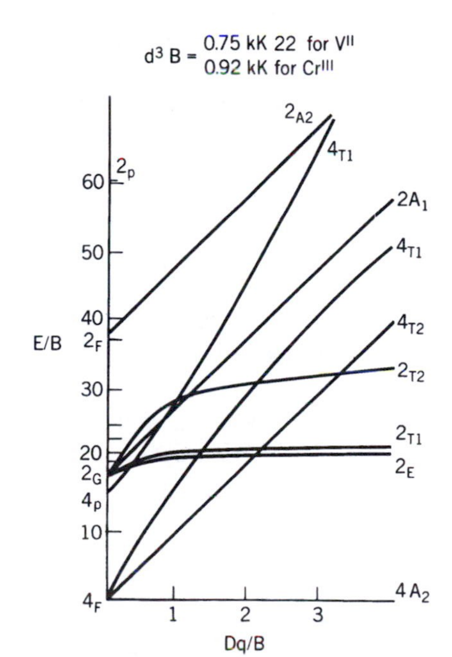

Figure 8.9. *Tanabe–Sugano diagram for* d³ *ion. (After Tanabe, Y.; Sugano, S.* J. Phys. Soc. Jpn. **1954,** 9, *753.)*

There are three main objectives for this experiment:

1. The investigation of the electronic spectra of several common Cr(III) complexes. The complexes $[Cr(H_2O)_6](NO_3)_3{\cdot}3H_2O$ and $[Cr(H_2O)_4Cl_2]\,Cl{\cdot}2H_2O$ are both commercially available. The complex $[Cr(en)_3]\,Cl_3{\cdot}3H_2O$ is prepared in this experiment and a Cr(III) compound, $[Cr(acac)_3]$, can also be prepared (see Experiment 22.A).

2. The determination of Δ_o for each compound and arranging the ligands in order of crystal field strength (spectrochemical series).
3. The interpretation of the changes in a series of spectra of $[Cr(H_2O)_4Cl_2]Cl \cdot 2H_2O$ obtained over an extended period of time.

Prior Reading and Techniques

Section 2.F: Reflux and Distillation

Section 5.D.3: Isolation of Crystalline Products (Suction Filtration)

Section 6.B: Visible Spectroscopy

Related Experiments

Chromium Chemistry: Experiments 22A, 23, and 33

EXPERIMENTAL SECTION

Preparation of Complexes

Hexaaquochromium(III) nitrate, $[Cr(H_2O)_6](NO_3)_3 \cdot 3H_2O$. This is a commercially available compound. It can be used without further purification.

Chromium(III) chloride hexahydrate, $CrCl_3 \cdot 6H_2O$. This compound is also commercially available. Its actual structure is $[Cr(H_2O)_4Cl_2]Cl \cdot 2H_2O$. It can also be used without further purification.

Tris(2,4-pentanedionato)chromium(III), $[Cr(acac)_3]$. This compound can be prepared (see Experiment 22.A).

PREPARATION OF TRIS(ETHYLENEDIAMINE)-CHROMIUM(III) CHLORIDE, $[Cr(en)_3]Cl_3$

Safety Recommendations

Methanol (CAS No. 67-56-1): This compound can be fatal if swallowed, and is harmful if inhaled or absorbed through the skin. ORL-HMN LDLo: 143 mg/kg, ORL-RAT LF50: 5628 mg/kg. The compound is flammable and highly volatile.

Zinc (CAS No. 7440-66-6): Zinc is not generally considered to be a dangerous material, however, the usual safety precautions (Section 1.A.3) should be taken.

Chromium(III) chloride hexahydrate (CAS No. 10060-12-5): Chromium compounds are considered mildly toxic. The reagent $CrCl_3 \cdot 6H_2O$ has MLD_{iv} in mice of 801 mg/kg. Certain of them have been listed as carcinogens by the EPA.

Chromium(III) nitrate nonahydrate (CAS No. 7789-02-8): Chromium compounds are considered mildly toxic. ORL-RAT LD50: 3250 mg/kg. The compound is harmful if inhaled or swallowed.

Ethylenediamine (CAS No. 107-15-3): This compound is harmful if swallowed, inhaled, or absorbed through the skin. ORL-RAT LD50: 500 mg/kg. It has an irritating ammonia odor (vapor pressure is 10 mm at 20 °C), so it should only be used in the **HOOD**.

CHEMICAL DATA

Compound	FW	Amount	mmol	bp (°C)	mp (°C)	Density
$CrCl_3 \cdot 6H_2O$	266.5	266 mg	1.0		83	1.760
Ethylenediamine	60.10	1 mL	15.0	118	8.5	0.899

Required Equipment

Magnetic stirring hot plate, magnetic stirring bar, 5-mL round-bottom flask, water condenser, Keck clip, calibrated Pasteur pipet, forceps, Hirsch funnel, clay tile.

Time Required for Experiment: 1.5 h for synthesis, 1.5 h additional time for spectral acquisition

EXPERIMENTAL PROCEDURE

Place 50–100 mg of mossy, granular zinc (previously briefly washed with 6*M* HCl to remove any surface ZnO) and 266 mg (1.0 mmol) of $CrCl_3{\cdot}6H_2O$ in a 5-mL round-bottom flask containing a magnetic stirring bar. Add 1 mL (15 mmol) of ethylenediamine, followed by 1 mL of methanol (calibrated Pasteur pipet). Attach a water condenser with a Keck clip. Place the assembly in a sand bath set on a magnetic stirring hot plate, and heat the mixture, with stirring, at reflux for 1 hr.

Isolation of Product

Cool the solution to room temperature. Collect the yellow crystalline product by suction filtration using a Hirsch funnel. Remove any unreacted zinc using forceps. Wash the filter cake with 500-μL portions of 10% ethylenediamine in methanol until the washings are colorless. Follow this with a 500-μL portion of ether and dry the product on a clay tile.

Characterization of Products

Prepare solutions of each complex in the solvents designated below. The molarity of the solutions must be such that $I/I_0 = 0.5$ at long wavelength absorbtion maxima. Water is used as the solvent for all the complexes except the insoluble $Cr(acac)_3$ complex, for which toluene is used. Calculate the molar extinction coefficient, E, for each solution, using the formula

$$E = \frac{\log(I/I_0)}{cl}$$

where I_0 is the intensity of the incident light, I is the intensity of the transmitted light, the cell length (l) is 1 cm, and c is the concentration in moles per liter. Obtain the Visible spectrum of each of these solutions separately, saving $CrCl_3{\cdot}6H_2O$ for last. Collect a spectrum of the chromium(III) chloride hexahydrate solution every hour for 3 h; obtain an additional spectrum after 6 and 24 h.

For each of the eight spectra obtained, determine the longest wavelength absorbtion maximum in *units of nanometers* (*nm*). Convert the wavelengths (which correspond to Δ_o) into frequency units of reciprocal centimeters (cm^{-1}) using the following relationship:

$$\Delta_o = \nu = \frac{1}{\lambda \text{ (in nm)}} (1 \times 10^7) \text{ cm}^{-1}$$

Other energy units for the absorption may be obtained using the following conversion factors:

$$1 \text{ cm}^{-1} = 1.24 \times 10^{-4} \text{ eV} = 0.01196 \text{ kJ·mol}^{-1}$$

Arrange the various ligands in order of increasing Δ_o. Compare this series with the spectrochemical series. Tabulate your data.

QUESTIONS

1. The Visible spectrum of the $Cr(acac)_3$ is significantly different from the others. Why?
2. Does the order of ligands obtained in this experiment correspond to the established order in the spectrochemical series? Explain any deviations.
3. Account for the changes in the highest wavelength maximum for $CrCl_3 \cdot 6H_2O$ with time. What reaction is occurring in solution?
4. Weak bands, with extinction coefficients ~1% of normal are observed at long wavelength for many Cr(III) complexes. What transitions do these bands correspond to? Why are these bands so weak?
5. Mn(II) and Fe(III) are examples of transition metal ions that are usually much more weakly colored than "normal" transition metal ions. Why are they so weakly colored?
6. Unlike most manganese(II) complexes, $[Mn(CN)_6]^{4-}$ is highly colored. Why? (*Hint:* The ligand is not colored and has a large Δ_o splitting.)

REFERENCES

1. Gillard, R. D.; Mitchell, P. R. *Inorg. Syn.* **1972,** *13,* 184.
2. Sawyer, D. T.; Heineman, W. R.; Beeke, J. M. *Chemical Experiments for Instrumental Methods,* Wiley: New York, 1984, p. 163.
3. Skoog, D. A. *Principles of Instrumental Analysis,* 3rd ed., Saunders: Philadelphia, 1985, Chapter 7.
4. Weissberger, A., Ed., *Physical Methods of Organic Chemistry,* Vol. II, in *Techniques of Organic Chemistry,* Interscience: New York, 1946, Chapter 17.
5. Silverstein, R. M.; Bassler, C. C.; Morrill, T. C. *Spectrometric Identification of Organic Compounds,* 4th ed., Wiley: New York, 1981, Chapter 6.

GENERAL REFERENCE

Lever, A. B. P., *Inorganic Electronic Spectroscopy,* Elsevier: Amsterdam, 1968.

Experiment 30

Preparation and Study of a Cobalt(II) Oxygen Adduct Complex

Part A: Preparation of *N,N'*-Bis(salicylaldehyde)ethylenediimine, $salenH_2$

Part B: Preparation of Co(salen)

Part C: Determination of Oxygen Absorption by Co(salen)

Part D: Reaction of the Oxygen Adduct with Chloroform

INTRODUCTION

A number of heme proteins contain a transition metal that can coordinate molecular oxygen. Some well-known examples are the iron containing myoglobin and hemoglobin proteins, copper containing hemocyanin and vanadium containing hemovanadin. Such metal proteins are involved in oxygen transport, necessary for life. A number of coordination compounds were investigated as "model compounds" that can mimic the biological systems. One such compound, which was extensively studied as an oxygen carrier, is a cobalt(II) complex of *N,N'*-bis(salicylaldehyde)ethylenediimine ($salenH_2$) of formula Co(salen).[1–3]

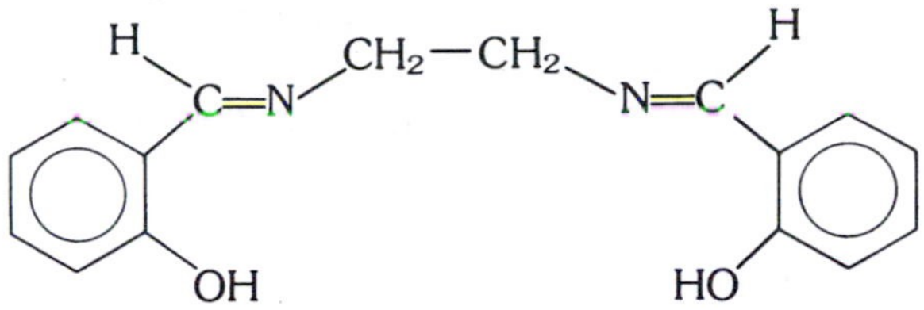

salenH_2

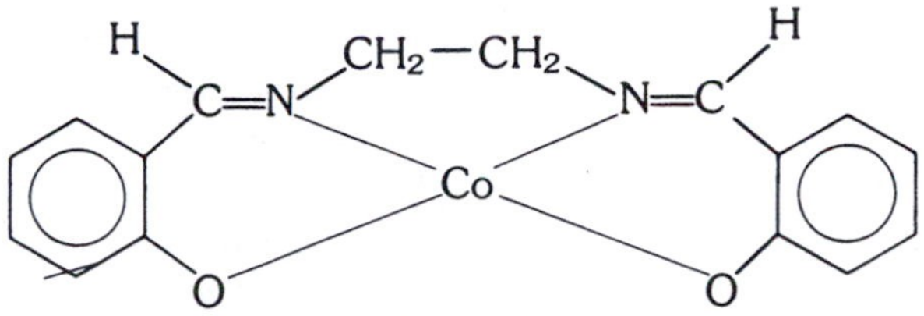

Co(salen)

The salen chelating ligand has two atoms (O, N) as the donor sites and belongs to a special class of ligands known as Schiff's bases.

Co(salen) exists in two forms: One is the **active** species that absorbs oxygen, and the other is the **inactive** variety that does not react with oxygen, both of which are dimeric. A solution of Co(salen) in a donor solvent (e.g., DMSO and DMF) readily absorbs oxygen, forming an adduct with a peroxo-bridged oxygen ligand.

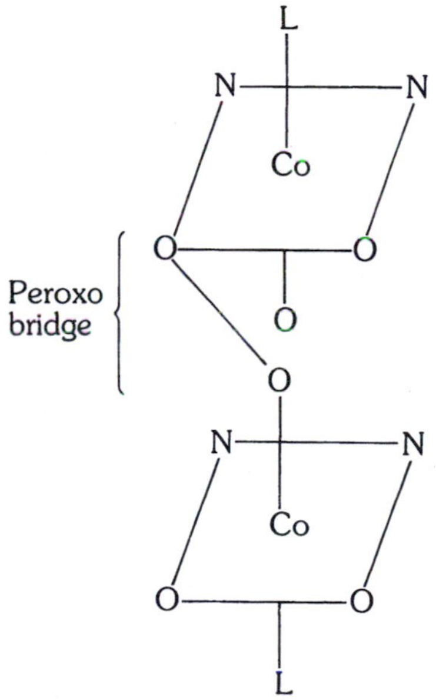

Peroxo bridge

In these diamagnetic compounds, the cobalt is present in three oxidation states.

In this experiment, the organic ligand salenH_2 is prepared, and subsequently reacted with hydrated cobalt acetate to prepare Co(salen). The oxygen up-take capacity of the compound is also investigated.

Prior Reading and Techniques

Section 5.C.2: Purging with an Inert Gas

Section 5.D.3: Isolation of Crystalline Products (Suction Filtration)

Related Experiments

Cobalt Complexes: Experiments 7B, 17, 26, 27, 35, and 47B

EXPERIMENTAL SECTION

Part A: Preparation of *N,N'*-Bis(salicylaldehyde)ethylenediimine, salenH_2

Safety Recommendations

Salicylaldehyde (CAS No. 90-02-8): This compound is harmful if swallowed, inhaled, or absorbed through the skin. ORL-RAT LD50: 520 mg/kg. It has been shown to have effects on fertility.

Ethylenediamine (CAS No. 107-15-3): This compound is harmful if swallowed, inhaled, or absorbed through the skin. ORL-RAT LD50: 500 mg/kg. It has an irritating ammonia odor (vapor pressure is 10 mm at 20 °C), so it should only be used in the **HOOD**.

CHEMICAL DATA

Compound	FW	Amount	mmol	mp (°C)	bp (°C)	Density
Salicylaldehyde	122.12	500 mg	4.09	1	197	1.146
Ethylenediamine	60.10	120 mg	2.00	8.5	118	0.899

Required Equipment

Magnetic stirring hot plate, sand bath, 10-mL Erlenmeyer flask, magnetic stirring bar, ice–water bath, Hirsch funnel, clay tile.

Time Required for Experiment: 0.5 h.

EXPERIMENTAL PROCEDURE

Place 5.0 mL of 95% ethanol in a 10-mL Erlenmeyer flask equipped with a magnetic stirring bar. Put the flask in a sand bath, set atop a magnetic stirring hot plate, and heat the ethanol to boiling. Immediately, with stirring, add 450 μL (4.09 mmol) of salicylaldehyde. To the boiling solution, add 140 μL (2.0 mmol) of ethylenediamine.

Isolation of Product

Stir the solution for 3–4 minutes, and cool the solution in an ice–water bath. Collect the yellow crystals by suction filtration using a Hirsch funnel. Wash the product with 2 drops of ethanol and dry on a clay tile. Determine the melting point and calculate a percentage yield.

Part B: Preparation of Co(salen)

Additional Safety Recommendations

Cobalt(II) acetate tetrahydrate (CAS No. 6147-53-1): This compound is harmful if inhaled or swallowed. ORL-RAT LD50: 708 mg/kg.

CHEMICAL DATA

Compound	FW	Amount	mmol	mp (°C)	Density
salenH_2	268.32	230 mg	0.56		
Cobalt(II) acetate tetrahydrate	249.09	200 mg	0.80		1.705

Required Equipment

Magnetic stirring hot plate, magnetic stirring bar, water bath, 25-mL two-necked flask, pressure equalizing addition funnel, water condenser, T joint, nitrogen source, oil bubbler, Hirsch funnel, vacuum desiccator.

Time Required for Experiment: 2 h.

EXPERIMENTAL PROCEDURE

Place 230 mg (0.86 mmol) of salenH_2 into a 25-mL two-necked flask equipped with a magnetic stirring bar. Attach a pressure equalizing addition funnel, a water condenser fitted with a T joint for N_2 flow, and an oil bubbler as shown in Figure 8.10. Set the apparatus in a water bath on top of a magnetic stirring hot plate. Add 12 mL of 95% ethanol through the funnel. Stir the solution and flush the apparatus with N_2 gas. Maintain a steady flow of N_2 (1 bubble/s). Heat the water bath to 70–80 °C.

Prepare a solution of 200 mg (0.80 mmol) cobalt acetate tetrahydrate, in 1.5 mL of water in a 10-mL beaker. Add this solution dropwise through the addition funnel to the salenH_2 solution in the flask. At first, a brown gelatinous precipitate will form. After 1 h of heating and stirring, this precipitate becomes bright red.

Isolation of Product

Discontinue the heat and N_2 flow and immerse the flask in cold water. Filter the product crystals by suction filtration using a Hirsch funnel. Wash the residue with a few drops of water, followed by several drops of 95% ethanol. Dry the product in a vacuum desiccator. Determine the percentage yield and the decomposition point for the complex.

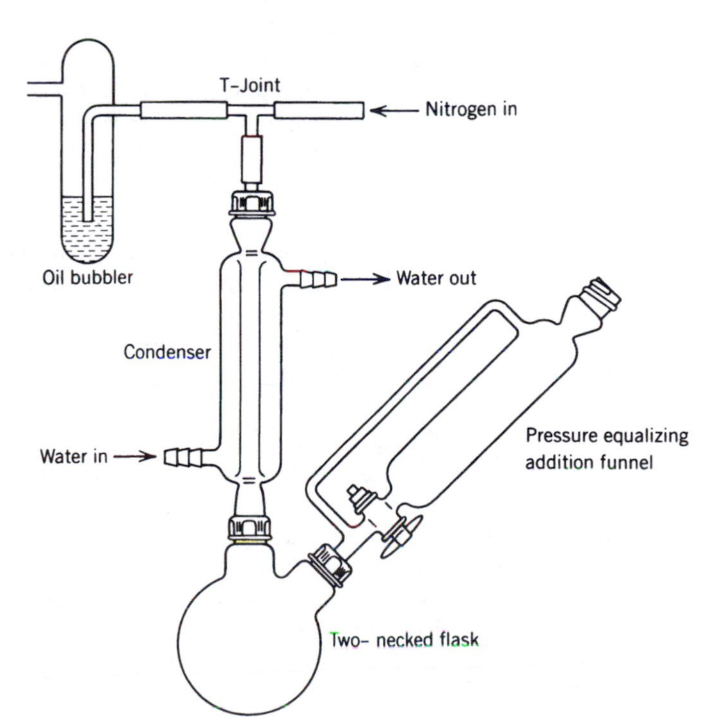

Figure 8.10. *Apparatus for Experiment 30.*

Part C: Determination of Oxygen Absorption by Co(salen)

Additional Safety Recommendations

Dimethyl sulfoxide (CAS No. 67-68-5): This compound is harmful if swallowed, inhaled, or absorbed through the skin. ORL-RAT LD50: 14,500 mg/kg.

Required Equipment

Graduated cylinder, 2- × 18-cm side arm test tube with rubber stopper, 1- × 7.5-cm test tube, forceps, two 10-mL burets, Tygon tubing.

Time Required for Experiment: 1.5 h.

EXPERIMENTAL PROCEDURE

Place 50–100 mg of Co(salen) (from Part 30.B) at the bottom of a 2- × 18-cm side arm test tube. Using a graduated cylinder, transfer 5 mL of DMSO to a 25-mL beaker, and saturate the complexing solvent by bubbling O_2 gas through it. Fill the 1- × 7.5-cm test tube with oxygen-saturated DMSO to a level 2 cm from the rim. Using a pair of forceps, carefully lower the smaller test tube into the side arm test tube. Make sure that no DMSO spills out from the test tube. In case DMSO spills out, dry the side arm test tube, and try again.

Using Tygon tubing, connect the bottom of a 10-mL buret (henceforth called a measuring buret) to the bottom of a second 10-mL buret (henceforth called a movable tube), which will act as a water reservoir. Secure the arrangement with clamps to a strong ring stand. Fill the movable tube with water, and allow water to drain so that the Tygon tubing is filled with water, and water begins to enter the measuring buret. Attach the top of the measuring buret to the side arm of the larger test tube. Adjust the height of the movable tube so that the water level remains near the bottom of the measuring buret. This apparatus is shown in Figure 8.11.

Flush the side arm test tube through its mouth for a few minutes with oxygen gas. Tightly seal the mouth of the side arm test tube with a rubber stopper. Adjust the height of the movable tube so that the water in the measuring buret and in the water reservoir tube are at the same levels. Note the initial water level from the measuring buret. Carefully invert the side arm tube in such a way that DMSO pours out from the smaller test tube, allowing it to react with the Co(salen) at the bottom of the side arm test tube. **Make certain that DMSO does not enter into the Tygon tubing!** Gently shake the side arm test tube so as to dissolve the Co(salen) in the DMSO. As oxygen is absorbed by this solution, the water level in the measuring buret will start rising. Continue shaking the side arm tube until the water level no longer changes. The process should not require more than 20 min. Readjust the height of the movable tube so that the water levels in both the tubes are the same. Note the final water level in the measuring buret.

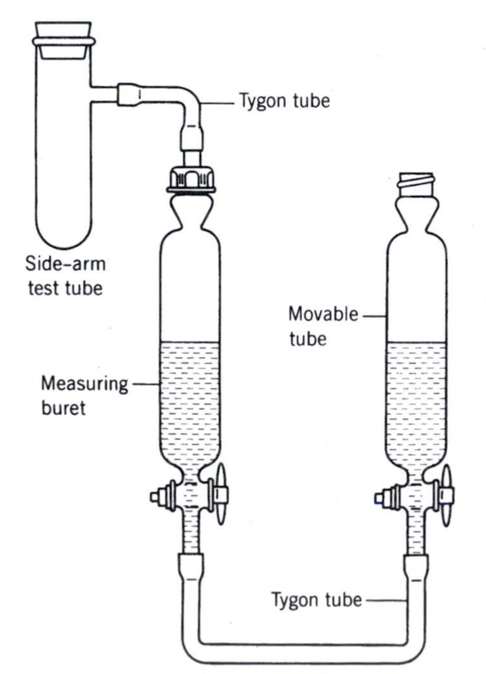

Figure 8.11. *Movable tube apparatus.*

Calculations

Calculate the decrease in the volume of water caused by the absorption of oxygen gas by the Co(salen) in DMSO. This is equivalent to the volume of oxygen uptake by Co(salen). Note the atmospheric pressure from a barometer. Record the room temperature. Using the vapor pressure of water at that temperature, calculate the number of moles of O_2 gas absorbed using the ideal gas law equation.

$$n = (P - f)V/RT$$

where f is the vapor pressure of water at room temperature.

Part D: Reaction of the Oxygen Adduct with Chloroform

Time Required for Experiment: 0.25 h.

EXPERIMENTAL PROCEDURE

Transfer the dark brown solid from Part 30.C into a centrifuge tube. Centrifuge the mixture and remove the supernatant liquid. Layer the residue with a few drops of chloroform and observe what happens. An evolution of oxygen gas should be seen as the complex deoxygenates.

QUESTIONS

1. In terms of the chemistry of Fe(II) describe the triggering mechanism in hemoglobin.
2. Why is CO so deadly to human respiration? (*Hint:* Consider the bonding to the iron in hemoglobin.)
3. There is a reasonable correlation between the O—O bond length in metal (dioxygen) complexes and the reversibility of the addition reaction. Discuss.
4. Cobalt is also found in other molecules of biological interest. Give an example and discuss its biological properties.
5. Certain porphyrin complexes of cobalt(II) can be substituted into globin to form cobalt analogs to hemoglobin. Discuss the similarities and differences between these complexes. A useful initial literature reference is Hoffman, B. M., Petering, D. H. *Proc. Natl. Acad Sci. USA* **1970,** *67,* 637.

REFERENCES

1. Appleton, T. G. *J. Chem. Educ.* **1977,** *54,* 443.
2. Klevan, J.; Peone, J.; Madan, S. K. *J. Chem. Educ.* **1973,** *50,* 670.
3. Basolo, F.; Hoffman, B. M.; Ibers, J. A. *Acct. Chem. Res.* **1975,** *8,* 384.

Experiment 31 Preparation of Dichloro-1,3-bis(diphenylphosphino)propanenickel(II)

INTRODUCTION

The nickel family of elements (nickel, palladium, and platinum) sees extensive use as catalysts for organic reactions. The compound being synthesized in this laboratory experiment, $Ni(dppp)Cl_2$, is used as a catalyst in carrying out Grignard coupling reactions to give optically pure products.

$$(C_6H_5\text{—Te})(H)C{=}C(H)(C_6H_5) + C_6H_5MgBr \xrightarrow[\text{THF}]{Ni(dppp)Cl_2} (C_6H_5)(H)C{=}C(H)(C_6H_5)$$

dppp = diphenylphosphinopropane

100% cis

It can also be used to catalyze cross coupling of Grignard reagents with aryl and alkenyl halides.[1]

The Ni^{2+} ion has a d^8 electron configuration. As Ni^{2+} is rather small, relatively few octahedral complexes of this metal are known. Many complexes of d^8 configuration are square planar, many are tetrahedral, and some are intermediate between these two geometries. The Ni(dppp)Cl_2 complex, shown below, exhibits an equilibrium between these forms in solution.[2]

$$\begin{array}{c} Cl \quad\quad Cl \\ C_6H_5 \quad \diagdown \; Ni \; \diagup \quad C_6H_5 \\ C_6H_5—P \quad\quad\quad P—C_6H_5 \\ H_2C \quad\quad CH_2 \\ CH_2 \end{array}$$

The complex is prepared by the reaction of 1,3-bis(diphenylphosphino)-propane, a bidentate ligand, with nickel(II) chloride hexahydrate in a mixed solvent of 2-propranol and methanol. The phosphorus atoms in the 1,3-bis(diphenylphosphino)propane have lone pairs available for donation and are therefore Lewis bases. The Ni^{2+} functions as a Lewis acid. The overall reaction is therefore simply a Lewis acid–base reaction.

$$NiCl_2{\cdot}6H_2O + (C_6H_5)_2P(CH_2)_3P(C_6H_5)_2 \rightarrow Ni(dppp)Cl_2 + 6H_2O$$

An example of an organic reaction using this product may be found in Ref. 1.

Prior Reading and Techniques

Section 5.D.3: Isolation of Crystalline Products (Suction Filtration)

Section 6.C: Infrared Spectroscopy

Related Experiments

Use of Inorganic Products in Organic Reactions: Experiments 32, 34, and 35

EXPERIMENTAL SECTION

Safety Recommendations

Nickel(II) chloride hexahydrate (CAS No. 7791-29-0): This compound is harmful if swallowed, inhaled, or absorbed through the skin. ORL-RAT LD50: 175 mg/kg. It is classified as a carcinogen.

1,3-Bis(diphenylphosphino)propane (CAS No. 6737-42-4): Toxicity data for this compound is not available. It would be prudent to observe the usual precautions (Section 1.A.3).

CHEMICAL DATA

Compound	FW	Amount	mmol	mp (°C)	Density
$NiCl_2{\cdot}6H_2O$	237.71	95 mg	0.40		
1,3-Bis(diphenylphosphino)-propane	412.46	45 mg	0.11	58–60	

Required Equipment

Magnetic stirring hot plate, 10-mL Erlenmeyer flask, side armed drying tube, 1-dram vial, magnetic stirring bar, Pasteur pipet, Hirsch funnel, heat lamp.

Time Required for Experiment: 2 h.

EXPERIMENTAL PROCEDURE

NOTE: *If it is desired to obtain solution magnetic moments or ^{1}H NMR spectra, nickel(II) bromide hydrate should be used in place of the nickel(II) chloride hexahydrate because of its greater solubility.*

Place 95 mg (0.4 mmol) of nickel(II) chloride hexahydrate and 2 mL of a 5:2 (v/v) mixture of 2-propanol and methanol in a 10-mL Erlenmeyer flask containing a magnetic stirring bar. With stirring, warm the mixture on a hot plate until just below the boiling point.

To this warm solution, add dropwise (Pasteur pipet) a hot solution of 1,3-bis(diphenylphosphino)propane (145 mg, 0.35 mmol) dissolved in 2.0 mL of 2-propanol. Heat the resulting mixture, which contains a flaky, red-brown precipitate, to near boiling while stirring for approximately $\frac{1}{2}$ h. Allow the solution to cool to room temperature.

Isolation of Product

Collect the red colored product by suction filtration using a Hirsch funnel. Wash the crystals with three $\frac{1}{2}$-mL portions of cold methanol and place them in a tared 1.0-dram vial, which in turn is placed in a side armed drying test tube. Dry the product under vacuum using a heat lamp. Weigh the dry product and determine the percentage yield.

Characterization of Product

Determine the melting point using a Fisher–Johns apparatus. Obtain an IR spectrum of the product and compare it to 1,3-bis(diphenylphosphino)propane.

Determine the magnetic moment of the solid product. Is the product square planar or tetrahedral? Dissolve the product in a minimum amount of methylene chloride and obtain the Visible spectrum. The square planar complex absorbs at approximately 500 nm, while the tetrahedral complex absorbs at 825 nm. Which complex is predominant in solution? If the bromo complex was prepared, determine the magnetic moment of the product in solution.

If the bromo complex was prepared, obtain the ^{1}H NMR spectrum from -5 to $+20$ ppm. This wide chemical shift range is necessary because of the paramagnetic moment of the nickel. The chemical shifts also show a marked temperature dependency.[2] The chloro complex could also be used for this purpose, but its solubility is low, so an FT NMR will be needed to obtain a high-resolution spectrum.

QUESTIONS

1. Similar complexes with Pd or Pt as the metal are always square planar and do not exhibit the square planar–tetrahedral equilibrium that the nickel does. Explain why.
2. A slight excess of the nickel(II) chloride hexahydrate is used in this reaction. Suggest a possible reason for this.
3. Draw the structure of the product you would obtain by the reaction of $NiCl_2 \cdot 6H_2O$ with 1,3-bis(dimethylphosphino)ethane. Name this compound.
4. Some nickel complexes are octahedral, some tetrahedral, and others square planar. Many can easily convert from one to another, as seen in this experiment. Using the literature, explain why all three geometries are stable for nickel. Give examples of each type.

REFERENCES

1. Kumada, M.; Tamao, K.; Sumitani, K. *Org. Syn.* **1978,** *58,* 127; also found in *Organic Syntheses,* Col. Vol. VI., Wiley: New York, 1988, p. 407.
2. Van Hecke, G. R.; Horrocks, W. D., Jr. *Inorg. Chem.* **1966,** *5,* 1968.

GENERAL REFERENCES

McAuliffe, C. A., "Phosphorus, Arsenic, Antimony and Bismuth Ligands" in *Comprehensive Coordination Chemistry,* G. Wilkinson, Ed., Pergamon: Oxford, 1987, Vol. 2, Chapter 14, p. 990.

LeVason, W.; McAuliffe, C. A., "Transition Metal Complexes Containing Bidentate Phosphine Ligands" in *Advances in Inorganic Chemistry and Radiochemistry,* H. J. Emeleus and A. G. Sharpe, Eds., Academic Press: New York, 1972, Vol. 14, p. 220.

Experiment 32 Preparation of Iron(II) Chloride (Use of $FeCl_3$ as a Friedel–Crafts Chlorination Source)

INTRODUCTION

Iron(II) chloride occurs naturally in the mineral Lawrencite. It can be obtained as white, very hygroscopic crystals by sublimation at 700 °C in a stream of hydrogen chloride. The compound finds uses as a reducing agent, a mordant in the dyeing industry, in pharmaceutical preparations, and in the metallurgical field.

The present experiment illustrates the preparation of iron(II) chloride using a unique approach. Iron(III) chloride is used as a chlorinating agent for an aromatic compound, and in turn is reduced to the iron(II) state.[1,2] While the use of $FeCl_3$ as a catalyst in Friedel–Crafts organic reactions is well known, $FeCl_3$ can act as the halogenating agent on its own ($FeCl_3$ in this case, of course, is no longer catalytic).

$$C_6H_5Br + 2FeCl_3 \rightarrow 2FeCl_2 + C_6H_4BrCl + HCl$$

The inorganic product, solid iron(II) chloride, is easily isolated from the reaction mixture in yields usually over 95%. It is difficult to prepare pure $FeCl_2$ by other methods. The other products of the reaction are the respective chlorinated aromatic compounds. This organic mixture can be isolated and the individual species separated and identified using HPLC.

Prior Reading and Techniques

Section 5.D.3: Isolation of Crystalline Products (Suction Filtration)

Section 5.D.9: Drying Techniques

Section 5.G.4: Liquid Chromatography: High-Performance Liquid Chromatography

Related Experiments

Use of Inorganic Products in Organic Reactions: Experiments 31, 34, and 35

Iron Chemistry: Experiments 40 and 45

EXPERIMENTAL SECTION

Safety Recommendations

Ferric Chloride (Anhydrous) (CAS No. 7705-08-0): This compound is harmful if inhaled, swallowed, or absorbed through the skin. It is extremely destructive to the tissue of the mucous membranes. ORL-RAT LD50: 1872 mg/kg.

Bromobenzene (CAS No. 108-86-1): This material is damaging to the mucous membranes and upper respiratory tract. ORL-RAT LD50: 2699 mg/kg.

CHEMICAL DATA

Compound	FW	Amount	mmol	mp (°C)	bp (°C)	Density
$FeCl_3$	162.21	162 mg	1.0	306		2.898
Bromobenzene	157.02	314 mg	2.0	−31	156	1.491

Required Equipment

Magnetic stirring hot plate, 3- and 5-mL conical vials, magnetic spin vane, 10-mL graduated cylinder, water condenser, sand bath, Pasteur pipet, Hirsch funnel, Pasteur pipet, silica gel column.

Time Required for Experiment: 3 h.

EXPERIMENTAL PROCEDURE

Place 162 mg (1 mmol) of anhydrous ferric chloride followed by 210 μL (314 mg, 2 mmol) of bromobenzene in a 3-mL conical vial containing a magnetic spin vane.

> **NOTE:** ***Weigh the $FeCl_3$ as quickly as possible and cap the vial immediately. This compound is very hygroscopic. An alternative is to use a glovebag that was flushed with nitrogen. The bromobenzene should be freshly distilled.***

Attach the vial to a water condenser and place the assembly in a sand bath or on an aluminum heating block on a magnetic stirring hot plate that is at 125 °C. Stir the mixture for 30 min while allowing the temperature to rise to 140 °C during this period.

Isolation of Product

Remove the resulting dark mixture from the sand bath and allow it to cool to room temperature. Add 2 mL (Pasteur pipet) of methylene chloride through the top of the condenser. Remove the vial and stir the mixture with a microspatula after removing the spin vane with forceps. Collect the solid by suction filtration using a Hirsch funnel. **Save the filtrate.** The tan-mustard colored $FeCl_2$ is washed with three 2-mL portions of additional methylene chloride. Rinse the reaction vial with the first portion of these washings. Combine all washings with the original filtrate.

Dry the solid product of iron(II) chloride under reduced pressure (16 mm) for 10 min at ~100 °C. Weigh this material and calculate the percentage yield.

Purification of Organic Product

Transfer the filtrate (in portions) using a Pasteur pipet to a Pasteur filter pipet chromatography column containing 500 mg (~1 in.) of silica gel followed by 250 mg (~$\frac{1}{2}$ in.) of powdered charcoal. Collect the eluate in a tared 5-mL conical vial containing a boiling stone. Wash the column with an additional 5 mL of methylene chloride. To assist the flow of eluent, a small rubber bulb may be used to apply a slight pressure to the column. This utilizes the essential aspects of flash chromatography.[3]

Concentrate the nearly colorless eluate on a warm sand bath **(HOOD)** under a gentle stream of N_2 gas. Weigh the liquid residue and carry out an HPLC analysis to determine the percentage composition of the mixture of chlorinated bromobenzenes.

High-Performance Liquid Chromatographic Analysis

Refer to the introduction to HPLC in Section 5.G.4. In the present analysis, add 2 drops of the liquid residue (Pasteur pipet) to 300 μL of methanol in a small vial. A sample of 2–3 μL of this solution is then injected into the HPLC and the analysis carried out under the following conditions.

The instrument used in the example below is a Beckman, Model 110A HPLC equipped with an octadecylsilane C_{18} column. The solvent used is a methanol–water (3:1) mixture with a flow rate of 0.9/0.3, respectively, which is 1.2 mL·min^{-1}.

Average retention times for the expected components are summarized in Table 8.1. The percentage composition of the mixture is determined by calculation of the area under the curves.

Table 8.1 Summary of Data as an Average of Six Runs

Compound	Retention Time (min)	Composition (%)
Bromobenzene	5.8	64.0
o-Chlorobromobenzene	7.5	4.1
p-Chlorobromobenzene	8.3	28.6
m-Chlorobromobenzene	9.7	2.7

QUESTIONS

1. A solution of iron(II) can be oxidized to iron(III) in acid solution by titration with dichromate ion. Under these conditions, the dichromate ion is reduced to chromic ion. If a 2.5-g sample of iron ore is dissolved in acid, 19.17 mL of 0.1*M* sodium dichromate solution is required to convert the iron(II) to iron(III). What percentage of iron is in the original ore?
2. Iron(II) can be distinguished from iron(III) using a thiocyanate solution. Iron(III) gives a blood red color but iron(II) gives a colorless solution. It is proposed that the structure of the species that produces the red color is $[Fe(H_2O)_5NCS]^{2+}$. Can you propose a structure for this species? Do you think that this color test could be used quantitatively to determine the amount of iron(III) in solution? If so, how would you do the measurements?
3. Iron(II) in aqueous solution can be oxidized, in the presence of molecular oxygen, to iron(III). In what ways is this reaction pH dependent? Write the balanced redox reactions in acidic and basic media.
4. From the literature, detail the utility of the Friedel–Crafts alkylation and acylation reactions. Give an example of each type of reaction, including the mechanistic sequence. What is the function of the $FeCl_3$? What other species can be used? What do they all have in common?

REFERENCES

1. Kovacic, P.; Brace, N. O. *J. Am. Chem. Soc.* **1954,** *76,* 5491.
2. Kovacic, P.; Brace, N. O. *Inorg. Syn.* **1960,** *6,* 172.
3. For example see:

a. Thompson, W. J.; Hanson, B. A. *J. Chem. Educ.* **1984,** *61*, 645.
b. Still, W. C.; Kahn, M.; Mitra, A. *J. Org. Chem.* **1978,** *43*, 2423.
c. Bell, L. W.; Edmondson, R. D. *J. Chem. Educ.* **1986,** *63*, 361.
Flash chromatographic equipment is commercially available from the Aldrich Chemical Co.

Experiment 33 Reaction of Cr(III) with a Multidentate Ligand: A Kinetics Experiment

INTRODUCTION

A detailed study of a chemical reaction depends on two important aspects of the reaction process: thermodynamics and kinetics. Thermodynamics is the study of changes in energy for a system, and determines the direction and spontaneity of a chemical change. While thermodynamics tells us *whether* a reaction will proceed as written, it does not tell us how quickly this will occur. Chemical kinetics is the study of the rates of chemical reactions. Such studies also help in understanding the reaction mechanisms of chemical reactions.

In any given reaction, the reactants, on mixing, begin forming the products. The products may also react to reform the reactants; such reactions are called equilibrium reactions. These two processes, the forward and reverse reactions, ultimately establish the equilbrium. This situation is represented by the general reaction

$$aA + bB \underset{\text{reverse}}{\overset{\text{forward}}{\rightleftharpoons}} cC + dD$$

The net reaction rate depends on how quickly the forward reaction takes place in comparison to the rate of the reverse reaction. Thus,

$$\text{net reaction rate} = \text{forward reaction rate} - \text{reverse reaction rate}$$

At equilibrium, the forward and reverse rates are equal, and thus, the net reaction rate is zero. When describing chemical reactions, the reaction rate is defined as the change in the concentration of any reagent per unit time, usually having units of moles per liter per minute ($\text{mol}\cdot\text{L}^{-1}\ \text{min}^{-1}$) or moles per liter per second ($\text{mol}\cdot\text{L}^{-1}\ \text{s}^{-1}$), and is expressed by the derivative $d[A]/dt$. A negative sign preceding the derivative indicates that the concentration of the reagent is decreasing. Rates may also be expressed in terms of the increase in concentration of a product, with a positive derivative term. Algebraic relationships may be established between the concentrations of A, B, C, and D and, hence, the reaction rates of all the species involved.

$$\frac{\text{mol A reacted}}{a} = \frac{\text{mol B reacted}}{b} = \frac{\text{mol C reacted}}{c} = \frac{\text{mol D reacted}}{d}$$

And also

$$-\left(\frac{1}{a}\right)\left(\frac{d[A]}{dt}\right) = -\left(\frac{1}{b}\right)\left(\frac{d[B]}{dt}\right) = \left(\frac{1}{c}\right)\left(\frac{d[C]}{dt}\right) = \left(\frac{1}{d}\right)\left(\frac{d[D]}{dt}\right)$$

As an example, consider the decomposition of hydrogen peroxide into oxygen and water.

$$2H_2O_2\ (\ell) \rightarrow O_2\ (g) + 2H_2O\ (g)$$

The stoichiometry indicates that 2 mol of H_2O_2 are consumed for every mole of oxygen formed. Thus, the concentration of O_2 formed is one half the concentration of H_2O_2 decomposed.

$$-\frac{d[H_2O_2]}{dt} = \frac{d[H_2O]}{dt} = 2\frac{d[O_2]}{dt}$$

The reaction rate for H_2O_2 decomposition is given by any of the above expressions.

The rate of any reaction may also depend on variables not explicitly given in the overall reaction formula, such as $[H^+]$, temperature, and the solvent. One of the goals of kinetics measurements is to determine the dependence of the rate of a reaction on the concentration of the reactants, explicit or otherwise. This relation is called the **rate law** and for the general reaction

$$aA + bB \rightarrow \text{products (P)}$$

the rate law can be written as follows:

$$\text{rate} = -(1/a)d[A]/dt = -(1/b)d[B]/dt = d[P]/dt = k[A]^m[B]^n$$

and

$$k = \text{rate}/[A]^m[B]^n$$

where k is the rate constant. The units of k are $(\text{mol}\cdot L^{-1})^{(1-m-n)}\ s^{-1}$. For a given reaction, the value of the rate constant k depends on temperature, not on the concentration of the reacting species.

In this expression, the exponents m and n are called the **reaction orders**.

NOTE: *The reaction orders m and n bear no relationship whatsoever to the overall reaction coefficients a and b.*

A reaction is said to have an order of m with respect to reagent A or of n with respect to reagent B. The sum, $m + n$, is the **overall order** of the reaction. The individual or overall orders of a reaction cannot be calculated from the reaction conditions: They must be determined experimentally.

Integrated Rate Laws: First Order

The rate laws above were expressed as differential quantities. An alternate way of writing them is as integrated rate laws. The differential equation for a first-order reaction is

$$-d[A]/dt = k[A]$$

Dividing both sides by [A] and multiplying by $-dt$,

$$d[A]/[A] = -k\,dt$$

At initial time t_0, the concentration of A would be $[A]_0$. At some later time t, the concentration of A would be $[A]_t$. Integrating,

$$\ln[A]_t/[A]_0 = -kt$$

or, equivalently

$$\ln[A]_t = -kt + \ln[A]_0 \qquad (33.1)$$

Solving for the concentration of A at any time t,

$$[A]_t = [A]_0 e^{-kt}$$

Note that Eq. 33.1 is the equation of a straight line. A plot of ln[A] (y axis) versus time (x axis) will therefore yield a straight line of slope $-k$ for a first-order reaction. The **half-life**, $t_{1/2}$, is defined as the time required for the concentration of A to decrease to one half its initial value, that is, from $[A]_0$ at time 0 to 0.5 $[A]_0$ at time $t_{1/2}$. Thus,

$$t_{1/2} = 0.693/k$$

The half-life of a first-order reaction is proportional to the rate constant and is independent of the concentration of the reactant.

Integrated Rate Laws: Second Order

A reaction is second order if the rate is proportional to the square of the reagent concentration, or to the product of two reagent concentrations.

$$\text{rate} = -d[A]/dt = k[A]^2$$

or

$$\text{rate} = k[A][B]$$

The first form is much easier to integrate. Dividing both sides by $[A]^2$ and multiplying by dt,

$$-d[A]/[A]^2 = k\,dt$$

Integrating, between time $t = 0$ and t,

$$1/[A]_t - 1/[A]_0 = kt$$

Rearranging,

$$1/[A] = kt + 1/[A]_0$$

If the inverse of [A] (y axis) is plotted against time (x axis), a straight line of slope k will be obtained. The other second-order form is not needed for this experiment.

Complex Reactions

Most reactions proceed via a fairly complex sequence of **elementary steps**. The order of reaction for any given elementary step is simply the coefficient of the reactant. Consider the following common reaction sequence:

$$A = M + C \qquad (33.2)$$

$$M + B \rightarrow D \qquad (33.3)$$

Overall reaction

$$A + B \rightarrow C + D$$

In the first step, A reversibly decomposes forming intermediate M and product C. The intermediate then reacts with a second reagent, B, forming product D. Let k_1 be the forward rate constant for Reaction 33.2, k_{-1} be the reverse rate constant, and k_2 be the rate constant for Reaction 33.3.

It is customary not to write rate laws in terms of intermediates. We must therefore solve for [M] and eliminate it from any rate law. From the previous reaction steps, we see that [M] increases when [A] increases, and [M] decreases when [B] and [C] increase. Thus,

$$d[\mathrm{M}]/dt = k_1[\mathrm{A}] - k_{-1}[\mathrm{M}][\mathrm{C}] - k_2[\mathrm{M}][\mathrm{B}]$$

When equilibrium is reached, the concentration of M does not change, that is, $d[\mathrm{M}]/dt = 0$. Thus, the above equation rearranges to

$$[\mathrm{M}] = \frac{k_1[\mathrm{A}]}{k_{-1}[\mathrm{C}] + k_2[\mathrm{B}]}$$

We also see from the elementary steps that

$$-d[\mathrm{A}]/dt = -d[\mathrm{B}]/dt = k_2[\mathrm{M}][\mathrm{B}]$$

Substituting for [M],

$$\text{rate} = -\frac{d[\mathrm{A}]}{dt} = -\frac{d[\mathrm{B}]}{dt} = \frac{k_1k_2[\mathrm{A}][\mathrm{B}]}{k_{-1}[\mathrm{C}] + k_2[\mathrm{B}]}$$

This rather formidable looking rate law can be drastically simplified depending on the nature of the reaction and how the reaction is carried out.

Case 1: $k_{-1}[\mathrm{C}] \gg k_2[\mathrm{B}]$

If the equilibrium step in Reaction 33.2 is fast (k_1 and k_{-1} large), and Reaction 33.3 is slow (k_2 small—Reaction 33.3 would be the **rate-determining step**), $k_{-1}[\mathrm{C}]$ will be much larger than $k_2[\mathrm{B}]$. The rate law then simplifies to

$$\text{rate} = \frac{k_1k_2[\mathrm{A}][\mathrm{B}]}{k_{-1}[\mathrm{C}]}$$

A reaction following this rate law is

$$\mathrm{I^-} + \mathrm{OCl^-} \rightarrow \mathrm{OI^-} + \mathrm{Cl^-}$$

The experimental rate law for this reaction is

$$\text{rate} = k\frac{[\mathrm{I^-}][\mathrm{OCl^-}]}{[\mathrm{OH^-}]}$$

The denominator of the rate law corresponds to C in the general mechanism, which was produced in the equilibrium Reaction 33.2. The OH^- must be forming from water, which must be reacting either with OCl^- or I^-. Since I^- is far too weak a base to undergo hydrolysis, the reaction must be

$$H_2O + OCl^- = HOCl + OH^-$$

The intermediate M, in this case, is HOCl. Reaction 33.3 must be between the intermediate and the other reactant, I^-, and form the products, OI^- and Cl^-.

$$HOCl + I^- \rightarrow OI^- + Cl^- + H^+$$

The overall reaction is obtained by adding the two elementary steps

$$HOCl + H_2O + OCl^- + I^- \rightarrow HOCl + OH^- + H^+ + OI^- + Cl^-$$

Canceling out terms found on both sides, and recognizing that H^+ and OH^- will immediately form water,

$$OCl^- + I^- \rightarrow OI^- + Cl^-$$

which is the overall reaction that was given. Note that although OH^- does not appear in the overall reaction, it does appear in one of the elementary steps, as does H^+. This reaction will therefore have a pH dependency.

Case 2: $k_{-1}[C] \ll k_2[B]$

Suppose that the equilibrium step of the general mechanism is slow and the second step is fast. In this case, $k_{-1}[C] \ll k_2[B]$. An alternate way of having this condition apply is to flood the reaction with reagent B, thereby increasing its concentration. This is the condition that applies in this experiment. Thus,

$$\text{rate} = \frac{k_1k_2[A][B]}{k_{-1}[C] + k_2[B]} = \frac{k_1k_2[A][B]}{k_2[B]} = k_1[A]$$

The reaction therefore appears to be only first order in A and is thus called a **pseudo-first-order** reaction.

Case 3: $k_{-1}[C] \approx k_2[B]$

The most difficult case, kinetically, is when both elementary steps occur at similar rates. An example of this is in the complex reaction

$$[Co(CN)_5(H_2O)]^{2-} + I^- \rightarrow [Co(CN)_5I]^{3-} + H_2O$$

where the observed rate equation is

$$\text{rate} = \frac{k'[Co(CN)_5(H_2O)]^{2-}[I^-]}{k'' + k'''[I^-]}$$

Note that there is no concentration term with k''. This indicates that the $k_{-1}[C]$ term is approximately constant and thus equal to k''. Only the solvent concentration does not change much in a reaction, so C is probably water. A reasonable mechanism is therefore

$$[Co(CN)_5(H_2O)]^{2-} = [Co(CN)_5]^{2-} + H_2O$$

$$[Co(CN)_5]^{2-} + I^- \rightarrow [Co(CN)_5I]^{3-}$$

The rate equation for this sequence would be

$$\text{rate} = \frac{k_1 k_2[\text{Co(CN)}_5(\text{H}_2\text{O})][\text{I}^-]}{k_{-1}[\text{H}_2\text{O}] + k_2[\text{I}^-]}$$

Since the $[H_2O]$ is essentially constant, the above rate law is the same as that experimentally observed.

Experimental Determination of Rate Laws and Reaction Orders

Any measurable property, which represents the progress of a reaction, may be used to monitor the rate of that reaction with time. Thus, the angle of optical rotation, the absorbance at a fixed wavelength, the volume of the liquid phase, the peak area in an NMR spectrum, the conductivity of a solution, and the pressure of a gas are just a few of the physical properties that may be followed as a function of time. Letting P represent such a property, then P_0 represents $[A]_0$ and P_t represents $[A]_t$. We can then use the equation:

$$\frac{P_t - P_\infty}{P_0 - P_\infty} = e^{-kt}$$

Thus, a plot of either $\log(P_t - P_\infty)$ or of $\log(P_\infty - P_0)$ (y axis) versus time (x axis) gives a straight line of slope $-k$ for a first-order reaction. The equation for second order is

$$P_t = P_0 - k[A]_0 t(P_t - P_\infty)$$

There are three common methods for determining the order of a reaction:

1. **Integrated Rate Laws.** As discussed above, if the reaction is first order, a plot of log[A] versus t yields a straight line of slope $-k$. In the case of a second-order reaction, a plot of 1/[A] versus t will provide a linear graph of slope k. In both cases, it is important to collect data out to at least 75% of the time needed for the reaction to reach completion.
2. **Fractional Lifetime (Half-Life) Method.** The fractional time method is based upon the concept that the time required to use up a given fraction of the limiting reagent is characteristic of the rate equation. The use of half-lives is generally recommended. A comparison of successive half-lives will reveal the order of reaction.

   The temperature dependence of the rate constant k is given by the Arrhenius equation

   $$\ln k = \ln A - E_a/RT$$

   where A is the preexponential or frequency factor and E_a is the activation energy. As the equation implies, reaction rates increase with temperature. A plot of ln k (y axis) versus $1/T$ gives a straight line of slope $-E_a/R$.
3. **Method of Initial Rates.** The method of initial rates[1] is often used in conjunction with the isolation method, where the concentrations of all reagents except one are in large excess, and thus essentially constant. In this method, the rate is measured at the beginning of the reaction for various initial concentrations of the variable reactant. The rate law in this case is

   $$(\text{rate})_0 = k[A]_0^m$$

Taking the logarithm of both sides,

$$\log(\text{rate})_0 = \log k + a \log[\text{A}]_0$$

Thus, a plot of $\log(\text{rate})_0$ (y axis) versus $\log[A]_0$ (x axis) gives a straight line of slope a, which is the order with respect to A.

The study of reaction kinetics is a vast field. It helps us in understanding the mechanism of a reaction, the effects of a catalyst, the effect of temperature, and so on, with regard to any reaction.

The following experiment deals with a number of features relating to the kinetic analysis of a chemical reaction.

The Kinetics of the Reaction of Cr(III) with EDTA

The reaction of Cr(III) with ethylenediaminetetraacetic acid (EDTA) has been studied via spectroscopic methods.[2] When a solution of a Cr(III) salt is added to a solution of the disodium salt of EDTA (in large excess, 0.1*M*, at a pH from 4 to 5), no reaction is immediately apparent. After some time, the light green color of the solution changes to deep purple, indicating that a reaction has taken place. The purple colored species is a complex of Cr(III) with the ligand EDTA. The relatively slow reaction rate makes this an attractive example for kinetic study.

The reaction leading to complex formation involves several slow steps. The first step is first order with respect to Cr(III) and inversely proportional to $[\text{H}^+]$ in the solution. The rate of the reaction is studied at several pH values within the range 3.5–5.5, where the purple complex is the main product species. The rate expression for the reaction is thus

$$\text{rate} = -\frac{d[\text{Cr(III)}]}{dt} = \frac{d[\text{Cr(III)L}]}{dt} = k[\text{Cr(III)}]^m[\text{H}^+]^n$$

where L = EDTA and m and n are the reaction orders with respect to Cr(III) and $[\text{H}^+]$. Since a large excess is used, the rate of the complex formation does not depend on the concentration of EDTA (as discussed previously).

Prior Reading and Technique

Section 6B: Visible Spectroscopy

Related Experiments

Chromium Chemistry: Experiments 22A, 23, and 29

Safety Recommendations

Chromium(III) nitrate nonahydrate (CAS No. 7789-02-8): Chromium compounds are considered mildly toxic. ORL-RAT LD50: 3250 mg/kg. The compound is harmful if inhaled or swallowed.

Ethylenediaminetetraacetic acid, disodium salt. (CAS No. 6381-92-6): This compound may be harmful if inhaled, ingested, or absorbed through the skin. ORL-RAT LD50: 2000 mg/kg.

CHEMICAL DATA

Compound	FW	Amount	mmol	bp (°C)	mp (°C)	Density
$Cr(NO_3)_3 \cdot 9H_2O$	400.15	0.24 g	0.6		60	
EDTA, disodium salt	374.28	9.306 g	25		>300	

Required Equipment

pH meter with combination electrode, visible spectrophotometer, 4 cuvettes, volumetric flasks (250 and 50 mL), four 50-mL beakers, 250-mL beaker, four 25-mL Erlenmeyer flasks, 10-mL graduated pipet, 25-mL pipet, Pasteur pipets, magnetic stirring hot plate.

Time Required for Experiment: Two 3-h laboratory periods

EXPERIMENTAL PROCEDURE

NOTE: *All solutions should be prepared ahead of time to keep this experiment within a 3-h laboratory period.*

Prepare a 0.1*M* stock solution of EDTA by dissolving 9.306 g (25 mmol) of the disodium salt of EDTA in a 250-mL beaker, using hot water. Allow the solution to cool to room temperature. Transfer the solution *quantitatively* to a 250-mL volumetric flask, and fill to the mark using deionized water.

Prepare a 0.012*M* $Cr(NO_3)_3 \cdot 9H_2O$ solution by dissolving 0.240 g (0.6 mmol) of the solid in a 50-mL volumetric flask.

While preparing the EDTA solution, set up a pH meter and calibrate it using buffer solutions (pH = 4.0 and 10) as shown by your instructor.

The experiment is carried out at four different solution pH values. Transfer four 25-mL aliquots of the 0.1*M* disodium EDTA solution to four labeled 50-mL beakers. Measure the solution pH. The pH of 0.1*M* disodium EDTA solution is ~4.5. Using either 6*M* NaOH or 6*M* HCl, adjust the pH of each EDTA solution in such a way that two of the solutions have pH values from 3.5 to 4.5 and the other two solutions have pH values from 5.0 to 5.8.

Label four 25-mL Erlenmeyer flasks in the same manner as the beakers. Transfer a 5.0-mL sample of each of the above EDTA solutions from the beaker to the similarly labeled flask. (Only one sample should be in each flask.) Pipet a 5.0-mL aliquot of Cr(III) solution into each of the four EDTA solutions in the flasks. Record the flask label and the pH values for each solution. Note that the solutions are now 0.006*M* in Cr(III).

Characterization of Reaction

As soon as the solutions are mixed, obtain the Visible spectrum of each solution from 320 to 650 nm. The absorbance maximum occurs from 545 to 570 nm. Obtain the absorbances (A_t) of each solution at the maximum wavelength at 10-min intervals for 3 h. Record the appropriate times and absorbances. Also note the color changes of the solutions with time, as well as which solution shows the most rapid color change.

At the end of the reaction, place all the solutions in a 100 °C water bath for 10–15 min. Cool the solutions and obtain the absorbance of each solution for the final time. This absorbance (A_∞) represents the absorbance of the reaction at infinite time. The difference ($A_\infty - A_t$) is a measure of the amount of unreacted Cr(III) at any time *t*. Obtain the Visible spectrum (320–650 nm) of one of the samples.

Analysis of Data

The use of a personal computer graphing program (such as Cricket Graph: Cricket Software, 30 Valley Stream Pkwy., Malvern, PA 19355) is most convenient to analyze the data.

In order to obtain the order of the reaction with respect to chromium(III), use the tabulated data to plot each of the following (on the *y* axis) versus time (*x* axis) for each pH.

$(A_\infty - A_t)$, which would be linear if zero order.

$\mathrm{Log}(A_\infty - A_t)$, which would be linear if first order.

$1/(A_\infty - A_t)$, which would be linear if second order.

The most linear plot determines which order is correct and this order should then be used in subsequent analyses.

All four pH plots should be on a single sheet of paper and should all be linear. Once the correct order is determined, obtain the slope of all four pH plots. The slope is the apparent rate constant, k', for that value of pH. A graph of log k' (y axis) versus pH (x axis) should give a straight line of slope a, where a is the order of the reaction with respect to H^+.

QUESTIONS

1. Derive the integrated rate law for a second-order reaction of type

$$\text{rate} = k[\text{A}][\text{B}]$$

2. Chromium(III) is an especially good species to use in investigating reaction kinetics. Why is it so good?
3. With respect to Question 2, why would chromium(II) be a poor choice?
4. From the literature, locate the reference to the first preparation of the EDTA complex of chromium(III). Compare this synthesis with that outlined in Ref. 2.

REFERENCES

1. Most physical chemistry textbooks will have a discussion of this method. For example, see: Atkins, P. W., *Physical Chemistry,* 4th ed., Freeman: New York, 1990, p. 783.
2. Hamm, R. C. *J. Am. Chem. Soc.*, **1953,** *75,* 5670.

GENERAL REFERENCES

Twigg, M. V., Ed., *Mechanisms of Inorganic and Organometallic Reactions,* Plenum: New York, 1983–present. Six volumes are currently available.

Rorbacher, D. B.; Endicott, J. F., Eds., *Mechanistic Aspects of Inorganic Reactions,* ACS Symposium Series No. 198, American Chemical Society: Washington, DC, 1982.

Benson, D., *Mechanisms of Inorganic Reactions in Solution,* McGraw-Hill: London, 1968.

Experiment 34 Organometallic Compounds and Catalysis: Synthesis and Use of Wilkinson's Catalyst

Part A: Synthesis of $RhCl(PPh_3)_3$, Wilkinson's Catalyst

Part B: Substitution of the Chloro Ligand in Wilkinson's Catalyst

Part C: Reaction of Wilkinson's Catalyst with Aldehydes

Part D: Reaction of Wilkinson's Catalyst with Ethylene

Part E: Absorption of Hydrogen by Wilkinson's Catalyst

Part F: Catalytic Hydrogenation of Olefins in the Presence of Wilkinson's Catalyst

INTRODUCTION

A catalyst is defined as a substance that accelerates the rate of achieving chemical equilibrium, and which can be recovered unchanged at the end of a reaction. In general, a chemical reaction between reactants A and B forming products X and Y can be considered to be in a state of equilibrium, where the relative amounts of the reactants and products formed is determined by the value of the equilibrium constant *K*.

$$A + B \overset{K}{\rightleftharpoons} X + Y$$

where

$$K = \frac{[X][Y]}{[A][B]}$$

The progress of a reaction at equilibrium depends on its thermodynamic and kinetic characteristics. The equilibrium constant *K* is related to the standard free energy change, $\Delta G°$, by the relation

$$\Delta G° = -RT \ln K$$

When $\Delta G°$ is negative, K is >1, indicating that the formation of the products is *thermodynamically* favorable. A reaction, though thermodynamically feasible, may not readily yield products if it is kinetically a very slow process.

The catalyst increases the rate of reaction by lowering the energy of activation in the rate-determining step. An uncatalyzed reaction would have a very high activation energy barrier, which prevents most collisions between reacting molecules from leading to a reaction. For example, the activation energy E_1 for the uncatalyzed decomposition reaction of hydrogen peroxide

$$2H_2O_2\ (aq) \rightarrow 2H_2O + O_2\ (g)$$

is 76 $kJ \cdot mol^{-1}$. In the presence of a catalyst (such as iodide ion), the reaction goes faster and has an activation energy of 57 $kJ \cdot mol^{-1}$. The catalyzed reaction proceeds at 2000 times the rate of the uncatalyzed one.

Catalytic processes can be broadly divided into two categories: heterogeneous and homogeneous catalysis. The classification is based upon the solubility and the type of catalyst. A heterogeneous catalysis is a process where the catalysts and reactants are in different phases, and the reaction occurs at a phase boundary. Heterogeneous catalysts are usually solids and the reactants are liquids or gases. An effective catalyst is one that has a large active surface area per unit volume. Thus, powders and porous solids are ideal candidates for heterogeneous catalysis. Metallic catalysts usually consist of a solid support (silica, alumina, or zeolites) onto which a layer of metal or metallic compound is deposited.

Homogeneous catalysis is a process where a catalyst and reactants remain in the same phase. If the reaction is carried out in the liquid phase, then the homogeneous catalyst must be soluble in the reaction medium. A homogeneous catalyst may be as simple as an anion (iodide ion in the previous example) or as complex as an organometallic cluster or a complex biological molecule such as an enzyme.

Organometallic compounds are extensively used as catalysts (heterogeneous and homogeneous) in industrial chemistry. Some specific examples include

Hydroformylation or oxo Process: An olefin reacts with CO and H_2 in the presence of a metal carbonyl to form aldehydes.

Wacker Process: An olefin is oxidized to an aldehyde or ketone in the presence of a soluble palladium salt, $[PdCl_4]^{2-}$.

Ziegler–Natta Process: Olefins are polymerized using an organoaluminum–titanium catalyst to form stereoregular polymers.

Wilkinson Process: An olefin is hydrogenated in the presence of a soluble catalyst like Wilkinson's catalyst, $RhCl(PPh_3)_3$.

As an example of homogeneous catalysis, we now consider an industrially important process—the manufacture of acetic acid. Acetic acid can be manufactured starting from ethylene by the **Wacker process**, where the olefin is oxidized to acetaldehyde in the presence of $[PdCl_4]^{2-}$. The sequence of reactions is

$$[PdCl_4]^{2-} + C_2H_4 + H_2O \rightarrow CH_3CHO + Pd^0 + 2HCl + 2Cl^-$$

$$Pd^0 + 2CuCl_2 + 2Cl^- \rightarrow [PdCl_4]^{2-} + 2CuCl$$

$$2CuCl + \tfrac{1}{2}O_2 + 2HCl \rightarrow 2CuCl_2 + H_2O$$

The overall reaction is

$$C_2H_4 + \tfrac{1}{2}O_2 \rightarrow CH_3CHO$$

The major mechanistic features of this reaction sequence can be shown by using what is known as a catalytic cycle or a **Tolman loop**, shown in Figure 8.12.

Most catalytic reactions for the hydrogenation of double bonds in organic compounds require high hydrogen pressures and high temperatures. It was found that some organometallic compounds can catalyze such hydrogenation reactions under mild reaction conditions. Two main types of hydrogenation catalysts include

1. The type containing at least one M—H bond, for example, $RhH(CO)(PPh_3)_3$.
2. The type having no M—H bonds, for example, $RhCl(PPh_3)_3$, which react with molecular hydrogen to form hydrides in solution.

The first successful homogeneous system developed for the reduction of olefins involved the use of $RhCl(PPh_3)_3$, called Wilkinson's catalyst. In solution the catalyst dissociates to a small extent.

$$RhCl(PPh_3)_3 = RhCl(PPh_3)_2 + PPh_3$$

$$2RhCl(PPh_3)_2 = [RhCl(PPh_3)_2]_2$$

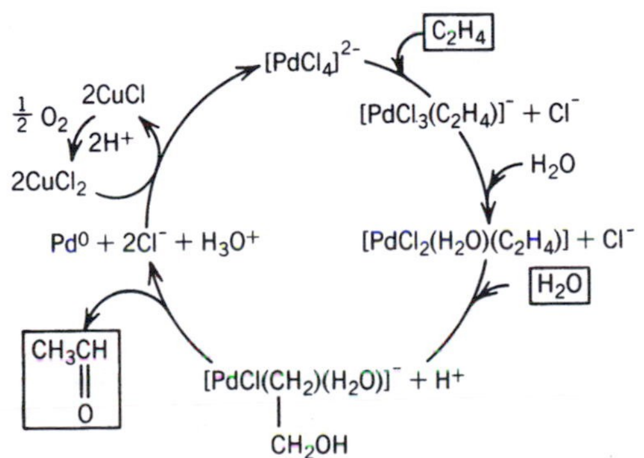

Figure 8.12. *Tolman loop for tthe oxidation of ethylene by Pd(II).*

Under these conditions, the bis(triphenylphosphine) species may dimerize as shown above. The dimer exists as an orange, halogen bridged species $(PPh_3)_2Rh(\mu\text{-}Cl)_2Rh(PPh_3)_2$.

Note that although $RhCl(PPh_3)_3$ is the precursor to the actual catalyst, the nature of the actual catalyst generated in solution is not yet established. The compound $RhCl(PPh_3)_3$ undergoes several interesting chemical reactions.

1. It reacts with hydrazine in the presence of PPh_3 to form $RhH(PPh_3)_4$.
2. It abstracts CO from aldehydes, formic acid, or acyl chlorides to form *trans*-$[RhCl(CO)(PPh_3)_2]$.
3. It absorbs ethylene forming the ethylene complex $RhCl(C_2H_4)(PPh_3)_2$. Here, ethylene substitutes for one PPh_3 ligand.
4. Thiocarbonyl compounds can be prepared by the reaction of CS_2 with $RhCl(PPh_3)_3$ in methanol.

Some of these complexes will be prepared in this laboratory experiment.

There are two convenient methods available for the preparation of Wilkinson's catalyst, $RhCl(PPh_3)_3$.

1. Reduction of hydrated rhodium trichloride by PPh_3 in ethanol results in the formation of a mildly air-sensitive, burgundy colored, highly crystalline product. The byproduct of the reaction is $Ph_3P{=}O$.
2. The more versatile method involves the reduction of hydrated rhodium chloride with C_2H_4 in an aqueous solution of methanol forming a chloro-bridged dimer, $(C_2H_4)_2Rh(\mu\text{-}Cl)_2Rh(C_2H_4)_2$. The dimer then reacts with a phosphine ligand, forming analogs of Wilkinson's catalyst.

The bromo and iodo analogs of Wilkinson's catalyst can be easily prepared (Part 34.B) by adding LiBr or LiI to the reaction mixture containing rhodium trichloride hydrate and triphenylphosphine in ethanol. The $RhCl(PPh_3)_3$ initially produced in the reaction mixture undergoes a substitution reaction, for example,

$$RhCl(PPh_3)_3 + LiBr \rightarrow RhBr(PPh_3)_3 + LiCl$$

Wilkinson's catalyst undergoes a wide variety of reactions. For example, CO and C_2H_4 can easily displace one PPh_3 ligand from $RhCl(PPh_3)_3$ forming the corresponding *trans*-bis(triphenylphosphine) complexes. Its affinity towards the CO group is so strong that it can abstract CO from an aldehyde (Part 34.C) or from formic acid.

$$RhCl(PPh_3)_3 + C_6H_5CHO \rightarrow \textit{trans}\text{-}Rh(CO)Cl(PPh_3)_2 + C_6H_6 + PPh_3$$

In Part 34.D, ethylene, like CO, also displaces one phosphine ligand from $RhCl(PPh_3)_3$, forming yellow crystals of $Rh(C_2H_4)Cl(PPh_3)_2$, bis(triphenylphosphine)ethylenechlororhodium(I).

$$RhCl(PPh_3)_3 + C_2H_4 \rightarrow \textit{trans}\text{-}Rh(C_2H_4)Cl(PPh_3)_2 + PPh_3$$

Isolation and study of compounds containing the hydrido group bound to a transition metal atom are important because in many catalytic hydrogenation reactions, hydrogen transfer is possible via M—H bonds. Wilkinson's catalyst, which is an excellent homogeneous hydrogenation catalyst, shows remarkable affinity towards hydrogen.

A dihydrido complex, $RhCl(PPh_3)_2H_2$, can be easily isolated from a solution of Wilkinson's catalyst in hydrogen saturated chloroform (Part 34.E). The best way to detect the presence of a hydride ligand is via IR or NMR spectroscopy. In the IR spectrum, the dihydrido complex shows two broad Rh—H stretching frequencies in the ~2000-cm^{-1} region, a characteristic region for transition metal–hydride bonds. The splitting of the band indicates that two hydride ligands are probably at cis positions relative to each other. The high field NMR spectrum of this compound in $CDCl_3$ shows three signals in the range from 18 to 30 ppm upfield from TMS.

When nitrogen is bubbled through a solution of the dihydride in chloroform containing excess PPh_3, the solution rapidly turns red, regenerating the parent species, $RhCl(PPh_3)_3$. The addition of hydrogen to Wilkinson's catalyst is, therefore, a reversible reaction, as can be confirmed by the presence or absence of the upfield NMR signal.

As mentioned previously, Wilkinson's catalyst and its analogs are excellent homogeneous hydrogenation catalysts for the reduction of nonconjugated olefins and acetylenes under normal temperature and pressure conditions. The hydrogenation rates[1] of the olefins compare well with those by heterogeneous catalysts. ^1H and ^{31}P NMR studies[2] indicate that when a solution of the catalyst in benzene or chloroform is saturated with H_2 gas, an octahedral cis dihydride species, Compound **A** (Fig. 8.13) is formed. This species then dissociates rapidly to a fluxional five-coordinate species, Compound **B** (Fig. 8.13), under the influence of the strong trans effect of H. It is this species that coordinates with an olefin, Compound **C** (Fig. 8.13). Note that the alkene coordinated complex has two cis PPh_3 groups and one of the two cis hydrogen ligands is at a trans position relative to one phosphine. This trans hydrogen becomes labilized and transfers itself to the olefin. It is also observed that the best situation for such a transfer is when the C=C bond and the M—H bond are coplanar, in a four-center transition state, which allows for the best overlap.

The catalytic alkene hydrogenation cycle involves the following steps, which are summarized in the Tolman loop in Figure 8.14.

1. The catalyst in toluene undergoes oxidative addition, forming a light yellow soluble five-coordinate dihydride complex (probably solvated at this stage) of rhodium(III) by absorbing H_2 gas.
2. The olefin coordinates to the active five-coordinate dihydride.
3. Hydrogen transfer occurs from the metal to the alkene. In the process, the alkyl group undergoes a π–σ shift.
4. A second hydrogen is transferred to the organic moiety, and the catalytic system undergoes reductive elimination, releasing alkane and generating the three-coordinate species $RhCl(PPh_3)_2$.
5. The $RhCl(PPh_3)_2$ complex reacts with a molecule of H_2, forming a five-coordinate species repeating the cycle again.

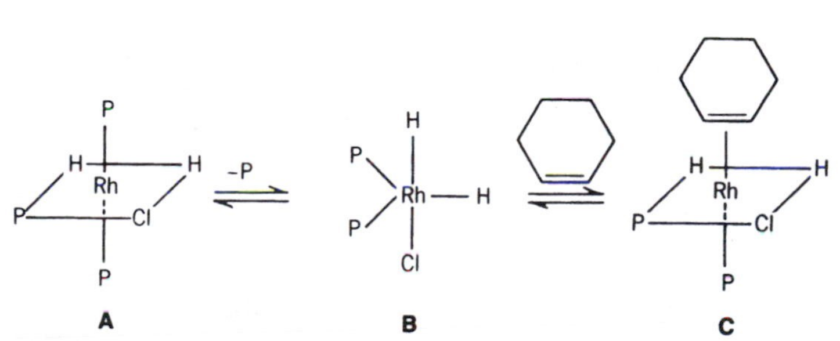

P H Rh H P Cl P -P P H Rh H P Cl P H Rh H P Cl P

A B C

Figure 8.13. *Hydrogenation reaction scheme.*

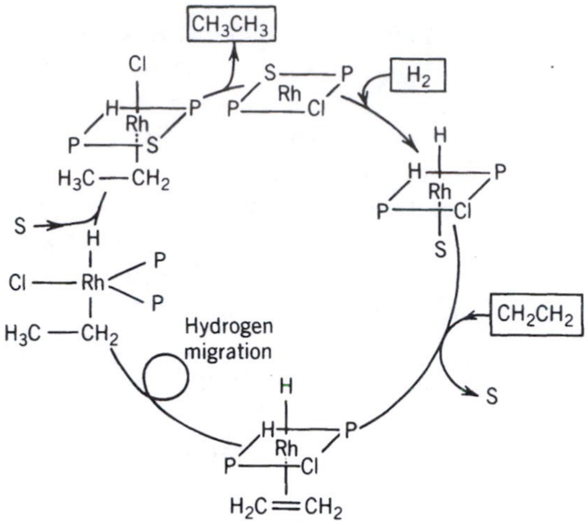

Figure 8.14. *Tolman loop for catalytic hydrogenation.*

Prior Reading and Techniques

Section 1.B.4: Compressed Gas Cylinders and Lecture Bottles
Section 2.F: Reflux and Distillation
Section 5.C.2: Purging with an Inert Gas
Section 5.G.3: Gas Chromatography
Section 6.C: Infrared Spectroscopy
Section 6.D: Nuclear Magnetic Resonance Spectroscopy

Related Experiments

Industrial Chemistry: Experiment 8
Metal Carbonyls: Experiments 42 and 43
Rhodium Chemistry: Experiments 21, 24, and 42

EXPERIMENTAL SECTION

Safety Requirements

Rhodium(III) chloride hydrate (CAS No. 20765-98-4): This compound is harmful if swallowed, inhaled, or absorbed through the skin. ORL-RAT LD50: 1302 mg/kg. It is a possible mutagen, although this has not been definitively established.

Triphenylphosphine (CAS No. 603-35-0): This compound is a mild lachrymator, but has a low vapor pressure, so use of small quantities is not a problem. The compound can cause skin irritation. ORL-RAT LD50: 700 mg/kg.

CHEMICAL DATA

Compound	FW	Amount	mmol	mp (°C)	Density
$RhCl_3 \cdot xH_2O$	209.26[a]	100 mg	0.48	Decomposes	
Triphenylphosphine	262.29	600 mg	2.29	79–81	

[a] Formula weight of the anhydrous material.

Part A: Synthesis of $RhCl(PPh_3)_3$, Wilkinson's Catalyst

$$RhCl_3 \cdot 3H_2O + P(C_6H_5)_3 \rightarrow RhCl(P(C_6H_5)_3)_3$$

Required Equipment

Magnetic stirring hot plate, 25-mL round-bottom flask, water condenser, magnetic stirring bar, sand bath, Hirsch funnel, 25-mL graduated cylinder.

Time Required for Experiment: 1.5 h.

EXPERIMENTAL PROCEDURE

Place 20 mL of absolute ethanol (graduated cylinder) in a 25-mL round-bottom flask equipped with a magnetic stirring bar. Attach a water condenser and place the apparatus in a sand bath upon a magnetic stirring hot plate. Heat the ethanol to just below its boiling point. Remove the condenser momentarily, and add 600 mg (2.29 mmol, a large excess) of triphenylphosphine to the hot ethanol and stir until dissolution is effected.

> **NOTE:** ***A small amount of solid may remain at this point. This is normal.***

Remove the condenser momentarily once again, and add 100 mg (0.48 mmol) of hydrated rhodium(III) chloride to the solution and continue to stir. Heat the solution to a gentle reflux. At first, a deep red-brown solution is obtained, which during the course of heating under reflux will slowly form yellow crystals. After ~20–30 min of reflux, the yellow crystals are converted into shiny burgundy-red crystals.

Isolation of Product

Collect the product crystals by suction filtration on a Hirsch funnel while the solution is hot. Wash the crystals with three 1-mL portions of ether. Dry the crystals on the filter by continuous suction. Calculate the percentage yield and determine the melting point of the product.

Characterization of Product

Obtain the IR spectrum and the ^{1}H NMR spectrum of the compound. Save these spectra for comparison with the various products prepared from Wilkinson's catalyst in subsequent steps.

Part B: Substitution of the Chloro Ligand in Wilkinson's Catalyst

$$RhCl_3 \cdot 3H_2O + P(C_6H_5)_3 + LiBr \rightarrow RhBr(P(C_6H_5)_3)_3 + LiCl$$

Safety Recommendations

Lithium bromide (CAS No. 7550-35-8): This compound is not normally considered dangerous. SCU-MUS LD50: 1680 mg/kg.

CHEMICAL DATA

Compound	FW	Amount	mmol	mp (°C)	Density
$RhCl_3 \cdot xH_2O$	209.26	25 mg	0.12	Decomposes	
LiBr	86.85	100 mg	1.15	547	3.464
Triphenylphosphine	262.29	150 mg	0.57	79–81	

Required Equipment

Same as required in Part 34.A.

Time Required for Experiment: 2 h.

EXPERIMENTAL PROCEDURE

The same experimental arrangement is utilized as in Part 34.A. Add 25 mg (0.12 mmol) of $RhCl_3 \cdot xH_2O$ to a hot solution of 150 mg (0.57 mmol) of PPh_3 in 5 mL of ethanol. The solution is held under reflux for about 5 min, whereupon the color of the solution becomes lighter. At this point, add 100 mg (1.15 mmol) of LiBr dissolved in 0.5 mL of hot ethanol (Pasteur pipet). Heat the mixture, with stirring, under reflux for an additional 1 h.

Isolation of Product

The orange crystalline product is collected by suction filtration using a Hirsch funnel, and washed with a few drops of ether. Calculate the percentage yield.

Characterization of Product

Obtain the IR spectrum of the product as a Nujol mull. Obtain the ^1H NMR spectrum in $CDCl_3$. Compare the spectra with the chloro analog prepared in Part 34.A.

Part C: Reaction of Wilkinson's Catalyst with Aldehydes

$$RhCl(P(C_6H_5)_3)_3 + CH_3(CH_2)_5CHO \rightarrow trans\text{-}Rh(CO)Cl(P(C_6H_5)_3)_2 + CH_3(CH_2)_4CH_3 + P(C_6H_5)_3$$

Additional Safety Recommendations

***n*-Heptanal** (CAS No. 111-71-7): The vapor of this compound is irritating to the eyes and mucous membranes. ORL-RAT LD50: 14 g/kg.

CHEMICAL DATA

Compound	FW	Amount	mmol	mp (°C)	bp (°C)	Density
n-Heptanal	114.19	50 μL	0.372	−43	153	0.818
$RhCl(PPh_3)_3$	925.23	50 mg	0.054	157		

Required Equipment

Magnetic stirring hot plate, 10-mL beaker, magnetic stirring bar, automatic delivery pipet, Pasteur pipet, Hirsch funnel, clay tile, or filter paper.

Time Required for Experiment: 1 h plus additional time for GC analysis

EXPERIMENTAL PROCEDURE

Place 50 mg [0.054 mmol of $RhCl(PPh_3)_3$, prepared in Part 34.A] in a 10-mL beaker equipped with a magnetic stirring bar. Add 1.0 mL of toluene (automatic delivery pipet).

Place the beaker on a magnetic stirring hot plate, and with stirring, warm the suspension to 40 °C. Using an automatic delivery pipet, add 50 μL (0.372 mmol) of *n*-heptanal to this mixture. The mixture rapidly turns yellow as stirring is continued, indicating formation of the trans complex.

Isolation of Product

Cool the solution in an ice–salt bath and add a few drops (5–6 drops) of ethanol to complete the crystallization. Filter the lemon yellow crystals by suction filtration on a Hirsch funnel. Collect and save the filtrate for GC analysis. Wash the solid product with 2–3 drops of ethanol followed by 500 μL of ether. Dry the crystals on a clay tile or on filter paper. Calculate the percentage yield.

Characterization of Product

Determine the melting point. Obtain the IR spectrum as a Nujol mull. Obtain the ^{1}H NMR spectrum. Compare these spectra with those of the starting material (Wilkinson's catalyst).

Perform a GC separation of the residual filtrate. The conditions are 8-ft silicone column, 125 °C, and He flow rate: 50 $mL \cdot min^{-1}$. The GC should show the presence of approximately 80% *n*-hexane, 10% 1-hexene, and small amounts of *cis*- and *trans*-2-hexene.

Part D: Reaction of Wilkinson's Catalyst with Ethylene

$$RhCl(P(C_6H_5)_3)_3 + C_2H_4 \rightarrow trans\text{-}Rh(C_2H_4)Cl(P(C_6H_5)_3)_2 + P(C_6H_5)_3$$

Safety Recommendations

Ethylene (CAS No. 74-85-1): This compound is a colorless, flammable, nonbreathable gas. Use it only in the **HOOD**.

CHEMICAL DATA

Compound	FW	Amount	mmol	mp (°C)	bp (°C)	Density
$PhCl(PPh_3)$	925.9	30 mg	0.0324			
C_2H_4	28.1	Sufficient amount	Sufficient amount	−169	−103	0.0013

Required Equipment

Hirsch funnel, 10-mL beaker, clay tile.

Time Necessary for Experiment: 0.25 h.

EXPERIMENTAL PROCEDURE

Prepare 2 mL of deoxygenated chloroform by bubbling N_2 gas through the chloroform for 2 min (**HOOD**). Place 30 mg (0.0324 mmol) of $RhCl(PPh_3)_3$ in a 10-mL beaker and add the deoxygenated chloroform. Bubble ethylene gas through the solution for a few minutes. The red color of the solution changes to yellow as reaction occurs. Concentrate the solution with a gentle stream of ethylene (**HOOD**). Yellow crystals of the desired compound will precipitate from solution.

Isolation of Product

Collect the product crystals by suction filtration using a Hirsch funnel. Dry the product on a clay tile or on filter paper.

Characterization of Product

Obtain an IR spectrum of the product as a Nujol mull. Obtain the ^{1}H NMR spectrum in $CDCl_3$ solution, and compare it to the spectrum of ethylene. What does the chemical shift tell you in terms of the ethylene donating electron density to the metal?

Part E: Absorption of Hydrogen by Wilkinson's Catalyst

$$RhCl(P(C_6H_5)_3)_3 + H_2 \rightarrow RhCl(P(C_6H_5)_3)_2H_2$$

Safety Recommendations

Hydrogen (CAS No. 1333-74-0): Hydrogen is an explosive gas. There must be no open flames when hydrogen is in use. It is a nonbreathable gas, so care should be exercised.

Required Equipment
Mercury bubbler, 10-mL side arm tube, Hirsch funnel, clay tile, or filter paper.

Time Required for Experiment; 1 h.

EXPERIMENTAL PROCEDURE

NOTE: *Perform Part 34.E in the HOOD.*

Assemble an apparatus as shown in Figure 8.15. Dissolve 50 mg (0.054 mmol) of $RhCl(PPh_3)_3$ (prepared in Part 34.A) in 1 mL of deoxygenated chloroform (see Part 34.D) in a 10-mL side arm tube. Bubble H_2 gas through the solution for a few minutes, whereupon the red solution will turn pale yellow. Concentrate the solution under the flow of H_2 gas. When the solution is sufficiently concentrated (0.2 mL), add deoxygenated ether dropwise until precipitation occurs.

Isolation of Product
Cool the flask in an ice–water bath and collect the light yellow crystals by suction filtration using a Hirsch funnel. Calculate the percentage yield.

Characterization of Product
Obtain the IR and ^1H NMR (in $CDCl_3$) spectra. Assign the IR and NMR bands. Remember to check the upfield region of the NMR spectrum.

Reversibility of Reaction
In order to study the reversibility of the absorption of H_2 gas, after obtaining the NMR spectrum, pass N_2 gas through the solution in the NMR tube with the help of a long syringe needle (see cannula technique for manipulations under N_2 gas). When the solution changes its color from light yellow to red, obtain the NMR spectrum once again. Compare the two NMR spectra. What changes have occurred?

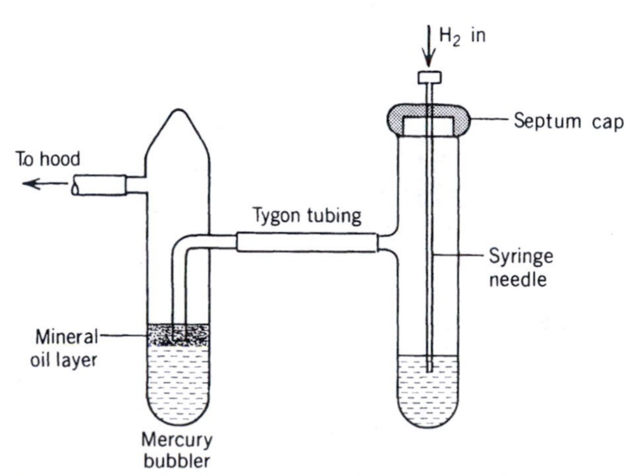

Figure 8.15. *Apparatus for Experiment 34.*

Part F: Catalytic Hydrogenation of Olefins in the Presence of Wilkinson's Catalyst

$$RhCl(P(C_6H_5)_3)_2H_2 + c\text{-}C_6H_{10} \rightarrow RhCl(P(C_6H_5)_3)_2 + c\text{-}C_6H_{12}$$

Additional Safety Recommendations

Cyclohexene (CAS No. 110-83-8): This compound may be harmful if inhaled, ingested, or absorbed through the skin. OSHA Standard-AIR: TWA 300 ppm.

CHEMICAL DATA

Compounds	FW	Amount	mmol	mp (°C)	bp (°C)	Density
$RhCl(PPh_3)_3$	925.23	25 mg	0.027			
C_6H_{10}	82.15	1 mL	9.87	−104	83	0.811

Required Equipment

Magnetic stirring hot plate, 10-mL side arm flask, 25-mL side arm flask, 10-mL graduated cylinder, calibrated Pasteur pipet, magnetic stirring bar, septum, long syringe needle, mercury bubbler.

Time Required for Experiment: 2 h.

EXPERIMENTAL PROCEDURE

NOTE: ***Carry out this experiment in a HOOD.***

Set up the apparatus as shown in Figure 8.15. Assemble a 25-mL side arm flask equipped with a magnetic stirring bar, and rubber septum at the mouth. Insert a long syringe needle through the septum. The side arm of the flask acts as the exit port for H_2 and should be connected using Tygon tubing to a mercury bubbler. Flush the flask with H_2 gas. Add 10 mL of toluene (graduated cylinder) to this flask and saturate it by bubbling H_2 gas (10 min) through it. Momentarily remove the septum and with stirring, dissolve 25 mg (0.027 mmol) of Wilkinson's catalyst in this solvent.

NOTE: ***The catalyst is relatively insoluble in toluene, but in the presence of H_2 gas it dissolves fairly rapidly due to the formation of the more soluble dihydrido species. The resulting solution is pale yellow.***

Discontinue the stirring, remove the septum, and with a calibrated Pasteur pipet, dropwise, add 1 mL of freshly distilled cyclohexene. As soon as the alkene is added, the solution turns deep red-brown in color. Reattach the septum. On rapid stirring in an atmosphere of H_2, the color of the solution lightens to pale yellow.

NOTE: ***If the H_2 gas flow is stopped and stirring is discontinued even for a short while (30 s), the solution again turns deep red-brown.***

Discontinue the hydrogen flow and discontinue stirring.

Characterization of Product

Remove a small sample of the solution for GC analysis. The presence of cyclohexane formed as a result of the hydrogenation of cyclohexene should be evident in the chromatogram. (Conditions: Silicone column, 8 ft length, flow rate: 50 mL·min^{-1} He, temperature 80–85 °C.) Calculate the percentage conversion of cyclohexene to cyclohexane from the relative areas of the peaks.

QUESTIONS

1. Draw the Tolman loop for the hydroformylation of ethylene.
2. Wilkinson's catalyst can be used to synthesize optically active amino acids. Discuss this, showing the mechanism. A useful reference is Chan, A. S. C. et al. *J. Am. Chem. Soc.* **1980,** *102,* 5952.
3. Show the mechanism for the hydrogenation of cyclohexene using Wilkinson's catalyst.
4. Alkene hydrogenation rates by Wilkinson's catalyst are enhanced by the presence of trace amounts of oxygen. Why is this true? A useful reference is Kushi, K. et al., *Chem. Lett.* **1972,** 593.
5. Based on a study of the literature, compare and contrast the catalytic ability and mechanism of the Ziegler–Natta catalyst with Wilkinson's catalyst.

REFERENCES

1. Young, J. F.; Osborn, J. A.; Jardine, F. H.; Wilkinson, G. *Chem. Commun.* **1965,** 131.
2. Brown, J. M.; Lucy, A. R. *Chem. Commun.* **1984,** 914.
3. James, B. R., "Addition of Hydrogen and Hydrogen Cyanide to Carbon–Carbon Double and Triple Bonds" in *Comprehensive Organometallic Chemistry,* G. Wilkinson, Ed., Pergamon: Oxford, 1982, Vol. 8, Chapter 51, p. 285.

GENERAL REFERENCE

Gavens, P. D.; Bottrill, M.; Kelland, J. W.; McMeeking, J., "Ziegler–Natta Catalysts" in *Comprehensive Organometallic Chemistry,* G. Wilkinson, Ed., Pergamon: Oxford, 1982, Vol. 3, Chapter 20, p. 89.

Experiment 35 Synthesis and Reactions of Cobalt Phenanthroline Complexes

Part A: Preparation of *Tris*(1,10-phenanthroline)cobalt(II) Bromide

Part A (Alternate): Preparation of *Tris*(1,10-phenanthroline)cobalt(II) Antimonyl-*d*-tartrate

Part B: Preparation of *Tris*(1,10-phenanthroline)cobalt(III) Tetrafluoroborate

Part C: Preparation of *Tris*(1,10-phenanthroline-5,6-quinone)cobalt(III) Hexafluorophosphate

Part D: Isolation of 1,10-Phenanthroline-5,6-quinone

INTRODUCTION

The reactivity of an organic molecule is markedly changed when it becomes coordinated to a metal. One specific example of enhanced reactivity of a coordinated ligand is the ready oxidation of 1,10-phenanthroline (phen) bonded

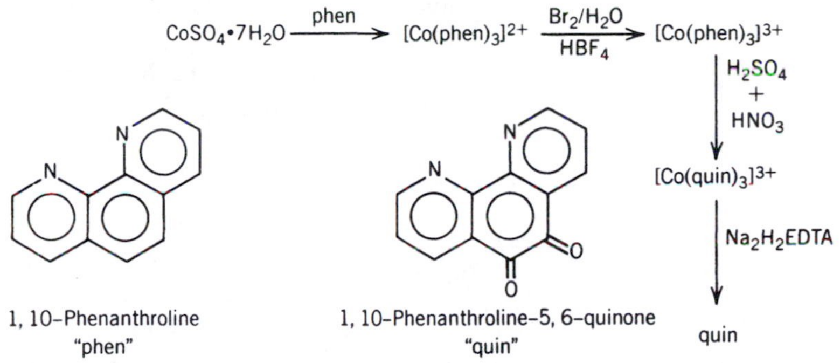

$CoSO_4 \cdot 7H_2O \xrightarrow{phen} [Co(phen)_3]^{2+} \xrightarrow[HBF_4]{Br_2/H_2O} [Co(phen)_3]^{3+} \xrightarrow{H_2SO_4 + HNO_3} [Co(quin)_3]^{3+} \xrightarrow{Na_2H_2EDTA} quin$

1, 10-Phenanthroline "phen"

1, 10-Phenanthroline-5, 6-quinone "quin"

Figure 8.16. *Reaction scheme and structures.*

to Co(III) to yield 1,10-phenanthroline-5,6-quinone (quin). The structure of these compounds is shown in Figure 8.16. This oxidation is normally quite difficult to accomplish. Using a nitric–sulfuric acid mixture, only a 1% yield of quin is obtained from free phen. However, phen can be converted into quin by oxidizing phen coordinated to Co(III), using a mixture of nitric and sulfuric acids. The process takes place in four steps.

1. Preparation of $[Co(phen)_3]^{2+}$, a phen complex of Co(II) as a tetrafluoroborate salt.
2. Oxidation of the Co(II) to Co(III), where the coordinated phen does not undergo oxidation. This step yields optically pure product D(+)-$[Co(phen)_3]^{3+}$.
3. Oxidation of the coordinated phen to metal bonded quin by the mixture of nitric and sulfuric acids in the presence of bromide ion. This conversion takes place without the loss of optical activity, and optically pure D(+)-$[Co(quin)_3]^{3+}$ is isolated.
4. The reaction of D(+)-$[Co(quin)_3]^{3+}$ with the disodium salt of ethylenediaminetetraacetic acid, Na_2H_2EDTA, which liberates the pure free quin.

Figure 8.16 summarizes the reaction process.

Prior Reading and Techniques

Section 2.F: Reflux and Distillation

Section 5.D.3: Isolation of Crystalline Products (Suction Filtration)

Section 5.D.4: The Craig Tube Method

Section 6.C: Infrared Spectroscopy

Section 6.D: Nuclear Magnetic Resonance Spectroscopy

Related Experiments

Cobalt Chemistry: Experiments 7A, 17, 26, 27, 30, and 47B

EXPERIMENTAL SECTION

Part A: Preparation of Tris(1,10-phenanthroline)cobalt(II) Bromide

Safety Recommendations

1,10-Phenanthroline hydrate (CAS No. 5144-89-8): This compound may be harmful, although its toxicological effects have not been extensively investigated. IPR-MUS LD50: 75 mg/kg. The normal precautions should be taken (Section 1.A.3).

Cobalt(II) sulfate heptahydrate (CAS No. 60459-08-7): This compound is harmful if inhaled or swallowed. ORL-RAT LD50: 768 mg/kg.

Potassium bromide (CAS No. 7758-02-3): This compound is not normally considered harmful, but the normal precautions should be taken (Section 1.A.3).

CHEMICAL DATA

Compound	FW	Amount	mmol	mp (°C)	Density
$CoSO_4 \cdot 7H_2O$	280.93	60 mg	0.214		2.019
1,10-Phenanthroline hydrate	198.23	120 mg	0.610	100	
KBr	119.01	50 mg	0.420	734	2.750

Required Equipment

Magnetic stirring bar, 10-mL Erlenmeyer flask, Pasteur pipet, magnetic stirring hot plate, Hirsch funnel, clay tile.

Time Needed for Experiment: 0.5 h.

EXPERIMENTAL PROCEDURE

Place 60 mg (0.214 mmol) $CoSO_4 \cdot 7H_2O$ and 120 mg (0.610 mmol) of 1,10-phenanthroline in a 10-mL Erlenmeyer flask equipped with a magnetic stirring bar. Using a calibrated Pasteur pipet, add 1 mL of deionized water, and stir to form a suspension. Add 50 mg (0.420 mmol) of KBr to this mixture. Stir the mixture for 5–10 minutes or until precipitation is complete.

Isolation of Product

Collect the yellow-brown precipitate by suction filtration using a Hirsch funnel. Wash the $[Co(phen)_3]Br_2$ crystals twice with 200-μL portions of water and dry them on a clay tile. Determine the percentage yield.

Characterization of Product

Obtain an IR spectrum of the product as a Nujol mull, and compare it with phenanthroline and published sources.[1]

Alternate Part A: Preparation of Optically Pure Tris(1,10-phenanthroline)cobalt(II) Antimonyl-*d*-tartrate

NOTE: ***The optically pure d isomer of tris(1,10-phenanthroline)cobalt(II) may be prepared by the following procedure. It can then be substituted for the bromide salt from Part 35.A in the reaction in Part 35.B, generating a stable, soluble, optically pure Co(III) tetrafluoroborate complex.***

EXPERIMENTAL PROCEDURE

Place 100 mg (0.51 mmol) of 1,10-phenanthroline monohydrate in 10 mL of water at 70–80 °C in a 25-mL beaker containing a magnetic stirring bar. Vigorously stir the solution, being careful not to heat it over 80 °C.

NOTE: *The phenanthroline does not dissolve well and will form a two layer system. It is important to stir vigorously so that complete reaction will be effected.*

In a 10-mL Erlenmeyer flask, prepare a solution of 24 mg (0.1 mmol) $CoCl_3 \cdot 6H_2O$ in 1 mL of water. Heat this solution to 80°C. In a separate 10-mL Erlenmeyer flask, prepare a solution of 251 mg (0.75 mmol) of potassium antimonyl-*d*-tartrate (for structure, see Experiment 27) in 1 mL of water, also at 80°C.

Transfer the cobalt solution to the phenanthroline solution using a Pasteur pipet. Rinse the flask with an additional few drops of water and add this to the phenanthroline solution as well. The solution will immediately turn yellow. Quickly add the potassium antimonyl-*d*-tartrate solution. The *d* isomer will immediately precipitate.

Isolation of Product

Cool the solution to room temperature and then in an ice–water bath for 10 min. Collect the green-yellow product by suction filtration using a Hirsch funnel. Wash the filter cake with a 100-μL portion of ice–water. Air dry the product for 15 min. In Part 35.B, this product can be substituted for the starting material, resulting in the optically pure tris(1,10-phenanthroline)cobalt(III) tetrafluoroborate being formed. The optical rotation of this compound can be measured following the procedure of Ref. 2, if desired.

Part B: Preparation of Racemic Tris(1,10-phenanthroline)cobalt(III) Tetrafluoroborate

Additional Safety Recommendations

Bromine (CAS No. 7726-95-6): This compound is an extremely corrosive liquid. **Only handle it while wearing gloves, in the HOOD.** It may be fatal if swallowed, inhaled, or absorbed through the skin. It is extremely destructive to the mucous membranes and skin. ORL-HMN LDLo: 14 mg/kg, IHL-HMN LCLo: 1000 ppm.

Fluoroboric acid (CAS No. 16872-11-0): This compound is harmful if swallowed, inhaled, or absorbed through the skin. It is extremely destructive to the mucous membranes and skin. It is also a lachrymator (makes you cry). **Use only in the HOOD while wearing gloves.**

Required Equipment

Gloves, 25-mL beaker, 10-mL round-bottom flask, calibrated Pasteur pipet, magnetic stirring bar, water condenser, magnetic stirring hot plate, Keck clip, plastic dropper, Hirsch funnel, clay tile.

Time Required for Experiment: 1 h.

EXPERIMENTAL PROCEDURE

Prepare a solution of saturated bromine water by adding 10 mL of water to 2–4 drops of liquid bromine (**HOOD**) in a 25-mL beaker. Use gloves when handling liquid bromine.

Place a mixture of 50 mg (0.0658 mmol) of $[Co(phen)_3]Br_2$ in 1.0 mL of the bromine water in a 10-mL round-bottom flask equipped with a magnetic stirring bar. Using a calibrated Pasteur pipet, add 500 μL of water to the flask, and attach a water condenser using a Keck clip. Heat the solution under mild reflux for 25 min.

Isolation of Product
Add 4–6 drops of 48% aqueous HBF_4 (use a plastic dropper) to the hot solution. Cool the flask to 0 °C in an ice–salt bath. The yellow product is collected by suction filtration using a Hirsch funnel, air dried by suction, and then dried on a clay tile. Calculate the percentage yield.

Characterization of Product
Obtain the IR spectrum as a Nujol mull. Compare it with the product from Part 35.A.

Part C: Preparation of Tris(1,10-phenanthroline-5,6-quinone)cobalt(III) Hexafluorophosphate

Additional Safety Recommendations
Sodium bromide (CAS No. 7647-15-6): This compound is not normally considered dangerous. ORL-RAT LD50: 3500 mg/kg. The normal precautions should be observed (Section 1.A.3).

Potassium hexachlorophosphate (CAS No. 17084-13-8): Toxicity data is not available for this compound. The compound is harmful if swallowed, inhaled, or absorbed through the skin. It is extremely destructive to the mucous membranes and skin.

CHEMICAL DATA

Compound	FW	Amount	mmol	mp (°C)	Density
$[Co(phen)_3](BF_4)_3$	859.3	26 mg	0.030		
NaBr	102.90	18 mg	0.175	755	3.21
KPF_6	184.07	67 mg	0.36	575	

Required Equipment
Ice–water bath, 10-mL round-bottom flask, water condenser, Keck clip, magnetic stirring bar, magnetic stirring hot plate, sand bath, 5-mL conical vial, Hirsch funnel, clay tile.

Time Required for Experiment: 3 h.

EXPERIMENTAL PROCEDURE
Add 26 mg (0.030 mmol) of $[Co(phen)_3](BF_4)_3$, 18 mg (0.175 mmol) of NaBr, and 6 drops of H_2SO_4 to a 10-mL round-bottom flask equipped with a magnetic stirring bar. Cool the mixture in an ice–water bath and add 6 drops of HNO_3 to the ice cold mixture. Attach a water condenser using a Keck clip, and transfer the apparatus into a sand bath set atop a magnetic stirring hot plate. Stir and heat the mixture gently for 40 min.

Isolation of Product
Prepare a solution of 67 mg of KPF_6 (0.36 mmol) in 8 drops of water in a 5-mL conical vial. Transfer the hot mixture from the reflux system to this vial. Stir the mixture thoroughly and cool it in an ice–salt bath (or in refrigerator overnight) until yellow crystals form (2 h). Collect the product crystals by suction filtration using a Hirsch funnel and wash them with 2 drops of ice cold water. Dry the crystals by suction and then on a clay tile. Calculate the percentage yield.

Characterization of Product
Obtain the IR spectrum of the product as a Nujol mull and compare it to products from the previous steps.

Part D: Isolation of 1,10-Phenanthroline-5,6-quinone

Additional Safety Recommendations

EDTA, disodium salt (CAS No. 6381-92-6): This compound is not normally considered dangerous, but the usual safety precautions should be observed (Section 1.A.3). ORL-RAT LD50: 2000 mg/kg.

Required Equipment

Magnetic stirring hot plate, magnetic stirring bar, 10-mL round-bottom flask, calibrated Pasteur pipet, Pasteur pipet, universal indicating pH paper, Keck clip, water condenser, Hirsch funnel, Craig tube.

Time Required for Experiment: 2 h.

EXPERIMENTAL PROCEDURE

Dissolve 30 mg of tris(1,10-phenanthroline-5,6-quinone)cobalt(III) hexafluorophosphate from Part 35.C and 30 mg of Na_2H_2EDTA in 1 mL of warm water (calibrated Pasteur pipet) in a 10-mL round-bottom flask containing a magnetic stirring bar. Adjust the pH of the solution to 5.5 by adding solid sodium bicarbonate (determine the pH by touching a Pasteur pipet to the solution and then to universal indicating pH paper). Attach a water condenser with a Keck clip and transfer the apparatus to a sand bath on a magnetic stirring hot plate. Reflux the solution for 1 h or until the color of the solution changes from yellow to red.

Isolation of Product

Cool the solution to room temperature, and then in an ice bath for 10 min. Collect the yellow solid, which is the impure quinone, by suction filtration using a Hirsch funnel. Recrystallize (Craig tube) the impure quinone from a minimum amount of hot methanol. Calculate the percentage yield of the quinone. Determine the melting point of the pure, yellow needles of product.

Characterization of Product

Obtain an IR spectrum of the product as a KBr pellet. Compare the spectrum to that of published sources.

QUESTIONS

1. What bands in the IR and in the ^1H NMR spectrum would be expected to change upon oxidation of 1,10-phenanthroline to the quinone?
2. When measuring the optical activity of the optically active tris(1,10-phenanthroline)cobalt(III) tetrafluoroborate, low values are frequently obtained. Offer a possible explanation for this result.
3. Racemic $[Co(phen)_3]^{3+}$ catalyzes the racemization of the optically active [*d*-$Co(phen)_3]^{3+}$. Propose a mechanism by which this can occur.
4. In the separation of optical isomers, what is the function of the antimonyl-*d*-tartrate? Based on the literature, what other materials may be used for this purpose?

REFERENCES

1. Gillard, R. D.; Hill, R. E.; Maskill, R. J. *J. Chem. Soc. (A)*, **1970,** 1447.
2. Hunt, H. R. *J. Chem. Educ.* **1977,** *54,* 710.

GENERAL REFERENCE

McWhinnie, W. R.; Miller, J. D., "The Chemistry of Complexes Containing 2,2′ Bipyridyl, 1,10-Phenanthroline or 2,2′,6′,2″ Terpyridyl as Ligands" in *Advances in Inorganic Chemistry and Radiochemistry,* H. J. Emeleus and A. G. Sharpe, Eds., Academic Press: New York, 1969, Vol. 12, p. 135.

Experiment 36 Preparation of Tetrakis(triphenylphosphine)platinum(0)

INTRODUCTION

One of the more unusual aspects of transition metal chemistry is the ability of many metals to form complexes where the metal is formally in the 0 oxidation state. The ligands commonly seen in these extremely low oxidation state complexes are CO, phosphines, and various organic (usually aromatic) groups.

Platinum forms a large variety of complexes in both the II and IV oxidation states. The Pt(II) complexes (d^8) are usually square planar, while Pt(IV) complexes (d^6) are octahedral, in accordance with crystal field theory. There are also a small number of Pt(0) complexes (d^{10}), nearly all containing tertiary phosphines as ligands.

The Pt(0) complexes are made by reaction of platinum dihalides or $[PtX_4]^{2-}$ complexes with phosphines in the presence of reducing agents, such as alcoholic potassium hydroxide or hydrazine, as shown in the unbalanced reaction

$$K_2PtCl_4 + P(C_6H_5)_3 \rightarrow Pt[P(C_6H_5)_3]_4$$

The platinum(0) phosphine complexes are important synthetic intermediates, as the phosphine ligands are readily lost. In solution at 25 °C, there is substantial dissociation of the tetrakis(triphenylphosphine)platinum(0) to tris(triphenylphosphine)platinum(0), but formation of the bis complex is slight. The complexes easily undergo oxidative addition reactions, resulting in the formation of Pt(II) complexes.[1]

Prior Reading and Techniques

Section 5.D.3: Isolation of Crystalline Products (Suction Filtration)

Section 6.C: Infrared Spectroscopy

Related Experiments

Platinum Chemistry: Experiments 37, 38, and 48

EXPERIMENTAL SECTION

Safety Recommendations

Potassium tetrachloroplatinate(IV) (CAS No. 10025-99-7): This compound is harmful if swallowed, inhaled, or absorbed through the skin. It is classified as an anticancer agent. IPR-MUS LD50: 45 mg/kg.

Potassium hydroxide (CAS No. 1310-58-3): This compound is highly corrosive and very hygroscopic. Ingestion will produce violent pain in the throat. ORL-RAT LD50: 1.23 g/kg. If contacted with the skin, wash with large quantities of water.

Triphenylphosphine (CAS No. 603-35-0): This compound is a mild lachrymator, but has a low vapor pressure, so use of small quantities is not a problem. The compound can cause skin irritation. ORL-RAT LD50: 700 mg/kg.

CHEMICAL DATA

Compound	FW	Amount	mmol	mp (°C)	Density
K_2PtCl_4	415.11	100 mg	0.24		
KOH	56.11	30 mg	0.53	380	
$P(C_6H_5)_3$	262.28	300 mg	1.14	80.5	1.194

Required Equipment

Magnetic stirring hot plate, 10-mL Erlenmeyer flask, magnetic stirring bar, sand bath, thermometer, Pasteur pipet, 10 mL beaker, Hirsch funnel, side arm flask.

Time Required for Experiment: 1 h.

EXPERIMENTAL PROCEDURE[2]

Place 2 mL of absolute ethanol in a 10-mL Erlenmeyer flask equipped with a magnetic stirring bar. Heat the flask to 65 °C in a sand bath using a magnetic stirring hot plate. When the ethanol has reached the desired temperature, add and dissolve 300 mg (1.14 mmol) of triphenylphosphine. Maintain the solution at 65 °C with stirring.

Prepare a solution of 30 mg of KOH in a mixture of 700 μL of ethanol and 150 μL of water in a 10-mL beaker. Add this, with stirring, to the above solution. Then, with continued stirring, dropwise (Pasteur pipet) add 100 mg (0.24 mmol) K_2PtCl_4 dissolved in 1 mL of water. After ~20 min the addition should be complete and a pale yellow precipitate should begin to separate.

Isolation of Product

Allow the solution to cool to room temperature. Recover the product, tetrakis(triphenylphosphine)platinum(0), by suction filtration using a Hirsch funnel. Wash the product crystals with 0.5 mL of cold water, followed by 0.5 mL of cold ethanol, and dry the crystals on a clay tile for a short while. Quickly, determine the weight of the product and the percentage yield. Do not leave the product exposed to air, as it will slowly decompose. For storage over extended periods of time, the $Pt(PPh_3)_4$ should be tightly sealed in a screw-cap vial that was flushed with N_2 gas.

Characterization of Product

Prepare a KBr pellet of the product and acquire the IR spectrum. Compare it to the spectrum for triphenylphosphine.

QUESTIONS

1. Does the product obey the 18-electron rule? Explain.
2. How can the platinum bond to the PPh_3 ligands when it is in a formal 0 oxidation state?
3. What multiplicity would the ^{195}Pt NMR ($I = \frac{1}{2}$) spectrum of the tetrakis(triphenylphosphine)platinum(0) be expected to have? Assume no coupling to ^1H occurs.
4. Transition metal compounds are found in low oxidation states with various ligands such as CO, NO, PPh_3, and aromatic rings. Give examples from the literature of each compound and how they are prepared. Why are they stable in low oxidation states with these ligands and not with such ligands as O^{2-} and F^-?

REFERENCES

1. Pierpont, C. G.; Downs, H. H. *Inorg. Chem.* **1975,** *14,* 343.
2. Ugo, R.; Cariati, F.; La Monica, G. *Inorg. Syn.* **1968,** *11,* 105.

GENERAL REFERENCE

Roundhill, D. M., "Zero-Valent Platinum Complexes," in *Comprehensive Coordination Chemistry,* G. Wilkinson, Ed., Pergamon: Oxford, 1987, Vol. 5, Section 52.5.3.1, p. 440.

Experiment 37 Platinum(II) Complexes—the Trans Effect

Part A: Preparation of *cis*-Dichloro(dipyridine)platinum(II)

Part B: Preparation of *trans*-Dichloro(dipyridine)platinum(II)

INTRODUCTION

A wide variety of Pt(II) complexes are known. There are various types, such as complex anions, $[PtX_4]^{2-}$; neutral complexes, $[PtL_2X_2]$, where L is a neutral ligand such as a phosphine; hydride complexes, $[PtHXL_2]$ and $[PtH_2L_2]$; and cationic species, $[PtL_4]^{2+}$. The great majority are of square planar geometry, but five- and six-coordinate Pt(II) complexes are also known.

Cis and trans isomerization is possible with square planar geometry and numerous cases have been reported for the Pt(II) system. In the general reaction

$$[PtX_3L]^- + Y^- = [PtX_2YL]^- + X^-$$

there are two possible reaction products: the cis and the trans products relative to the ligand L. The proportion of cis to trans isomer varies drastically with the nature of the ligand. In some cases, nearly 100% cis product is isolated, while in others, nearly 100% trans product is obtained. An extensive series of ligands was investigated and arranged according to their ability to direct the incoming group Y to the position trans to themselves. This phenomenon is known as the trans effect. The trans directing order of some common ligands is as follows:

$$H_2O, OH^-, NH_3, py < Cl^-, Br^- < C_5H_5^- < PR_3 < CN^-, CO$$

For example, reacting pyridine (py) with $[PtCl_4]^{2-}$ results first in $[PtCl_3py]^-$. A second pyridine ligand can either substitute trans to the first pyridine or trans to one of the chlorines. Since chloride is the better trans director, the second pyridine substitutes trans to a chloride ligand (cis to the other pyridine), and the cis isomer of $[Pt(py)_2Cl_2]$ is obtained.

Conversely, reacting chloride with $[Pt(py)_4]^{2+}$ results first in $[Pt(py)_3Cl]^+$. A second chloride ligand can substitute either trans to one of the pyridines or trans to the chloride. Since chloride is the better trans director, the second chloride substitutes trans to the first, and the trans isomer of $[Pt(py)_2Cl_2]$ is obtained.

In this experiment, both *cis*- and *trans*-dichlorodipyridineplatinum(II) are synthesized.

Prior Reading and Techniques

Secton 5.D.3: Isolation of Crystalline Products (Suction Filtration)
Section 5.D.4: The Craig Tube Method
Section 6.C: Infrared Spectroscopy

Related Experiments

Isomerism: Experiments 26, 27, 46, and 49
Platinum Chemistry: Experiments 36, 38, and 48

EXPERIMENTAL SECTION

Safety Recommendations

Potassium tetrachlloroplatinate(IV) (CAS No. 10025-99-7): This compound is harmful if swallowed, inhaled, or absorbed through the skin. It is classified as an anticancer agent. IPR-MUS LD50: 45 mg/kg.

Pyridine (CAS No. 110-86-1): Pyridine is harmful if swallowed, inhaled, or absorbed through the skin. It has a noxious smell and is a general anaesthetic. Dispense it only in the **HOOD**. Wash all utensils in contact with the pyridine in the **HOOD** with acetone. ORL-RAT LD50: 891 mg/kg.

CHEMICAL DATA

Compound	FW	Amount	mmol	mp (°C)	bp (°C)	Density
K_2PtCl_4	415.26	100 mg	0.24			
Pyridine	79.10	40 μL	0.5	−42	115	0.978

Part A: Preparation of *cis*-Dichloro(dipyridine)platinum(II)

Required Equipment

Magnetic stirring hot plate, magnetic spin vane, calibrated Pasteur pipet, automatic delivery pipet, 5-mL conical viral, Hirsch funnel, clay tile, or filter paper.

Time Necessary for Experiment: 2 h.

EXPERIMENTAL PROCEDURE[1]

Place 100 mg (0.24 mmol) of potassium tetrachloroplatinate(II) in a 5-mL conical vial containing a magnetic spin vane. Add 1 mL of water (calibrated Pasteur pipet) and stir at room temperature on a magnetic stirring hot plate until the solid dissolves.

Using a Pasteur pipet, slowly add a previously prepared solution of 40 μL (0.5 mmol) of freshly distilled pyridine (**HOOD**) in 250 μL of water, while stirring. Cap the vial. In a few minutes a light creamy yellow precipitate begins to form. Continue stirring for 1 h. Cool the solution in an ice–salt bath for 30 min to complete the precipitation of the product.

Isolation of Product

Collect the solid product by suction filtration using a Hirsch funnel. Wash the light creamy yellow crystals three times with 0.5 mL of ice–water to remove any KCl. Dry the solid on a clay tile or on filter paper. One half of the solid obtained in this reaction should be used for preparation of the trans product in Part 37.B. Calculate a percentage yield and determine the decomposition point.

Characterization of Product

Obtain the IR spectrum of the product as a KBr pellet. Compare it to the trans material to be prepared in Part 37.B and that of pyridine.

Part B: Preparation of *trans*-Dichloro(dipyridine)platinum(II)

Required Equipment

Magnetic stirring hot plate, two 10-mL beakers, calibrated Pasteur pipet, automatic delivery pipet, Pasteur filter pipet, sand bath, Hirsch funnel, clay tile, or filter paper.

Time Required for Experiment: 2 h.

EXPERIMENTAL PROCEDURE[1]

NOTE: *This reaction should be run in the HOOD.*

Place 40 mg (0.094 mmol) of *cis*-dichloro(dipyridine)platinum(II) (prepared in Part 37.A) in a 10-mL beaker containing a magnetic stirring bar, and add 0.5 mL of freshly distilled pyridine (automatic delivery pipet) and 1 mL of water (calibrated Pasteur pipet). Place the beaker in a sand bath set atop a magnetic stirring hot plate.

Stir the solution and heat it to just below boiling. The solids should dissolve to yield a colorless solution of tetrapyridineplatinum(II) chloride, $[Pt(py)_4]Cl_2$. Any undissolved solid should be removed at this point by drawing the solution into a Pasteur filter pipet and transferring it to a clean 10-mL beaker.

Add a boiling stone and gently heat the solution to dryness. Add 1 mL of concentrated HCl (calibrated Pasteur pipet), with stirring, to the white residue.

NOTE: *Concentrated HCl is quite corrosive. Handle it with care.*

The solution formed on heating is again evaporated to dryness as before (**HOOD**), resulting in a pale yellow-cream colored solid. If any white solid is present, an additional 0.5 mL of HCl should be added, and the evaporation repeated.

Isolation of Product

Transfer the yellow-cream crystals (scrape) to a Hirsch funnel and wash them with three 0.5-mL portions of ice–water to remove any excess HCl. Further wash the product with a 1-mL portion of ethanol and a 1-mL portion of ether. Dry the product on a clay tile or on filter paper.

Purification of Product

Recrystallization may be effected by dissolving the product crystals (Craig tube) in a minimum amount of boiling chloroform, reducing the volume by 50% by evaporation using a gentle stream of nitrogen, and cooling the mixture in an ice bath.

Characterization of Product

Obtain the IR spectrum of the product, and compare it to the trans product prepared in Part 37.A and to that of pyridine.

QUESTIONS

1. Other than spectroscopy, suggest a physical technique that would allow one to distinguish between the *cis*- and *trans*-platinum isomers.
2. Why is platinum the element of choice to study the trans effect?
3. Platinum(II) complexes are usually square planar. Why is this geometry especially preferred by this oxidation state?
4. Based upon a study of the literature, discuss Grinberg's polarization theory, as it applies to the rationalization of the trans effect. Contrast it with the theory (for CO, phosphines, and olefins) which invokes a five-coordinate intermediate.

REFERENCE

1. Kauffman, G. B. *Inorg Syn.* **1964,** *7,* 249.

GENERAL REFERENCES

Basolo, F.; Pearson, R. C., "The Trans Effect in Metal Complexes" in *Progress in Inorganic Chemistry,* F. A. Cotton, Ed., Interscience: New York, 1962, Vol. 4, p. 381.

Roundhill, D. M., "The Trans Effect" in *Comprehensive Coordination Chemistry,* G. Wilkinson, Ed., Pergamon: Oxford, 1987, Section 52.9.1.1

Chapter 9
Chemistry of Organometallic and Related Compounds

Experiment 38: Organoplatinum(II) Complexes: Preparation of η^4-$C_8H_{12}PtCl_2$

Experiment 39: NMR Investigation of Molecular Fluxionality: Synthesis of Allylpalladium Complexes

Experiment 40: Preparation and Use of Ferrocene

Experiment 41: Preparation of Organopalladium Complexes

Experiment 42: Synthesis of Metal Carbonyls

Experiment 43: Sunlight Photochemistry: Preparation of Dicarbonyl-(η^5-methylcyclopentadienyl)triphenylphosphine manganese(0)

Experiment 44: Synthesis of Metal Nitrosyl Complexes

Experiment 45: ^{13}C NMR Analysis of Cyclopentadienylirondicarbonyl Dimer

Also see the following organo-main group element experiments:

Experiment 6: Synthesis of Dichlorophenylborane

Experiment 7: Synthesis and Reactions of Carboranes

Experiment 15: Synthesis of Trichlorodiphenylantimony(V) Hydrate

Experiment 38 Organoplatinum(II) Complexes: Preparation of η^4-$C_8H_{12}PtCl_2$

INTRODUCTION

The earliest known organotransition metal complex was discovered by Zeise in 1845. Zeise's salt, $K[Pt(C_2H_4)Cl_3]$, was prepared by heating a mixture of platinum(II) chloride and platinum(IV) chloride in ethanol, evaporating the solvent, and treating the residue with KCl. Over 100 years after it was first discovered, the complex was shown to have the following structure.

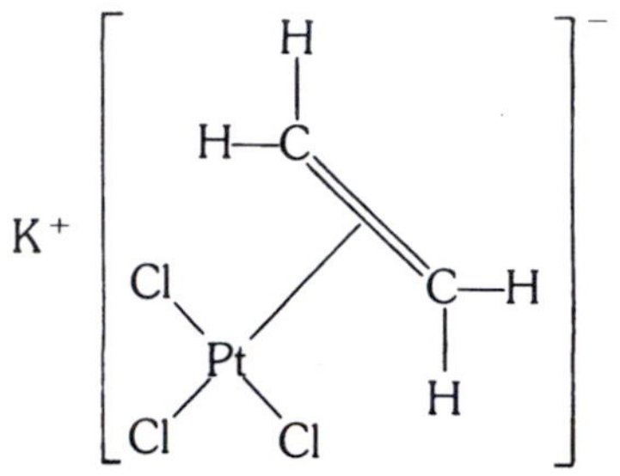

Zeise's salt is an example of a π **complex,** wherein a transition metal is complexed with an unsaturated molecule. This type of bonding was first described by M. J. S. Dewar as consisting of two aspects.

1. The donation of electrons from the filled olefin π orbital to an empty metal σ orbital.
2. The donation from filled metal *d* orbitals to empty olefin π^* orbitals.

This "give-and-take" arrangement is termed synergistic bonding and is shown in Figure 9.1. Since the metal donates electrons into the empty π^* orbital of the olefin, the bond order in the olefin is reduced. The degree to which the metal donates is easily followed in the IR spectrum, where a shift of the C=C bond stretch to lower frequency is observed.

Organoplatinum(II) complexes can also be made via the reduction of Pt(IV) with a cyclic olefin (which consequently is partially oxidized). In these reactions, first described by Kharasch and Ashford,[1] a solution of glacial acetic acid and chloroplatinic acid is mixed with the appropriate olefin. The olefin used in this experiment is 1,5-cyclooctadiene. The diene is a chelating ligand and occupies two coordination sites on the platinum(II) square plane. The diene interacts with the platinum(II) using both double bonds in a π fashion, thus donating four electrons (indicated by the notation η^4), as shown in Figure 9.2 (only the carbon framework is shown).

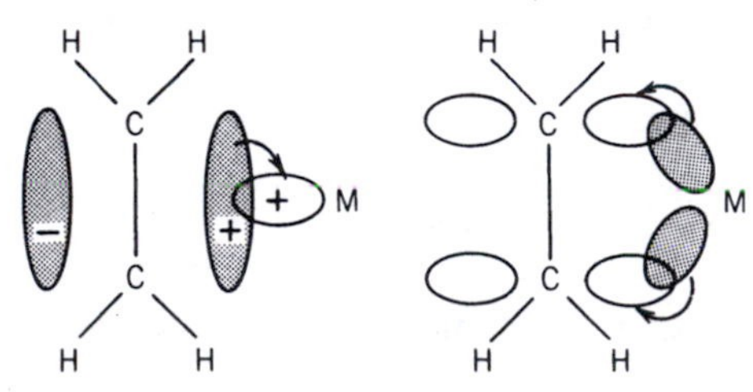

Figure 9.1. *Synergistic bonding between a metal and olefin.*

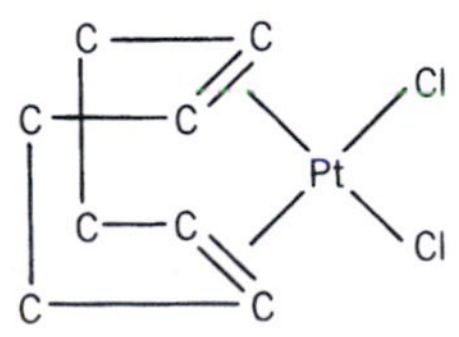

Figure 9.2. *Structure of* η^4-$C_8H_{12}PtCl_2$.

The product is a white, air-stable solid. It is prepared by the reaction of 1,5-cyclooctadiene with chloroplatinic acid.

$$C_8H_{12} + H_2PtCl_6(H_2O)_x \rightarrow (\eta^4\text{-}C_8H_{12})PtCl_2$$

As is usually the case for square planar complexes, the product obeys the 16-electron rule.[2]

Prior Reading and Techniques

Section 5.D.3: Isolation of Crystalline Products (Suction Filtration)

Section 5.F.2: Evaporation Techniques

Related Experiments

Platinum Chemistry: Experiments 36, 37, and 48

EXPERIMENTAL SECTION

Safety Recommendations

Chloroplatinic acid (CAS No. 26258-7): This compound is harmful if swallowed, inhaled, or absorbed through the skin. IVN-RAT LD50: 49 mg/kg.

1,5-Cyclooctadiene (CAS No. 111-78-4): This compound is harmful if swallowed, inhaled, or absorbed through the skin. The vapor pressure at room temperature is fairly high, 6.8 mm at 25 °C, and the compound has a disagreeable odor. Use it only in the **HOOD.**

CHEMICAL DATA

Compound	FW	Amount	mmol	mp(°C)	bp(°C)	Density
$H_2PtCl_6 \cdot xH_2O$	409.82	100 mg	0.24	60		2.430
1,5-Cyclooctadiene	108.18	200 mg	0.96	−69	149	0.882

Required Equipment

Magnetic stirring hot plate, 10-mL Erlenmeyer flask, sand bath, automatic delivery pipet, Hirsch funnel, side arm flask.

Time Required for Experiment: 2 h.

EXPERIMENTAL PROCEDURE[3]

Place 100 mg (0.24 mmol) of hydrated chloroplatinic acid and 1.0 mL of glacial acetic acid in a 10-mL Erlenmeyer flask. Heat the mixture, with swirling, to 75 °C using a sand or water bath set at this temperature.

Using an automatic delivery pipet, add 226 μL (200 mg, 0.96 mmol) of 1,5-cyclooctadiene to the warm solution (**HOOD!**). Mix the black solution by swirling, and then allow it to cool to room temperature. Add 500 μL of water and allow the dark solution to stand at room temperature for approximately 1 h.

Isolation of Product

Collect the grayish precipitate that forms during this period by suction filtration using a Hirsch funnel. Approximately 50 mg is obtained.

Purification of Product

Transfer the solid product to a 10-mL Erlenmeyer flask and dissolve it in 1.5 mL of methylene chloride by warming on a hot plate. Transfer this solution by Pasteur pipet to a Pasteur filter pipet containing a 1-in. high column of silica gel (~500 mg). Collect the filtrate in a 5-mL conical vial containing a boiling stone. Wash the column with an additional 1 mL of methylene chloride and collect the filtrate in the same vial.

Concentrate the filtrate on a warm sand bath under a slow stream of nitrogen gas (**HOOD**) to the point where a white precipitate of the product appears. Collect the precipitate of cyclooctadienylplatinum(II) dichloride by suction filtration using a Hirsch funnel. Wash the white product with two 200-μL portions of hexane. Dry it on a clay tile or on filter paper.

Characterization of Product

Prepare a KBr pellet of the product and obtain the IR spectrum. If desired, prepare an NMR tube of the product in $CDCl_3$ and obtain the ^1H- and ^{13}C-NMR spectra. Compare the spectra to those of 1,5-cyclooctadiene.

QUESTIONS

1. Palladium and platinum can interact with cyclic polyolefins of higher order, such as 1,3,5,7-cyclooctatetraene. Draw structures showing two different ways that the metal can interact with this ligand. (*Hint:* In all cases, the ligand acts as 4-electron donor.)
2. What effect would you expect the bonding of an olefin to platinum(II) have on its ^1H chemical shift and on its IR absorption (C=C stretch) frequency?
3. Pi-bonded systems (alkenes and alkynes) can bond to platinum in other than the π fashion seen in this experiment. Describe the bonding in the complex $(Ph_3P)_2Pt(dpa)$, where dpa is diphenylacetylene. A useful reference is Fachinetti, G. et al. *J. Chem. Soc. Dalton Trans.* **1978,** 1398.
4. Palladium reacts readily with mesityl oxide to form a π-allylic system, much like the allyl palladium dimer prepared in Experiment 39. Platinum, however, gives a nonallylic, olefin complex product, where the platinum bonds to the mesityl oxide in two different ways. Describe the structure of both materials. A useful reference is Parshall, G. W.; Wilkinson, G. *Inorg. Chem.* **1962,** *1,* 896.

REFERENCES

1. Kharasch, M. S.; Ashford, T. *J. Am. Chem. Soc.* **1936,** *58,* 1733.
2. Tolman, C. A. *Chem. Soc. Rev.* **1972,** *1,* 337.
3. Drew, D.; Doyle, J. R. *Inorg. Syn.* **1972,** *13,* 48.

GENERAL REFERENCES

Mingos, D. M. P., "The Bonding of Unsaturated Organic Molecules to Transition Metals" in *Comprehensive Organometallic Chemistry,* G. Wilkinson, Ed., Pergamon: Oxford, 1982, Vol. 3, Chapter 19, p. 1.

Hartley, R. R., "Platinum" in *Comprehensive Organometallic Chemistry,* G. Wilkinson, Ed., Pergamon: Oxford, 1982, Vol. 6, Chapter 39, p. 471.

Experiment 39

NMR Investigation of Molecular Fluxionality: Synthesis of Allylpalladium Complexes

INTRODUCTION

Palladium(II) forms a large variety of square planar organometallic complexes with various olefinic organic groups. In the case of the reaction of $PdCl_2$ with allyl bromide, the allylpalladium bromide complexes shown in Figure 9.3 may be synthesized.

Complexes between a metal salt and an olefin have been known since 1827. In the palladium complexes, the olefin donates electron density from its filled π orbital to an empty palladium π symmetry orbital. The palladium, in turn, donates electron density from a filled σ orbital to the empty olefin π^* orbital. This results in a lowering of the C—C bond order and a consequent lowering of the olefin IR absorption frequency.

There is some difficulty in assigning the number of electrons donated by the allyl group. Viewing the allyl group as a neutral ligand (most convenient in this case), it would function as a 1 (*monohapto*)- or 3 (*trihapto*)-electron donor, depending on whether it were σ or π bound. If π bound, the allyl group is bidentate (occupies two coordination sites), while if σ bound, it is monodentate. Alternatively, the allyl group can be treated as an anion, where it functions as a 2- or 4-electron donor.

In noncoordinating solvents, the complex is found in the π form, where it is a 16-electron species,[1] the most stable electronic arrangement for square planar geometry. (The simplest electron count in this case is Pd^{2+} = 8 electrons, allyl anion = 4, 2 × chloride ion = 4, total = 16 electrons.) In a strongly complexing solvent, the dimer is cleaved, forming the monomeric species $[Pd(\eta^3\text{-}C_3H_5)Cl(DMSO)]$, also a 16-electron system. (Pd^{2+} = 8 electrons, allyl anion = 4, chloride = 2, DMSO = 2, total = 16 electrons.) Additional, reversible, interaction with the relatively basic DMSO solvent allows conversion from the π allylic to the σ bonded form, $[Pd(\eta^1\text{-}C_3H_5)Cl(DMSO)_2]$. ($Pd^{2+}$ = 8 electrons, $\sigma\text{-}C_3H_5$ anion = 2, chloride = 2, 2 × DMSO = 4, total = 16 electrons.) This reaction sequence is shown in Figure 9.3.

When the allyl group is π bound, the complex is stereochemically rigid. There are three types of nonequivalent hydrogen atoms, shown in Figure 9.4. Hydrogen c is clearly unique, being part of the only CH group. The b hydrogen atoms are syn to hydrogen c, and the a hydrogen atoms are anti to hydrogen c. The ^1H NMR spectrum would therefore show three signals. When the allyl group is σ bound, there is free rotation about the C—C single bond, thus rendering the a and b hydrogen atoms equivalent. The ^1H NMR spectrum would

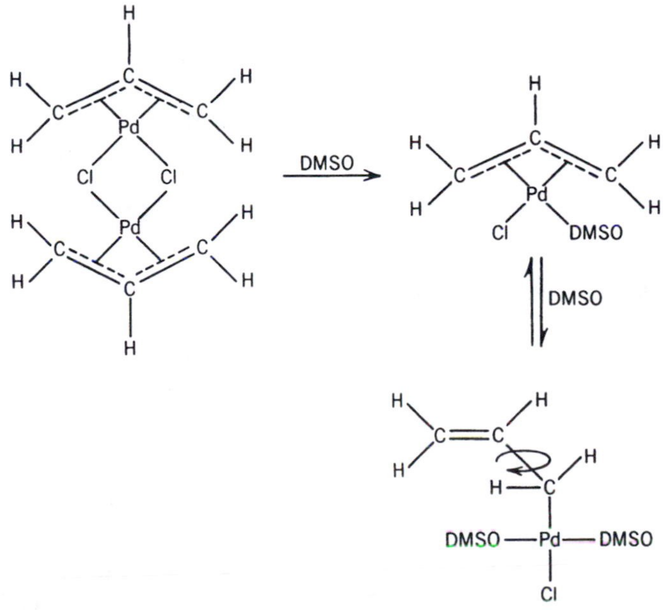

Figure 9.3. *Allylpalladium chloride dimer and DMSO cleavage.*

Figure 9.4. *Nonequivalent protons in the allyl group.*

therefore show only two signals. Molecules showing this kind of motion are said to be fluxional.

Prior Reading and Techniques

Section 5.D.4: The Craig Tube Method
Section 5.F.3: Removal of Solvent Under Reduced Pressure
Section 5.I.3: Extraction Procedures: Simple Extraction
Section 6.C: Infrared Spectroscopy
Section 6.D: Nuclear Magnetic Resonance Spectroscopy

Related Experiments

Molecular Fluxionality: Experiment 45
Palladium Chemistry: Experiments 20, 41, and 46

EXPERIMENTAL SECTION

Safety Recommendations

Palladium(II) chloride (CAS No. 7647-10-1): This compound may be fatal if swallowed, inhaled, or absorbed through the skin. It may be carcinogenic. ORL-RAT LD50: 2704 mg/kg.

Allyl bromide (CAS No. 106-95-6): This compound is a mild lachrymator and is flammable. It may be fatal if inhaled, swallowed, or absorbed through the skin. IHL-RAT LC50: 1000 mg/m^3/30 m.

Acetic acid (CAS No. 10908-8): Acetic acid is harmful if swallowed, inhaled, or absorbed through the skin. Concentrated acetic acid is very corrosive and has an unpleasant smell. It has been found to have effects on male fertility and to have behavioral effects on newborns. ORL-RAT LD50: 3530 mg/kg.

CHEMICAL DATA

Compound	FW	Amount	mmol	mp (°C)	bp (°C)	Density
$PdCl_2$	177.31	100 mg	0.56	500[a]		4.0
Allyl bromide	120.98	500 μL	3.69	−119	70	0.892

[a] Decomposes.

Required Equipment

Magnetic stirring hot plate, two 25-mL round-bottom flasks, magnetic stirring bar, 10-mL graduated cylinder, sand bath, Pasteur pipet, Pasteur filter pipet, Hirsch funnel, clay tile, Craig tube.

Time Required for Experiment: 3 h.

EXPERIMENTAL PROCEDURE[1]

Add 100 mg (0.56 mmol) of finely divided $PdCl_2$ to a 25-mL round-bottom flask equipped with a magnetic stirring bar. Add 3 mL of glacial acetic acid (graduated cylinder) and 3 mL of water. Attach a water condenser and place the apparatus in a sand bath on a magnetic stirring hot plate. Heat the mixture, with stirring, to 100 °C for 15 min.

> **NOTE:** ***At the end of this time, if all of the solid has not dissolved, filter the mixture by suction filtration on a Hirsch funnel, retaining the liquid. Recycle any solid obtained on the filter as directed below.***

Using an automatic delivery pipet, add 500 μL (3.69 mmol) of allyl bromide to the reaction solution through the top of the condenser. Heat the solution to 60 °C, with stirring, for 1 h.

> **NOTE: *Do not heat the solution over 60 °C or decomposition will occur.***

Isolation of Product

Cool the pale yellow mixture to room temperature. Add 3 mL of methylene chloride (graduated cylinder), swirl, and transfer the supernatant liquid into a clean 25-mL round-bottom flask using a Pasteur pipet. Repeat this extraction procedure two additional times if any solid remains. Combine the liquid extractions and dry them for 15 min over anhydrous $MgSO_4$.

Transfer the liquid from the drying agent using a Pasteur filter pipet to a 25-mL round-bottom flask, and rotary evaporate the solution to dryness. The resulting orange-yellow powder is the allylpalladium dimer, di-μ-chlorobis(η^3-allyl)dipalladium(II). Dry the product on a clay tile and determine a percentage yield. If desired, the product may be recrystallized (Craig tube) from a minimum amount of hot methanol.

CHARACTERIZATION OF PRODUCT

Infrared Spectrum Obtain the IR spectrum of the product as a KBr pellet and compare the spectrum to that of allyl bromide.

NMR Spectra Dissolve one half of the product (~20 mg) in a minimum amount of $CDCl_3$ and obtain the ^{1}H NMR spectrum. Dissolve the other one half of the product in a minimum amount of DMSO-d^6 and obtain the ^{1}H NMR spectrum. If equipment is available for variable temperature work, obtain the DMSO-d^6 spectrum at 0, 40, and 60 °C in addition to that at room temperature.[2]

Recycle of Palladium[3]

The starting material, palladium(II) chloride, is rather expensive, and is fairly easy to regenerate. Evaporate the NMR solvents from the product and combine the solid with any unreacted $PdCl_2$ that was filtered in an earlier step. Add the solids to a concentrated solution of sodium methoxide and reflux the mixture for 30 min. A solid mass of palladium should precipitate. Filter the solid and place it in a small amount of aqua regia solution (4:1 hydrochloric and nitric acid). Slow evaporation of all liquid (**HOOD**) gives crude $PdCl_2$. Dissolve the solids in concentrated warm hydrochloric acid, and again, evaporate to dryness. The remaining solid is $PdCl_2$, sufficiently pure for further use.

QUESTIONS

1. Account for the multiplicities in the ^{1}H NMR spectra of both products.
2. Write a mechanism showing the fluxionality of the monomer in DMSO. Be sure to show how the A and B protons can interconvert.
3. Why does the DMSO-d^6 spectrum change with temperature?
4. Provide reasoned arguments as to whether a Pd—Pd bond is present in the dimer.
5. Search the literature and detail two cases of organometallic molecular fluxionality not involving palladium.

REFERENCES

1. Maitlis, P. M.; Espinet, P.; Russell, M. J. H., "Allylic Complexes of Pd(II)" in *Comprehensive Organometallic Chemistry*, G. Wilkinson, Ed., Pergamon: Oxford, 1982, Vol. 6, Chapter 38.7, p. 385.

2. Lindley, J. *J. Chem. Educ.* **1980,** *57,* 671.
3. Bailey, C. T.; Lisensky, G. C. *J. Chem. Educ.* **1985,** *62,* 896.

GENERAL REFERENCES

Mingos, D. M. P., "The Bonding of Unsaturated Organic Molecules to Transition Metals" in *Comprehensive Organometallic Chemistry,* G. Wilkinson, Ed., Pergamon: Oxford, 1982, Vol. 3, Chapter 19, p. 1.

Mann, B. E., "Non-Rigidity in Organometallic Compounds" in *Comprehensive Organometallic Chemistry,* G. Wilkinson, Ed., Pergamon: Oxford, 1982, Vol. 3, Chapter 20, p. 89.

Experiment 40 Preparation and Use of Ferrocene

Part A: Preparation of Cyclopentadiene

Part B: Preparation of Ferrocene

Part C: Acetylation of Ferrocene

Part D: HPLC Analysis of the Acetylation Reaction

INTRODUCTION

In terms of importance in organometallic chemistry, the cyclopentadienyl ligand (C_5H_5, abbreviated Cp) is second only to the carbonyl ligand. For electron counting purposes, a π-bound C_5H_5 ligand may either be considered as a 6-electron donor anion, $C_5H_5^-$, or as a 5-electron donor neutral radical, C_5H_5. Both notations are in common use and the formal oxidation state of the metal must be adjusted accordingly. The cyclopentadienyl ligand can also be σ bonded, and act as 1-electron donor.

Metallocenes are bis(cyclopentadienyl)metal-type compounds, $(\eta^5\text{-Cp})_2M$. The earliest discovered and best known example is ferrocene, $(\eta^5\text{-Cp})_2Fe$.

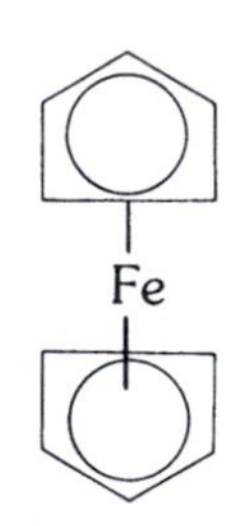

Other examples include manganocene, cobaltocene, and nickelocene. These metallocenes are also known as "sandwich" compounds, because the metal is sandwiched between two Cp rings. The aromatic nature of the Cp ring can be easily demonstrated because Cp undergoes Friedel–Crafts acylation when treated with acetyl chloride and aluminum chloride in the same manner as benzene. Many "half-sandwich" compounds are also known. Two examples of such compounds are $(\eta^5\text{-Cp})Fe(CO)_2X$ and $(\eta^5\text{-Cp})Mn(CO)_3$. Most of these compounds are coordinatively saturated and diamagnetic.

Cyclopentadienyl metallocenes are generally prepared in two steps.

1. Cyclopentadiene is generated by the thermal cracking of its commercially available dimer, followed by its conversion to a $C_5H_5^-$ salt by treating it with a base.
2. A metal salt is treated with the $C_5H_5^-$ salt in a suitable solvent.

An example of the metal–arene "sandwich" compounds is bis(benzene)-chromium, $Cr(C_6H_6)_2$, first prepared by E. O. Fischer in 1955. However, a more important class of compounds in this series are the arene tricarbonyl derivatives of the Group 6 (VIB) metals. These compounds have found a wide range of synthetic usage in organic chemistry.

The arene compounds are prepared by simply boiling the metal hexacarbonyls in benzene.

$$Cr(CO)_6 + C_6H_6 \rightarrow Cr(CO)_3(C_6H_6)$$

Ferrocene is a diamagnetic orange solid (mp 174 °C) that sublimes at 100 °C; it is stable in air and insoluble in water. Its preparation involves the two steps just outlined for cyclopentadienyl compounds: Cyclopentadiene is prepared by the thermal cracking of dicyclopentadiene and it is converted into the cyclopentadienide anion by treatment with KOH.

$$C_5H_6 + KOH \rightarrow K^+C_5H_5^- + H_2O$$

The use of KOH is preferable, because apart from being basic, it is also a good dehydrating agent. The potassium salt is very unstable in air. Its preparation must, therefore, be carried out under an inert atmosphere. The potassium salt is then reacted with hydrated iron(II) chloride, resulting in the formation of ferrocene.

$$2K^+C_5H_5^- + FeCl_2{\cdot}4H_2O = (\eta^5\text{-}C_5H_5)_2Fe + 2KCl + 4H_2O$$

Ferrocene has extensive chemistry. An interesting reaction of ferrocene is the facile displacement of bromide from ferrocenyl bromide by a variety of nucleophiles. In addition, it very readily undergoes Friedel–Crafts acylation and alkylation reactions. In fact, ferrocene is 10^6 times more reactive than benzene towards acetylation. Thus, in the presence of a Friedel–Crafts catalyst (phosphoric acid in this experiment) ferrocene can be conveniently acetylated. The catalyst generates the electrophile $[CH_3CO]^+$ from acetic anhydride, which then attacks the cyclopentadienyl ring of ferrocene forming monoacetyl and diacetyl products. At the end of the acetylation reaction, one obtains a product mixture

$$(CH_3\overset{\overset{\displaystyle O}{\|}}{C})_2O + H_3PO_4 \rightarrow [CH_3CO]^+ + CH_3COOH + H_2PO_4^-$$

$$Fe + CH_3CO^+ \rightarrow Fe\ (\text{one ring}-\overset{O}{\overset{\|}{C}}CH_3) + Fe\ (\text{both rings}-\overset{O}{\overset{\|}{C}}CH_3)$$

of unreacted ferrocene, and the mono- and diacetylated ferrocenes. Thin-layer chromatography may be used to quickly identify how many components there are in the product mixture, and which solvent is the best for efficient separation of these components. Column chromatography is used to separate and purify the products. High-performance liquid chromatography may be used to follow the course of the reaction, as well as for the detection of reaction products.

Prior Reading and Techniques

Section 5.C: Vacuum and Inert Atmosphere Techniques
Section 5.G.2: Thin-Layer Chromatography
Section 5.H: Sublimation
Section 6.C: Infrared Spectroscopy
Section 6.D: Nuclear Magnetic Resonance Spectroscopy

Related Experiments

Iron Chemistry: Experiments 32 and 45

EXPERIMENTAL SECTION

Part A: Preparation of Cyclopentadiene

Safety Recommendations

Dicyclopentadiene (CAS No. 77-73-6): This compound is harmful if swallowed, inhaled, or absorbed through the skin. ORL-RAT LD50: 353 mg/kg.

Silicone oil (CAS No.: none): No toxicity data is available for this compound, but it is normally not considered dangerous. The usual precautions should be taken (Section 1.A.3).

CHEMICAL DATA

Compound	FW	Amount	mmol	mp (°C)	bp (°C)	Density
Dicyclopentadiene	132.21	2 mL	14.9	−1	170	0.986

Required Equipment

Magnetic stirring hot plate, 25-mL round-bottom side arm flask with stopcock, distillation head, water condenser, vacuum adapter, 10-mL round-bottom flask, sand bath, ice–water bath, magnetic stirring bar, calibrated Pasteur pipet, source of nitrogen, source of vacuum.

Time Required for Experiment: 1.5 h.

EXPERIMENTAL PROCEDURE

NOTE: *This part of the experiment should be carried out in the HOOD because of the strong odor of the dicyclopentadiene and cyclopentadiene.*

Freshly prepared cyclopentadiene is necessary to accomplish this synthesis. Assemble the inert atmosphere apparatus shown in Figure 9.5. Add 1 mL of silicone oil (Nujol or mineral oil may be used instead) to the 10-mL round-bottom side arm flask using a calibrated Pasteur pipet.

Purge the apparatus with N_2 gas for 15 min and maintain a slow nitrogen flow throughout the experiment (20–30 bubbles of N_2 gas per minute as measured with a bubbler). Heat the oil to about 60°C, and using a calibrated Pasteur pipet, add 2 mL (14.9 mmol) of dicyclopentadiene dropwise, through the thermometer port. Replace the thermometer and collect the fraction distilling at 42–45°C in the receiving flask. The vacuum adaptor may be connected to a vacuum

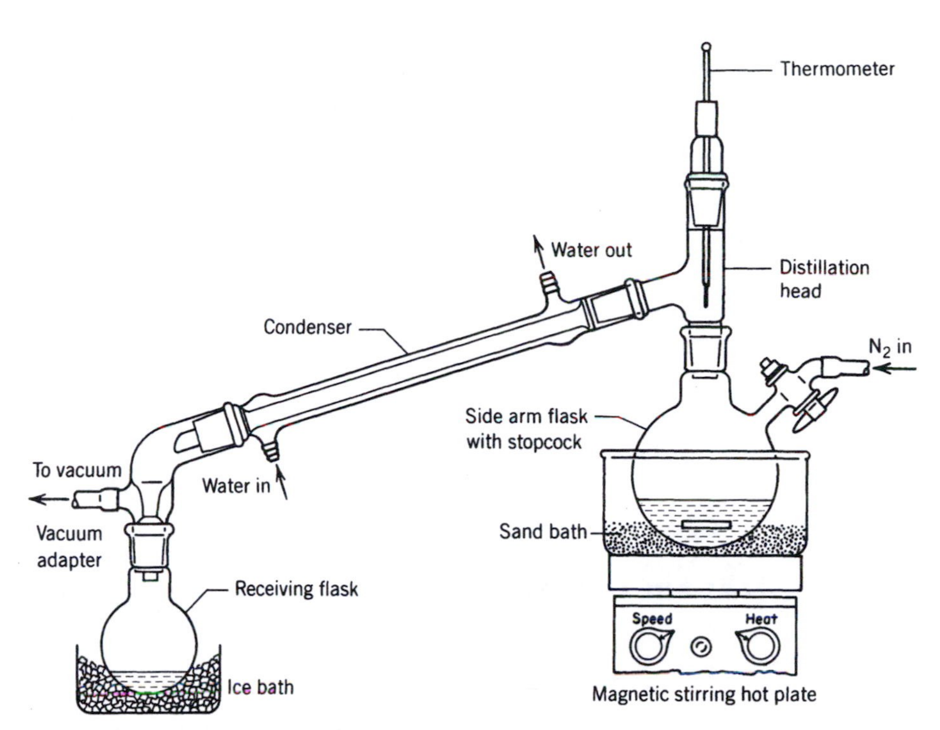

Figure 9.5. *Apparatus for Experiment 40.*

system, if desired, and the cyclopentadiene distilled under slightly reduced pressure. Otherwise, use the vacuum adaptor line as an exhaust to a **HOOD**.

> **NOTE:** ***While the distillation is in progress proceed to Part 40.B.***

Placing an ice–water bath around the receiving flask will assist the collection, by preventing loss of cyclopentadiene due to evaporation. The cyclopentadiene prepared must be used immediately, or else it will undergo a Diels–Alder reaction, reforming dicyclopentadiene. Alternatively, it must be stored at a temperature of −78 °C or below.

Part B. Preparation of Ferrocene

Additional Safety Recommendations

Ethylene glycol (CAS No. 107-21-1): Also known as common antifreeze. This compound is harmful if swallowed, inhaled, or absorbed through the skin. ORL-HMN LDLo: 398 mg/kg, ORL-RAT LD50: 4700 mg/kg.

Iron(II) chloride tetrahydrate (CAS No. 13478-10-9): This compound is harmful if swallowed. ORL-RAT LD50: 984 mg/kg.

Potassium hydroxide (CAS No. 1310-58-3): This compound is highly corrosive and very hygroscopic. Ingestion will produce violent pain in the throat. ORL-RAT LD50: 1.23 g/kg. If contacted with the skin, wash with large quantities of water.

Dimethyl sulfoxide (CAS No. 67-68-5): This compound is harmful if swallowed, inhaled, or absorbed through the skin. ORL-RAT LD50: 14,500 mg/kg.

CHEMICAL DATA

Compound	FW	Amount	mmol	mp (°C)	bp (°C)	Density
Cyclopentadiene	66.10	600 μL	7.26		42.5	0.80
$FeCl_2 \cdot 4H_2O$	198.81	750 mg	3.77			1.930
KOH	56.11	1.5 g	26.7			

Required Equipment

Mortar and pestle, two 10-mL side arm flasks, 10-mL graduated cylinder, syringe, source of nitrogen, magnetic stirring hot plate, magnetic stirring bar, Pasteur pipet, 25-mL graduated cylinder, sublimation apparatus (optional), Hirsch funnel, clay tile, filter paper.

Time Required for Experiment: 1 h.

EXPERIMENTAL PROCEDURE[1,2]

While the distillation proceeds, grind 1.5 g of KOH in a mortar as rapidly as possible.

> **CAUTION: *KOH is very corrosive and highly hygroscopic. Wear gloves.***

Quickly transfer the finely ground powder to a 10-mL side arm flask equipped with a magnetic stirring bar that was previously flushed with N_2 gas. Using a graduated cylinder, add 2.50 mL of ethylene glycol. Bubbling N_2 gas through the mixture helps in stirring the solution and maintaining a positive pressure of the gas. Stir the solution using a magnetic stirring hot plate.

Using a syringe, add 600 μL (7.26 mmol) of the freshly distilled cyclopentadiene prepared in Part 40.A directly to the mixture of KOH in ethylene glycol. Continue the slow passage of N_2 gas and stir the flask vigorously. The solution will turn brown in color because of the formation of potassium cyclopentadienide salt.

While the potassium cyclopentadienide is forming, prepare a solution of 750 mg (3.77 mmol) of $FeCl_2 \cdot 4H_2O$ in 2 mL (graduated cylinder) of DMSO in a 10-mL side arm flask under a purge of N_2 gas. Using a Pasteur pipet that was flushed with N_2, transfer the $FeCl_2 \cdot 4H_2O$ solution dropwise to the cyclopentadienide solution prepared above. The addition should be carried out slowly, over a period of approximately 10 min. Stir the solution continuously for 15–20 min to complete the reaction. Transfer the materials to a 25-mL Erlenmeyer flask.

Isolation of Product

Prepare a mixture of 10 g of ice and 8 mL of 6*M* HCl in a graduated cylinder, and add this mixture to the 25-mL flask. Stir the mixture thoroughly to neutralize any remaining KOH (if necessary add more HCl, test with pH paper). Filter the orange crystals of product using a Hirsch funnel and wash the crystals with two 100-μL portions of water. Draw air through the product on the Hirsch funnel for 5 min, dry the crystals between the folds of filter paper, and finally dry the product on a clay tile. Calculate the percentage yield and determine the melting point. This product, when dry, is quite satisfactory for subsequent reactions. If desired, it can be purified by sublimation.

Sublimation of Product

Sublimation may be carried out in the apparatus shown in Figure 5.40. Place a cold finger (a test tube of proper size or a centrifuge tube) in an adapter fitted tightly to a filtration flask. Place the dry product in the flask and fill the cold finger with ice water. Evaculate the flask using vacuum, while at the same time warming the flask on a hot plate. Ferrocene will sublime and collect on the outside surface of the cold finger. When sublimation is complete, disassemble the sublimation apparatus and carefully collect the crystals. Calculate the percentage yield and obtain the melting point of the purified product.

Characterization of Product

Obtain the IR spectrum (KBr pellet) of the ferrocene product and compare it to published data.[3] Obtain the NMR spectrum of the sample in $CDCl_3$ and compare the chemical shift to that of cyclopentadiene.

Part C: Acetylation of Ferrocene

Additional Safety Recommendations

Acetic anhydride (CAS No. 108-24-7): This compound is harmful if swallowed, inhaled, or absorbed through the skin. ORL-RAT LD50: 1780 mg/kg. It reacts with water to form acetic acid, and must be kept dry. It is also a mild lachrymator (makes you cry).

Iodine (CAS No. 7553-56-2): Iodine is harmful if swallowed, inhaled, or absorbed through the skin. It is a lachrymating agent (makes you cry). ORL-RAT LD50: 14 g/kg. Ingestion of 2–3 g has been fatal.

Sodium bicarbonate (CAS No. 144-55-8): This compound is not normally considered dangerous, but normal precautions (Section 1.A.3) should be taken. ORL-RAT LD50: 4220 mg/kg.

CHEMICAL DATA

Compound	FW	Amount	mmol	mp (°C)	bp (°C)	Density
Ferrocene	186.04	150 mg	0.81	174	249	
Acetic anhydride	102.09	500 μL	8.7	−73	138	1.082

Required Equipment

Magnetic stirring hot plate, 10-mL round-bottom flask, magnetic stirring bar, Pasteur pipets, automatic delivery pipet, air condenser, $CaCl_2$ drying tube, sand bath, ice–water bath, Hirsch funnel, clay tile, six TLC plates, silica gel, cotton, six 8-cm bottles with caps, two 10-mL Erlenmeyer flasks.

Time Required for Experiment: 1.5 h.

EXPERIMENTAL PROCEDURE

Place 150 mg (0.81 mmol) of ferrocene (prepared in Part 40.B) in a 10-mL round-bottom flask equipped with a magnetic stirring bar. Using a Pasteur pipet, add 2–3 drops of 85% phosphoric acid. Add 500 μL (0.525 g, 8.7 mmol) of acetic anhydride, using an automatic delivery pipet. Attach an air condenser and a $CaCl_2$ drying tube. Heat the mixture in a sand bath at 100 °C for 10 min. Add ~2 g of crushed ice to the reaction mixture. When the ice has all melted, neutralize the mixture by adding solid $NaHCO_3$ in *small quantities* (to avoid sudden overflow of the solution because of rapid evolution of CO_2 gas) to the flask until CO_2 evolution ceases.

Isolation of Product

Cool the flask in an ice–water bath for 20–30 min. A brown solid consisting of a mixture of ferrocene and its mono- and diacetylated derivatives will precipitate out from the orange solution. Collect the solid by suction filtration using a Hirsch funnel and wash the brown crystals with a 100-μL portion of water. Dry the product on a clay tile.

Separation of Products by Thin-Layer Chromatography (TLC)

Obtain six TLC plates (2.5 × 7.5 cm) with silica absorbent (see Section 5.G.2). Prepare a concentrated solution of the product prepared above in 2–3 drops of toluene. With a lead pencil, draw a line on the silica coating 1 cm from the 2.5-cm edge of a TLC plate. Dip one end of the fine capillary applicator in the product solution. Touch this end to just below the line drawn on the TLC plate. Two such spots should be made on each slide. Prepare five such plates, with two spots on each. Keep a sixth plate for later work (see below).

Select five small bottles with covers and having lengths slightly larger than 7.5 cm to serve as developing chambers. Add 2–3 mL of a different trial solvent [such as petroleum ether (bp 60–70 °C), diethyl ether, a mixture of petroleum ether and diethyl ether, ethyl acetate, and 10% ethyl acetate–90% petroleum ether by volume] to each bottle.

Insert one of the spotted TLC plates into each bottle, with the spotted end dipping in the solvent. Cap the bottles. Make sure that the pencil mark remains above the solvent level. Allow the solvents to rise along the plates until the solvent front has reached three quarters of the way to the top of the plate. Remove the plates from the developing chambers and allow them to dry in air. Develop the plates by placing them in a closed container containing few crystals of iodine for a few minutes. The iodine vapors will form brown spots where the components of the product mixture have moved on the plates. You will notice several spots on each plate, indicating that there are as many components as there are spots. By inspecting the slides and location of spots, determine which solvent gives the best separation of the products.

In order to determine which spot is due to unreacted ferrocene, make two spots on the sixth TLC plate—one each of the solution of the product mixture and of pure ferrocene in toluene. Place the plate in the developing chamber containing the solvent that gave the best separation of the products. Develop the plate as described before.

Column Chromatographic Separation of the Products

Prepare a microchromatographic column in a Pasteur pipet by pushing a small wad of cotton down to the tip of the pipet. Fill the pipet nearly to the top with silica gel, which serves as the solid phase. Begin dripping the solvent selected in the TLC analysis through the column, at the rate of one drop per second. Place 40–60 mg of product in the solvent selected and add this dropwise to the column. Continue eluting with the sovent. **Do not allow the column to run dry.**

The column may exhibit two colored bands because of the two principal components, which are ferrocene and monoacetylated ferrocene. Unreacted ferrocene travels at a faster rate than the other product(s). When the first band begins eluting, place a clean 10-mL Erlenmeyer flask under the column, and collect the eluent. When the fraction is completely eluted, collect the second fraction in a separate 10-mL Erlenmeyer flask. When separation is complete, evaporate the solvent from each fraction collected (**HOOD**) with a gentle stream of N_2 on a warm sand bath.

Characterization of Product(s)

Obtain IR spectra (KBr pellets) of both components. Compare the spectra with that of pure ferrocene and determine the identity of each fraction.

Part D: HPLC Analysis of the Acetylation Reaction

> **NOTE:** ***The equipment and safety recommendations for Part 40.D are identical to Part 40.C, with the following changes: additional equipment, for example, separatory funnel and HPLC unit.***

EXPERIMENTAL PROCEDURE

In a 10-mL round-bottom flask, dissolve 230 mg (1.24 mmol) of ferrocene in 6.5 mL of acetic anhydride. After cooling the solution in an ice bath, add 500 μL of 85% phosphoric acid. Transfer the flask to a sand bath maintained at 50 °C. After 60 min, withdraw a 2.0-mL portion of the reaction mixture and quench the reaction by adding the fraction directly onto 5 g of ice in a 25-mL beaker. Add solid $NaHCO_3$ to neutralize the acid present until evolution of CO_2 ceases. Extract the mixture with 5.0 mL of ether, using a separatory funnel. Discard the lower water layer. Collect the ether layer in a small Erlenmeyer flask containing a small amount of anhydrous Na_2SO_4 to dry the sample. Allow the ether layer to stand for a few minutes and then pipet it into a 5-mL conical vial. Evaporate the solvent by passing a stream of N_2 gas over the ether layer (**HOOD**). Extract the residue with exactly 2.5 mL of chlorofom. Save this solution. Repeat the above procedure at 60 and 75°C.

Inject an 8-μL sample of the $CHCl_3$ solutions onto the HPLC column. The experimental conditions are

Packing material: silica gel (10 μm)

Flow rate: 1.0 $mL{\cdot}min^{-1}$

Mobile phase: mixture of ether–methanol in 10:1 ratio

Pressure: 500 psi

Elution: at room temperature

Detector: UV detector at 254 nm

Obtain the chromatogram for each fraction. A typical chromatographic run for a reaction mixture at 60 °C after heating for 75 min and with flow rate of 4.0 $mL{\cdot}min^{-1}$ shows peaks with retention times of ~1.3, ~2.4, and ~4.7 min, due to ferrocene, 1-acetylferrocene, and 1,1′-diacetylferrocene, respectively. Determine at which temperature the yield of 1-acetylferrocene is maximized, and at what temperature the maximum amounts of the various products are formed.

A variation of this work[3] can be accomplished by using different reaction times at a fixed temperature (e.g., by collecting fractions at 50 °C, after time intervals of 15, 30, 45, 60, 75, and 90 min). Quenching and HPLC analysis is as described previously.

QUESTIONS

1. Draw the reaction mechanism for the acetylization of ferrocene.
2. What is the geometric orientation of the Cp rings in ferrocene? Propose an explanation.
3. The second acetylation of ferrocene invariably occurs on the second ring, never on the same ring as the first acetylation. Explain.

4. Many more cyclopentadienyl metal compounds are known than arene metal compounds. Explain why.
5. Cyclopentadiene can bond in other ways than seen here with iron. From the literature, describe the bonding in titanocene, a rather controversial subject. (*Hint:* More than one form of titanocene is known.) Can cyclopentadiene bond in other ways as well?

REFERENCES

1. Angelici, R. J., *Synthesis and Techniques in Inorganic Chemistry,* 2nd ed., Saunders: Philadelphia, 1977.
2. Bozak, R. E. *J. Chem. Educ.* **1966,** *43,* 73.
3. Haworth, D. T.; Liu, T. *J. Chem. Educ.* **1976,** *53,* 730.

GENERAL REFERENCES

Deeming, A. J., "Compounds with η^5-Carbon Ligands" in *Comprehensive Organometallic Chemistry,* G. Wilkinson, Ed., Pergamon: Oxford, 1982, Vol. 4, Chapter 31.3, Section 4, p. 475.

Birmingham, J., "Synthesis of Cyclopentadienyl Metal Compounds" in *Advances in Organometallic Chemistry,* F. G. A. Stone and R. West, Eds., Academic Press: New York, 1964, Vol. 2, p. 365.

Wilkinson, G.; Cotton, F. A., "Cyclopentadienyl and Arene Metal Compounds" in *Progress in Inorganic Chemistry,* F. A. Cotton, Ed., Interscience: New York, 1959, Vol. 1, p. 1.

Experiment 41 Preparation of Organopalladium Complexes

Part A: Preparation of Dichlorobis(benzonitrile)palladium(II)

Part B: Preparation of Di-μ-chlorodichlorodiethylenedipalladium(II)

INTRODUCTION

Palladium is a member of the platinum family of metals, the other members being ruthenium, rhodium, osmium, iridium, and platinum. It was first discovered in 1803 by Wollaston, who added potassium cyanide to a solution of native platinum in aqua regia. This yielded a yellow precipitate of palladium(II) cyanide. (Do not try this reaction in the laboratory!) Palladium is a silver-white metal of very high density (12.02 g·cm^{-3}). At room temperature, palladium has the unusual property of absorbing up to 900 times its own volume of hydrogen. Hydrogen readily diffuses though heated palladium and this provides a common way for purifying the gas. Palladium compounds find major use as catalysts in hydrogenation reactions and in catalytic converters in automobiles.

In this experiment, an organopalladium complex is prepared via a two step synthesis. Palladium(II) chloride is first reacted with benzonitrile (C_6H_5CN) to form dichlorobis(benzonitrile)palladium(II)

$$PdCl_2 + 2C_6H_5CN \rightarrow Pd(C_6H_5CN)_2Cl_2$$

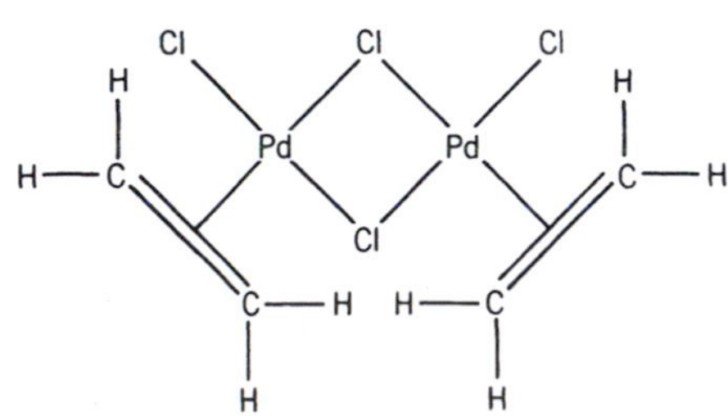

Figure 9.6. *Structure of di-μ-chlorodichlorodiethylenedipalladium(II).*

This material, in turn, is reacted with ethylene gas, forming di-μ-chlorodichlorodiethylenedipalladium(II), a palladium dimer (Fig. 9.6). The **di-μ-chloro** prefix indicates that two of the chlorines are bridging, and the **dichloro** that follows indicates that two chlorines are terminally attached. Each bridging chlorine is bonded directly to one palladium and uses a lone pair of electrons to act as a

Lewis base to the other palladium. Such bridging ligands are very common in inorganic chemistry.

Prior Reading and Techniques

Section 1.B.4: Compressed Gas Cylinders and Lecture Bottles

Section 5.D.3: Isolation of Crystalline Products (Suction Filtration)

Section 6.C: Infrared Spectroscopy

Related Experiments

Palladium Chemistry: Experiments 20B, 39, and 46

EXPERIMENTAL SECTION

Safety Recommendations

Palladium(II) chloride (CAS No. 7647-10-1): This material is moderately toxic and may cause heavy metal poisoning. Do not breathe the dust or get it on your hands. ORL-RAT LD50: 2704 mg/kg.

Benzonitrile (CAS No. 100-47-0): This compound has an overwhelming almond smell. Do not breathe the vapor. All utensils coming into contact with benzonitrile should be kept in the **HOOD** and not removed until washed. ORL-MUS LD50: 971 mg/kg.

Ethylene (CAS No. 74-85-1): This compound is a colorless, flammable, nonbreathable gas. Use it only in the **HOOD.**

CHEMICAL DATA

Compound	FW	Amount	mmol	bp (°C)	mp (°C)	Density
Palladium(II) chloride	177.31	130 mg	0.73		500[a]	4.000
Benzonitrile	103.12	1 mL	9.6	190.7	−13	1.010
Ethylene	28.05	Sufficient amount	Sufficient amount	−103.7	−169.15	0.0013

[a] Decomposes.

Part A: Preparation of Dichlorobis(benzonitrile)palladium(II)

Required Equipment

Air condenser, 3-mL conical vial, magnetic spin vane, automatic delivery pipet, sand bath, magnetic stirring hot plate, Hirsch funnel, 10-mL graduated cylinder, fritted glass filter, clay tile or filter paper.

Time Needed for Experiment: 2.0 h.

EXPERIMENTAL PROCEDURE

NOTE: *Part 41.A should be done entirely in the HOOD.*

Place 130 mg (0.73 mmol) of finely divided $PdCl_2$ in a 3-mL conical vial equipped with a magnetic spin vane and fitted with an air condenser. Add 1 mL of benzonitrile (9.6 mmol) using an automatic delivery pipet. A large excess of benzonitrile is used to force the reaction to completion. Heat the mixture gently in a sand bath set upon a magnetic stirring hot plate, until most of the

$PdCl_2$ has dissolved. Any insoluble material remaining at this point should be removed by suction filtration of the hot solution using a Hirsch funnel.

Isolation of Product

Cool the filtrate to room temperature and add 2 mL (graduated cylinder) of hexane to the reaction mixture. Collect the yellow crystals of dichlorobis(benzonitrile)palladium(II) which precipitate by suction filtration using a fritted glass filter (10–20 μ). Wash the crystals with a 1-mL portion of hexane and dry them on a clay tile. Weigh the product and calculate the percentage yield. One half of this product is retained for characterization and the remaining material is used in Part 41.B.

Characterization of Product

Obtain the melting point of the product. Prepare a KBr pellet of the material and obtain the IR spectrum. If the Nicolet–Aldrich FT IR computer data base searching program is available, search the major peaks of the benzonitrile complex. Compare the spectrum to that obtained for the product of Part 41.B.

Part B: Preparation of Di-μ-chlorodichlorodiethylenedipalladium(II)

Required Equipment

Magnetic stirring hot plate, 10-mL Erlenmeyer flask, magnetic stirring bar, 10-mL graduated cylinder, Hirsch funnel, ethylene lecture bottle and stand, fritted glass filter, clay tile, or filter paper.

Time Required for Experiment: 1 h.

EXPERIMENTAL PROCEDURE

Weigh out one half the dichlorobis(benzonitrile)palladium(II) product from Part 41.A, and place it in a 10-mL Erlenmeyer flask equipped with a magnetic stirring bar. Add 2 mL of toluene (graduated cylinder) and stir until dissolution is effected. Any insoluble material can be removed by suction filtration using a Hirsch funnel. In the **HOOD,** pass a slow stream of ethylene gas through the solution, until precipitation is complete (~15 min).

Isolation of Product

Collect the crystals that form by suction filtration using a fritted glass filter (10–20 μ) and wash them with a 1-mL portion of hexane. Dry the crystalline product on a clay tile and calculate the percentage yield.

Characterization of Product

Obtain the melting point of the product. Prepare a KBr pellet of the material and obtain the IR spectrum. If the Nicolet–Aldrich FT IR computer data base searching program is available, search the major peaks of the ethylene complex. Compare the spectrum to that obtained for the product of Part 41.A.

QUESTIONS

1. What is a Lewis base? Does the "normal" bond from the chlorine to the first palladium differ from the electron pair donation from the chlorine to the second palladium?
2. Compare the IR spectrum of the dichlorobis(benzonitrile)pallladium(II) with benzonitrile. Identify as many bands in both spectra as possible.
3. Compare the IR spectra of the two products from this experiment. Suggest a simple way of proving that the desired reaction has taken place.
4. Palladium has two stable oxidation states, II and IV. Why does this make it capable of being an oxidation catalyst?

5. Why should Wollaston's reaction not be tried in an introductory laboratory? What is aqua regia?
6. Search the literature to find at least three specific uses for palladium as a catalyst.

REFERENCE

1. Oberhansli, W. E.; Dahl, L. F. *J. Organomet. Chem.* **1965,** *3,* 43.

GENERAL REFERENCE

Maitlis, P. M.; Espinet, P.; Russell, M. J. H., "Monoolefin and Acetylene Complexes of Palladium" in *Comprehensive Organometallic Chemistry,* G. Wilkinson, Ed., Pergamon: Oxford, 1982, Vol. 6, Chapter 38.5, p. 351.

Experiment 42 Synthesis of Metal Carbonyls

Part A: Preparation of *trans*-Chlorocarbonylbis(triphenylphosphine)rhodium(I)

Part B: Preparation of *mer*-Carbonyltrichlorobis(triphenylphosphine)-rhodium(III)

Part C: Synthesis of an SO_2 Adduct of *trans*-Chlorocarbonylbis(triphenylphosphine)-rhodium(I)

INTRODUCTION

Compounds in which a metal atom is directly bonded to carbon are known as organometallic compounds. Organometallic compounds are heavily used in the area of organic synthesis and in industrial chemistry. Metals in organometallic compounds are generally found in low oxidation states, with the most common carbon ligands being CO (called a carbonyl ligand), alkenes, $C_5H_5^-$ (cyclopentadienyl anion, abbreviated Cp^-), and C_6H_6 (benzene). The bonding to π olefins in these compounds was first described[1] by M. J. S. Dewar as consisting of two aspects.

1. Electrons are donated from the filled olefin π orbital to an empty metal σ orbital.
2. Electrons are "back-donated" from filled metal *d* orbitals to empty olefin π^* orbitals.

This "give-and-take" arrangement is termed synergistic bonding. Bonding to the CO group is similarly synergistic. The carbonyl ligand donates the lone pair of electrons on the carbon to an empty metal σ orbital, and the metal "back-donates" electrons from the filled metal π orbital to the empty π^* orbital of the carbonyl as shown in Figure 9.7.

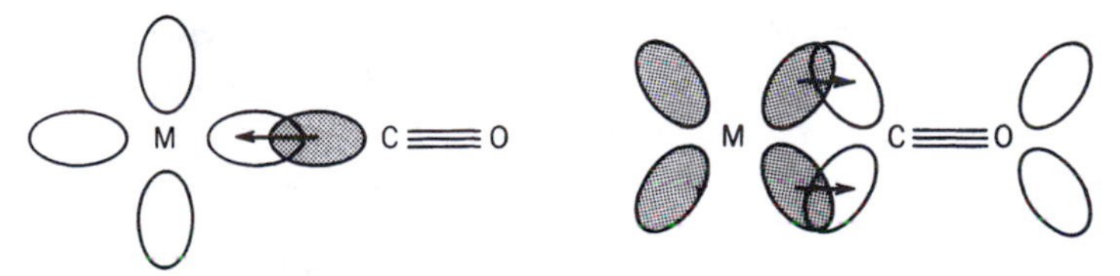

Figure 9.7. *Metal–carbonyl bonding.*

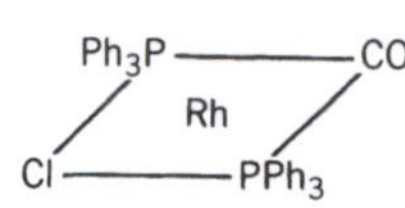

Figure 9.8. *Structure of* trans-*chlorocarbonylbis(triphenylphosphine)-rhodium(I).*

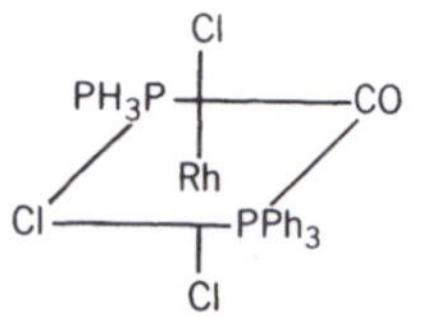

Figure 9.9. *Structure of* mer-*carbonyltrichlorobis(triphenylphosphine)-rhodium(III).*

The donation of electrons from the metal to the π^* orbitals of the carbonyl has a drastic effect on the IR frequency of the CO stretch. In free carbon monoxide, this stretch occurs at 2143 cm^{-1}. As the metal donates electron density to the π^* orbital, the bond order of the carbonyl will decrease (an antibonding orbital is being filled), and the IR stretch appears closer to that of C═O at 1700 cm^{-1}. Infrared spectroscopy is therefore a very sensitive indicator of the nature of bonding in metal carbonyls.

Metal carbonyls are most often prepared by the direct reaction of a metal with carbon monoxide gas. This reaction is quite dangerous, as CO will bind nonreversibly with hemoglobin, and is therefore extremely toxic. In this reaction, CO is generated *in situ* (within the reaction system) employing the much safer reagent DMF as the source of the CO group. A square planar triphenylphosphine complex (Fig. 9.8) is prepared in Part 42.A via the reaction of the starting material, rhodium(III) chloride hydrate, with triphenylphosphine and DMF (unbalanced).

$$RhCl_3 \cdot 3H_2O + 2(C_6H_5)_3P + HCON(CH_3)_2 \rightarrow RhCl(CO)(P(C_6H_5)_3)_2$$

Addition of chlorine to the square planar complex, above, results in a six-coordinate rhodium(III) species, *mer*-carbonyltrichlorobis(triphenylphosphine)-rhodium(III), shown in Figure 9.9.

Prior Reading and Techniques

Section 1.B.4: Compressed Gas Cylinders and Lecture Bottles

Section 2.F: Reflux and Distillation

Section 5.D.3: Isolation of Crystalline Products (Suction Filtration)

Section 5.F.3: Removal of Solvent Under Reduced Pressure

Section 6.C: Infrared Spectroscopy

Related Experiments

Metal Carbonyls: Experiments 34, 43, and 45

Rhodium Chemistry: Experiments 21, 24A, and 34

EXPERIMENTAL SECTION

Safety Recommendations

Rhodium(III) chloride hydrate (CAS No. 20765-98-4): This compound is harmful if swallowed, inhaled, or absorbed through the skin. ORL-RAT LD50: 1302 mg/kg. It is a possible mutagen, although this has not been definitively established.

Triphenylphosphine (CAS No. 603-35-0): This compound is a mild lachrymator, but has a low vapor pressure, so use of small quantities is not a problem. The compound can cause skin irritation. ORL-RAT LD50: 700 mg/kg.

N-N-Dimethylformamide (CAS No. 68-12-2): This compound is harmful if swallowed, inhaled, or absorbed through the skin. The vapor is irritating to the eyes and mucous membranes. DMF reacts violently with carbon tetrachloride. ORL-RAT LD50: 2800 mg/kg.

Chloroform (CAS No. 67-66-3): This compound is a narcotic agent (it knocks you out). Avoid breathing the fumes. Avoid contact with the skin. It is classified as a carcinogen. ORL-RAT LD50: 908 mg/kg. Use only in the **HOOD.**

Carbon tetrachloride (CAS No. 56-23-5): This compound is classified as a carcinogen. Avoid contact with the skin. Avoid breathing the fumes. Use only in the **HOOD.** ORL-RAT LD50: 2350 mg/kg.

Sodium bisulfite (CAS No. 7631-90-5): This compound is harmful if swallowed, inhaled, or absorbed through the skin. ORL-RAT LD50: 2000 mg/kg.

CHEMICAL DATA

Compound	FW	Amount	mmol	bp (°C)	mp (°C)	Density
Rhodium(III) chloride hydrate	209.26[a]	25 mg	0.119			
DMF	73.10	2 mL	25.8	153	−61	0.944
Triphenylphosphine	262.29	100 mg	0.381	377	79	

[a] Anhydrous material.

Part A: Synthesis of *trans*-Chlorocarbonylbis(triphenylphosphine)rhodium(I)

Required Equipment

Magnetic stirring hot plate, 10-mL round-bottom flask, magnetic stirring bar, water condenser, sand bath, Pasteur filter pipet, Hirsch funnel, clay tile.

Time Required for Experiment: 2.5 h.

EXPERIMENTAL PROCEDURE[2]

In a 10-mL round-bottom flask equipped with a magnetic stirring bar, place 2-mL DMF, followed by 25 mg of hydrated rhodium(III) chloride. Attach a water condenser to the flask, place the apparatus in a sand bath set upon a magnetic stirring hot plate, and stir the solution for 5 min. The solution should then be heated at reflux until the color changes from dark brown to lemon yellow (~20 min).

Cool the solution to room temperature. Remove any solids that may still be present by suction filtration using a Hirsch funnel. Transfer the solution to the funnel using a Pasteur filter pipet. Wash the funnel with a few drops of DMF to ensure that no rhodium(I) carbonyl adheres to it. Return the liquid to the reaction flask using a Pasteur pipet.

Place the reaction vessel in the **HOOD.** Cautiously, in small portions, add 100 mg of triphenylphosphine to the solution, until no further evolution of gas is evident.

> **CAUTION:** ***The gas being released is carbon monoxide, which is highly toxic.***

By the end of the addition, shiny yellow crystals of product should precipitate from the solution.

Isolation of Product

To complete the precipitation, add a few drops of absolute ethanol, and cool the solution in a rock salt–ice bath for 30 min. Suction filter the crystals using a Hirsch funnel, and wash them with a 0.5-mL portion of absolute ethanol and one of ether. Dry the crystals on a clay tile and determine the percentage yield.

Characterization of Product

Prepare a KBr pellet of the product and obtain an IR spectrum. Determine the stretching frequency of the CO band.

Part B: Synthesis of *mer*-Carbonyltrichlorobis(triphenylphosphine)rhodium(III)

Required Equipment

Automatic delivery pipet, 5-mL conical vial, magnetic spin vane, Hirsch funnel, and clay tile.

Time Required for Experiment: 30 min.

EXPERIMENTAL PROCEDURE

NOTE: *Students should work in pairs. The following reaction should be done in the HOOD.*

In a 5-mL conical vial equipped with a magnetic spin vane, place 25 mg of $Rh(CO)Cl(PPh_3)_2$ (prepared in Part 42.A) and 1.5 mL of chloroform (automatic delivery pipet). When the solid has completely dissolved to form a yellow solution, add 1 mL of chlorine saturated CCl_4 (automatic delivery pipet). The solution immediately turns red-brown.

NOTE: *The chlorine–CCl_4 solution may be easily prepared by bubbling chlorine gas (HOOD) through the CCl_4 for 30 s. The chlorine may be delivered directly from a lecture bottle or generated by attaching an addition funnel to a side arm flask containing solid $KMnO_4$. Dropwise, add concentrated HCl to the solid $KMnO_4$, forming chlorine gas, which is delivered to the CCl_4 through the side arm.*

Allow the mixture to stand for 10 min, whereupon precipitation of the product should begin. Attach the conical vial to a rotary evaporator and strip off all solvent in the **HOOD.** (Alternatively, pass a gentle stream of nitrogen over the solution until all solvent has evaporated.) The CCl_4–chloroform mixture stripped off by the rotary evaporator should be collected by the laboratory instructor for proper disposal.

Isolation of Product

Disperse the solid in 1 mL of ethanol, suction filter the resulting mixture using a Hirsch funnel, and wash the filter cake with 0.5-mL portions each of ethanol and ether. Dry the solid on a clay tile and determine the percentage yield.

Characterization of Product

Prepare a KBr pellet of the product and obtain the IR spectrum. Determine the stretching frequency of the CO band. How does the frequency compare to those of free carbon monoxide and $Rh(CO)Cl(PPh_3)_2$?

Part C: Synthesis of an SO_2 adduct of *trans*-Chlorocarbonylbis(triphenylphosphine)rhodium(I)

Required Equipment

SO_2 generator (see Fig. 9.10), 5-mL beaker, acetone-dry ice slush.

Time Required for Experiment: 1 h.

EXPERIMENTAL PROCEDURE

> **NOTE:** ***The entire procedure should be carried out in the HOOD. Sulfur dioxide (SO_2) is a toxic gas that attacks the mucous membranes. IHL-HMN LCLo: 1000 ppm/10 M.***

In a 5-mL beaker, dissolve 10 mg (0.014 mmol) of *trans*-$Rh(CO)Cl(PPh_3)_2$ (prepared in Part 42.A) in 1 mL of chloroform. Place this in the liquid SO_2 generator shown in Figure 9.10.

The generator consists of a wide-mouthed bottle (100 mL) fitted with a three-hole rubber stopper. A centrifuge tube, Pasteur pipet, and vent tube are fitted into the stopper as shown. The bottom of the bottle is covered with solid $NaHSO_3$. The centrifuge tube is filled with an acetone–dry ice slush.

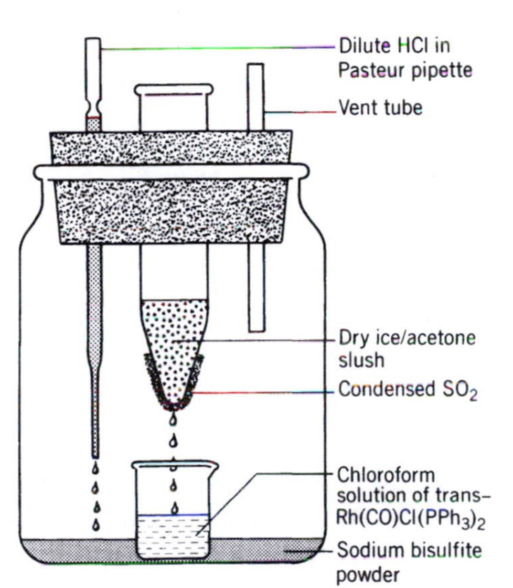

Figure 9.10. *Liquid SO_2 generator.*

> **NOTE:** ***Dry ice can cause severe burns if handled with bare hands. Use tongs or heavy gloves.***

Place the beaker in the center of the bottle so that the centrifuge tube is centered directly above it. The Pasteur pipet is filled with 6*M* HCl and inserted loosely into the third hole in the stopper. The HCl is slowly dripped onto the $NaHSO_3$ powder, generating SO_2 gas, which condenses onto the centrifuge tube. The condensed liquid SO_2, in turn, drips directly into the reaction beaker. The solution turns green upon addition of the SO_2.

Isolation of Product

When the reaction is complete (~2 min), remove the beaker and allow the solvent to evaporate under a gentle nitrogen flow, holding the beaker in an ice–salt bath. The green residue is the SO_2 adduct.

Characterization of Product

Immediately, obtain the IR spectrum as a Nujol mull. Compare the IR spectrum to those of the products from Parts 42.A and B.

The reversibility of the SO_2 adduct formation reaction can be easily demonstrated by warming the green adduct on a watch glass held over a water bath, regenerating the original yellow trans compound. This reaction can be monitored via IR spectroscopy.

QUESTIONS

1. What is the hybridization of the rhodium atom in each of the products?
2. The infrared CO stretching frequency is higher for $Rh(CO)Cl_3(PPh_3)_2$ than for $Rh(CO)Cl(PPh_3)_2$. Explain. (*Hint:* Consider the oxidation states of the Rh and the ability of the metal to back-donate electrons to the carbonyl group.) Where would you expect the CO band for $Rh(CO)Cl_2(PPh_3)_2$ to appear?

3. Write balanced redox equations for the reactions carried out in Parts 42.A and B.
4. From the literature, determine the major industrial usages of rhodium and its compounds.

REFERENCES

1. Dewar, M. J. S. *Bull. Soc. Chim. Fr.* **1959,** *18,* C79.
2. Singh, M. M.; Szafran, Z.; Pike, R. M. *J. Chem. Educ.* **1990,** *67,* A180.

GENERAL REFERENCES

Hughes, R. P., "Rhodium-Carbonyl Compounds" in *Comprehensive Organometallic Chemistry,* G. Wilkinson, Ed., Pergamon: Oxford, 1982, Vol. 5, Chapter 35, p. 277.

Griffith, W. P., "Carbonyls, Cyanides, Isocyanides and Nitrosyls" in *Comprehensive Inorganic Chemistry,* J. C. Bailar, et al., Eds., Pergamon: Oxford, 1973, Vol. 4, Chapter 46, p. 105.

Hieber, W., "Metal Carbonyls" in *Advances in Organometallic Chemistry,* F. G. A. Stone and R. West, Eds., Academic Press: New York, 1970, Vol. 8, p. 1.

Experiment 43 Sunlight Photochemistry: Preparation of Dicarbonyl(η^5-methylcyclopentadienyl) triphenylphosphinemanganese(0)

INTRODUCTION

The recent growth of interest in solar energy lends increasing significance to solar-activated chemistry. One example of this is photoactivated chemical synthesis, where solar radiation may be employed to carry out chemical reactions. Although many photochemical reactions are known, few are carried out using sunlight, partly because of the low intensity of atmospherically filtered radiation in the useful wavelength regions.

The present experiment proceeds under sunlight activation and involves the photosubstitution of a triphenylphosphine ligand to replace a carbonyl group in $(CH_3C_5H_4)Mn(CO)_3$.

$$(CH_3C_5H_4)Mn(CO)_3 + P(C_6H_5)_3 = (CH_3C_5H_4)Mn(CO)_2P(C_6H_5)_3 + CO$$

The manganese starting material is widely used as an octane booster in unleaded gasolines.

To increase yields and decrease photodecomposition, the reaction is usually carried out in a weakly coordinating solvent, such as THF or ether. The product, $(CH_3C_5H_4)Mn(CO)_2P(C_6H_5)_3$, is a stable crystalline solid, which precipates from the reaction mixture and can be recrystallized in high purity.

Inert atmosphere techniques are required for optimum yields. In the presence of air and light, the crystalline product is accompanied by large amounts of brown oxidation products. A second substitution can be accomplished by using larger excesses of triphenylphosphine.

Prior Reading and Techniques

Section 5.C: Vacuum and Inert Atmosphere Techniques

Section 5.D.3: Isolation of Crystalline Products (Suction Filtration)

Section 5.D.4: The Craig Tube Method

Section 6.C: Infrared Spectroscopy

Related Experiments

Manganese Chemistry: Experiment 22B

Metal Carbonyls: Experiments 34, 42, and 45

EXPERIMENTAL SECTION

Safety Recommendations

Methylcyclopentadienylmanganese tricarbonyl (CAS No. 12108-13-3): This compound may be fatal if inhaled, swalllowed, or absorbed through the skin. It is a possible carcinogen. Use it with care. ORL-RAT LD50: 50 mg/kg. Only use this compound in the **HOOD**.

Triphenylphosphine (CAS No. 603-35-0): This compound is a mild lachrymator, but has a low vapor pressure, so use of small quantities is not a problem. The compound can cause skin irritation. ORL-RAT LD50: 700 mg/kg.

CHEMICAL DATA

Compound	FW	Amount	mmol	mp (°C)	bp (°C)	Density
$(c\text{-}CH_3\text{—}C_5H_4)\text{-}Mn(CO)_3$	218.09	100 μL	0.63	−1	232	1.380
$P(C_6H_5)_3$	262.28	200 mg	0.76	80.5	377	1.194

Required Equipment

Hirsch funnel, 50-mL Erlenmeyer flask, stopper, clay tile or filter paper, Craig tube.

Time Necessary for Experiment: 30 min (first laboratory period), 1-week waiting period, 1 h (second laboratory period).

EXPERIMENTAL PROCEDURE[1–3]

In the **HOOD**, place 10 mL of hexane, 100 μL (0.63 mmol) of $(c\text{-}CH_3C_5H_4)Mn(CO)_3$, and 200 mg (0.77 mmol) of triphenylphosphine in a 50-mL Erlenmeyer flask. Stopper the flask, and shake it until all the triphenylphosphine has dissolved. A vortex mixer is useful for this purpose.

Remove the stopper and bubble N_2 gas through the solution for about 5 min. Restopper the flask and place the Erlenmeyer flask in a location where it will be in the presence of direct sunlight during the daytime. Allow the flask to remain there for 1 week. A crystalline precipitate should form during this period.

Isolation of Product

Collect the product by suction filtration using a Hirsch funnel. Recrystallize the golden-yellow needles of product (Craig tube method) by dissolving them in a minimum amount of hot acetone, cooling in an ice–water bath, and reprecipitating by adding 3–5 drops of distilled water. Dry the product on a clay tile or on filter paper. Determine the melting point and the percentage yield.

Characterization of Product

Obtain the IR spectrum of both the starting material and the product as KBr pellets. Compare the results.

> **NOTE: *If an absorption band is observed at 1836 cm^{-1}, an impurity is present, which is the disubstituted product, ($CH_3C_5H_4$)Mn(CO)(P(C_6H_5)$_3$)$_2$. The impurity is red in color.***

QUESTIONS

1. Why are metal carbonyls considered extremely toxic?
2. Vanadium hexacarbonyl, $V(CO)_6$, is fairly unstable. Why? On the other hand, the vanadium hexacarbonyl anion is fairly stable. Why?
3. The simple manganese carbonyl, $Mn(CO)_5$, does not exist; however, many derivatives of this compound do exist: $Mn(CO)_5X$ (X = H, Cl, Br, I, R, . . .). Explain. The dimer is also known. Describe its structure and explain its stability.
4. Give an example of an organic reaction that is photochemically activated.
5. From the literature, obtain the physical properties and preparations of the following metal carbonyl compounds: $Ni(CO)_4$, $Fe(CO)_5$, $Cr(CO)_6$, and $W(CO)_6$. Which obey the 18-electron rule?

REFERENCES

1. Calabro, D. C.; Lichtenberger, D. L. *J. Chem. Educ.* **1982,** *59,* 686.
2. Lewis, J.; Nyholm, R. S.; Osborne, A. G.; Sandher, S. S.; Stiddard, M. H. B. *Chem. Ind.,* **1963,** 1398.
3. Wrighton, M. S. *Acc. Chem. Res.* **1979,** *12,* 303.

GENERAL REFERENCES

Treichel, P. M., "Photochemistry" in *Comprehensive Organometallic Chemistry,* G. Wilkinson, Ed., Pergamon: Oxford, 1982, Vol. 4, Chapter 29, p. 1.

Kutal, C., "Photochemical Processes" in *Comprehensive Coordination Chemistry,* G. Wilkinson, Ed., Pergamon: Oxford, 1987, Vol. 1, Chapter 7.3, p. 385.

Griffith, W. P., "Carbonyls, Cyanides, Isocyanides and Nitrosyls" in *Comprehensive Inorganic Chemistry,* J. C. Bailar et al., Eds., Pergamon: Oxford, 1973, Vol. 4, Chapter 46, p. 105.

Hieber, W., "Metal Carbonyls" in *Advances in Organometallic Chemistry,* F. G. A. Stone and R. West, Eds., Academic Press: New York, 1970, Vol. 8, p. 1.

Experiment 44 Synthesis of Metal Nitrosyl Complexes

Part A: Preparation of Trichloronitrosylbis(triphenylphosphine)ruthenium(II)

Part B: Preparation of Dinitrosylbis(triphenylphosphine)ruthenium(-II)

INTRODUCTION

The nitric oxide molecule, NO, can readily lose its one π^* antibonding electron, to form the very stable nitrosyl cation, NO^+. This cation is isoelectronic with carbon monoxide and forms many similar complexes with transition metals. In electron counting, the ligand can be counted as the 2-electron donor NO^+ (or alternatively, as the 3-electron donor NO^0) when it is in a linear M—N—O geometry. When in a bent geometry, the NO ligand acts as a 1-electron donor. The ligand is also found as a bridging group.

Unlike metal carbonyls, few complexes with only NO as a ligand are known. Most complexes contain both NO and CO. In this experiment, a somewhat unusual complex containing no carbonyl ligands will be synthesized. If ruthenium(III) chloride is treated with *N*-methyl-*N*-nitroso-*p*-toluenesulfonamide (Diazald)® and triphenylphosphine, an octahedral nitrosyl product is obtained.

$$RuCl_3 + 2PPh_3 + CH_3C_6H_4SO_2N(CH_3)N{=}O \xrightarrow{\text{ethanol}} RuCl_3(NO)(PPh_3)_2$$

The product is a ruthenium(II) complex. The dinitrosyl complex can be synthesized in a similar manner.

$$RuCl_3 + 2PPh_3 + \text{Diazald}^{®} + (C_2H_5)_3N \rightarrow Ru(NO)_2(PPh_3)_2$$

The ruthenium is in a formal oxidation state of −II if the nitrosyl ligand is taken as having a charge of 1+.

The geometry adopted by the NO group can usually be inferred from the IR stretching frequency. Linear NO ligands generally have stretching frequencies from 1800–1900 cm^{-1}, while bent and bridging NO ligands have much lower frequencies: from 1300–1600 cm^{-1} for bridging groups and from 1525–1690 cm^{-1} for bent geometries.

Prior Reading and Techniques

Section 2.F: Reflux and Distillation

Section 5.D.3: Isolation of Crystalline Products (Suction Filtration)

Section 5.F.2: Evaporation Techniques

Section 6.C: Infrared Spectroscopy

Related Experiments

Metal Carbonyls: Experiments 34, 42, 43, and 45

EXPERIMENTAL SECTION

Part A: Preparation of Trichloronitrosylbis(triphenylphosphine)ruthenium(II)

Safety Recommendations

Ruthenium(III) chloride trihydrate (CAS No. 14898-67-0): This compound is harmful if swallowed, inhaled, or absorbed through the skin. IPR-RAT LD50: 360 mg/kg.

***N*-methyl-*N*-nitroso-*p*-toluenesulfonamide** (CAS No. 80-11-5): This compound is known by the trade name Diazald®. It is harmful if swallowed, inhaled, or absorbed through the skin, and causes severe irritation. ORL-RAT LD50: 2700 mg/kg.

Triphenylphosphine (CAS No. 603-35-0): This compound is a mild lachrymator, but has a low vapor pressure, so use of small quantities is not a problem. The compound can cause skin irritation. ORL-RAT LD50: 700 mg/kg.

CHEMICAL DATA

Compound	FW	Amount	mmol	mp (°C)	Density
$RuCl_3 \cdot xH_2O$	207.42	78 mg	0.30		
$P(C_6H_5)_3$	262.28	475 mg	1.81	80.5	1.194
Diazald®	214.24	125 mg	0.58	61	

Required Equipment

Magnetic stirring hot plate, 10-mL round-bottom flask, two 10-mL beakers, water condenser, 10-mL graduated cylinder, magnetic stirring bar, Keck clip, $CaCl_2$ drying tube, sand bath, glass funnel, Hirsch funnel, Pasteur pipets.

Time Required for Experiment: 1.5 h.

EXPERIMENTAL PROCEDURE

In separate 10-mL beakers, prepare solutions of 78 mg (0.30 mmol) of hydrated ruthenium(III) chloride dissolved in 6-mL of absolute ethanol (graduated cylinder) and of 125 mg (0.58 mmol) of Diazald® dissolved in 6-mL absolute ethanol.

Place 475 mg (1.81 mmol) of triphenylphosphine and 18 mL of absolute ethanol in a 50-mL round-bottom flask, equipped with a magnetic stirring bar. Attach a water condenser protected by a $CaCl_2$ drying tube with a Keck clip. Clamp the assembly in a sand bath atop a magnetic stirring hot plate. While stirring vigorously, bring the solution to a boil.

Momentarily remove the condenser, and using a small funnel, add **in rapid succession** the previously prepared solutions of $RuCl_3$ and Diazald®. Reflux the mixture for 15 min.

Isolation of Product

Allow the mixture to cool to room temperature. Green platelets of the product, trichloronitrosylbis(triphenylphosphine)ruthenium(II) will precipitate at this point. Collect the solid by suction filtration using a Hirsch funnel and wash the filter cake with 250-μL portions of ethanol, water, ethanol, and hexane, in that order. Allow the product to dry on a clay tile. Calculate the percentage yield of impure product.

Recrystallization of Product

Extract the impure green complex with five 20-mL portions of methylene chloride. Filter the extracts, add a boiling stone, and reduce the volume to 3–5 mL **(HOOD!)**. Cool the extract in an ice–water bath. Wash the resulting pure yellow-orange crystals with a 250-μL portion of hexane. (Further reduction of the extract volume provides a second crop of crystals.) Calculate the percentage yield and obtain the melting point.

Characterization of Product

Obtain an IR spectrum of the product as a KBr pellet. What is the NO stretching frequency? What does this tell you about the nature of the NO geometry in this complex? Compare the spectrum with that of the product prepared in Part 44.B. See Question 3 for an interesting additional characterization.

Part B: Preparation of Dinitrosylbis(triphenylphosphine)ruthenium(−II)

Additional Safety Recommendations

Triethylamine (CAS No. 121-44-8): This compound has a very disagreeable odor and should only be used in the **HOOD.** It is harmful if swallowed, inhaled, or absorbed through the skin. ORL-RAT LD50: 460 mg/kg.

CHEMICAL DATA

Compound	FW	Amount (mg)	mmol	mp (°C)	Density
$RuCl_3 \cdot xH_2O$	207.42	50	0.19		
$P(C_6H_5)_3$	262.28	310	1.18	80.5	1.194
Diazald	214.24	80	0.37	61	

Required Equipment

Magnetic stirring hot plate, 25-mL round-bottom flask, two 10-mL beakers, water condenser, magnetic stirring bar, Keck clip, sand bath, glass funnel, Hirsch funnel, Pasteur pipets.

Time Required for Experiment: 1 h.

EXPERIMENTAL PROCEDURE

NOTE: ***This experiment should be carried out in the HOOD.***

Place 310 mg (1.18 mmol) of triphenylphosphine in a 25-mL round-bottom flask containing a magnetic stirring bar. Add 12 mL of absolute ethanol and stir until dissolution is achieved. Attach a water condenser with a Keck clip and transfer the reaction apparatus to a sand bath set on a magnetic stirring hot plate. With stirring, heat the solution to boiling. While waiting for the solution to heat, proceed to the next step.

In a 10-ml beaker, prepare a solution of 50 mg (0.19 mmol) of ruthenium(III) chloride hydrate in 4 mL of ethanol. In a separate 10-mL beaker, prepare a solution of 80 mg (0.37 mmol) Diazald® in 4 mL of ethanol and in another 10 mL beaker, 1.25 mL of triethylamine.

Using a small glass funnel, add the ruthenium solution to the boiling triphenylphosphine solution through the condenser. Continue the heating and stirring. Dropwise, using a Pasteur pipet, add triethylamine (~1 mL) until a deep purple color develops. At this point, using the funnel, add the Diazald solution all at once to the reaction mixture. Heat the mixture at reflux for 5 min and allow it to cool to room temperature.

Isolation of Product

Collect the gray-colored precipitate of product under suction filtration using a Hirsch funnel. Wash the filter cake with successive 500-µL portions of ethanol, water, ethanol, and hexane. Obtain the melting point of the product.

Characterization of Product

Obtain an IR spectrum of the product as a KBr pellet. What is the NO stretching frequency? What does this tell you about the nature of the NO geometry in this complex? Compare the spectrum with that of the product prepared in Part 44.A.

QUESTIONS

1. Do the product complexes obey the 18-electron rule?
2. Draw the molecular orbital diagram for NO. What would the bond order be in NO, NO^+, and NO^- ?
3. Draw a diagram showing the bonding interactions between a linear nitrosyl ligand and a metal.
4. $[Co(en)_2Cl(NO)]^+$ has a CoNO bond angle of 121°, and an infrared NO frequency of 1611 cm^{-1}. By analogy to organic chemistry, in what manner is the nitrosyl group bonding in this complex? Does the complex obey the 18-electron rule? Treating the NO ligand as neutrally charged, how many electrons does it donate?
5. Ahmad et al. (Ref. 1, below) report the product of Part 44.A to be a Ru(II) complex, although through electron counting, one might have assumed it to be a Ru(III) complex with the nitrosyl group being a neutral three-electron donor. Suggest a way in which it could be determined whether the ruthenium is in the II or III oxidation state.

REFERENCE

1. Ahmad, N.; Levison, J. J.; Robinson, S. D.; Uttley, M. F. *Inorg. Syn.* **1974,** *15,* 51.

GENERAL REFERENCES

Griffith, W. P., "Carbonyls, Cyanides, Isocyanides and Nitrosyls" in *Comprehensive Inorganic Chemistry,* J. C. Bailar et al., Eds., Pergamon: Oxford, 1973, Vol. 4, Chapter 46, p. 105.

Johnson, B. F. G.; McCleverty, J. A., "Nitric Oxide Compounds of Transition Metals" in *Progress in Inorganic Chemistry,* F. A. Cotton, Ed., Interscience: New York, 1966, Vol. 7, p. 277.

Experiment 45 — ^{13}C NMR Analysis of the Cyclopentadienylirondicarbonyl Dimer

Part A: Preparation of $[(\eta^5\text{-}C_5H_5)Fe(CO)_2]_2$

Part B: Variable Temperature ^{13}C NMR Investigation of $[(\eta^5\text{-}C_5H_5)Fe(CO)_2]_2$

INTRODUCTION

The compound $[\eta^5\text{-}C_5H_5Fe(CO)_2]_2$ (Fig. 9.11) has some strikingly unusual structural features, not frequently found in other carbonyl compounds:

1. It is a dimeric compound.
2. The molecule posesses both terminal and bridging carbonyl groups.
3. It is a diamagnetic compound having a short Fe—Fe bond.

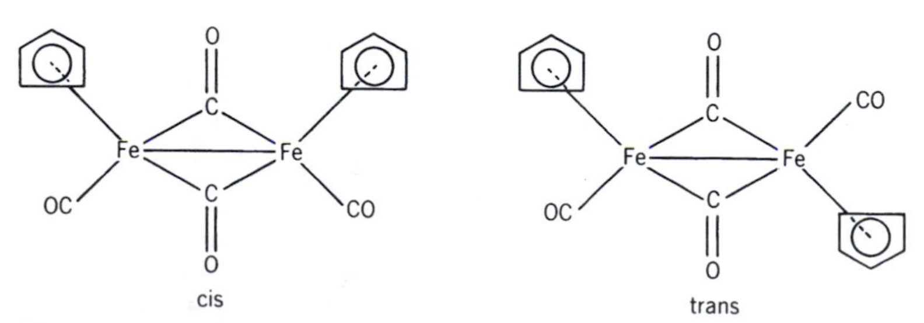

Figure 9.11. *Structure of cyclopentadienylirondicarbonyl dimer.*

4. The two Fe atoms and the two bridging CO groups are coplanar. One C_5H_5 ring and one terminal CO group lie above this plane, and the other C_5H_5 and CO are situated below this plane.
5. In solution, the compound exists mainly as cis and trans isomers, both being present in rapid equilibrium. Trace amounts of nonbridged species are also present in solution.
6. Because of the presence of the metal–metal bond, the compound is very sensitive to reduction. It can be easily reduced by sodium metal (sodium amalgam) to a very air-sensitive anionic species.

All these properties make this compound an ideal target for detailed investigation by both IR and NMR (both ^1H and ^{13}C). While the IR and ^1H NMR spectra of the compound help in determining the nature of C_5H_5 and CO groups, the ^{13}C NMR spectrum, combined with variable temperature studies, establishes the dynamic processes of the fluxional behavior of this molecule.

The title compound can be easily prepared by heating a mixture of iron pentacarbonyl and dicyclopentadiene. Iron pentacarbonyl, however, is quite toxic, so the synthesis is somewhat problematic. The cyclopentadienylirondicarbonyl dimer is commercially available, however, so that its spectrum may be studied without having to synthesize it.

Terminal CO groups show IR stretches in the range from 1850 to 2150 cm^{-1}, with the bridging carbonyl group stretches occurring from 1750 to 1850 cm^{-1}. This is quite close to the characteristic frequency range for organic carbonyl groups.

Prior Reading and Techniques

Section 2.F: Reflux and Distillation

Section 5.C.2: Purging with an Inert Gas

Section 5.D.3: Isolation of Crystalline Products (Suction Filtration)

Section 6.D: NMR Spectroscopy

Related Experiments

Metal Carbonyls: Experiments 34, 42, 43, and 45

Molecular Fluxionality: Experiment 39

EXPERIMENTAL SECTION

Part A: Preparation of $[(\eta^5\text{-}C_5H_5)Fe(CO)_2]_2$

Safety Recommendations

Dicyclopentadiene (CAS No. 77-73-6): This compound is harmful if swallowed, inhaled, or absorbed through the skin. ORL-RAT LD50: 353 mg/kg.

Iron pentacarbonyl (CAS No. 13463-40-6): This compound is an extremely toxic liquid. It may be fatal if inhaled, swallowed, or absorbed through the skin. ORL-RAT LD50: 40 mg/kg. Only use this compound in an efficient fume **HOOD.** Only handle this compound while wearing gloves.

Cyclopentadienylirondicarbonyl dimer (CAS No. 38117-54-3): No toxicity data is available for this compound. Most metal carbonyls are quite toxic, so great caution should be taken in handling this material.

CHEMICAL DATA

Compound	FW	Amount	mmol	mp (°C)	bp (°C)	Density
$Fe(CO)_5$	195.90	1.5 g	7.65	−20	103	1.490
Dicyclopentadiene	132.21	6.4 mL	47.7	−1	170	0.986

Required Equipment

Magnetic stirring hot plate, magnetic stirring bar, 25-mL side arm flask with stopcock, water condenser, Keck clip, $CaCl_2$ drying tube, mercury bubbler, sand bath, syringe, Hirsch funnel.

Time Required for Experiment: 3h in lab, overnight and one additional hour.

EXPERIMENTAL PROCEDURE[1–3]

> **NOTE: *This compound is commercially available. If so desired, it can be prepared in the laboratory in microscale quantities according to the following procedure.***

Because of the toxicity of iron pentacarbonyl, the entire reaction must be carried out in an efficient HOOD.

Attach the side arm of a 25-mL side arm stopcock flask containing a magnetic stirring bar to a nitrogen source. Connect a water condenser protected by a $CaCl_2$ drying tube, using a Keck clip to the flask. Finally, using Tygon tubing, attach the drying tube to a mercury bubbler, which acts as an outlet for the N_2 gas purge. The equipment is shown in Figure 9.12. Place the flask in a sand bath on a magnetic stirring hot plate. The temperature of the sand bath must be accurately controlled. Hang a thermometer in the sand bath to constantly

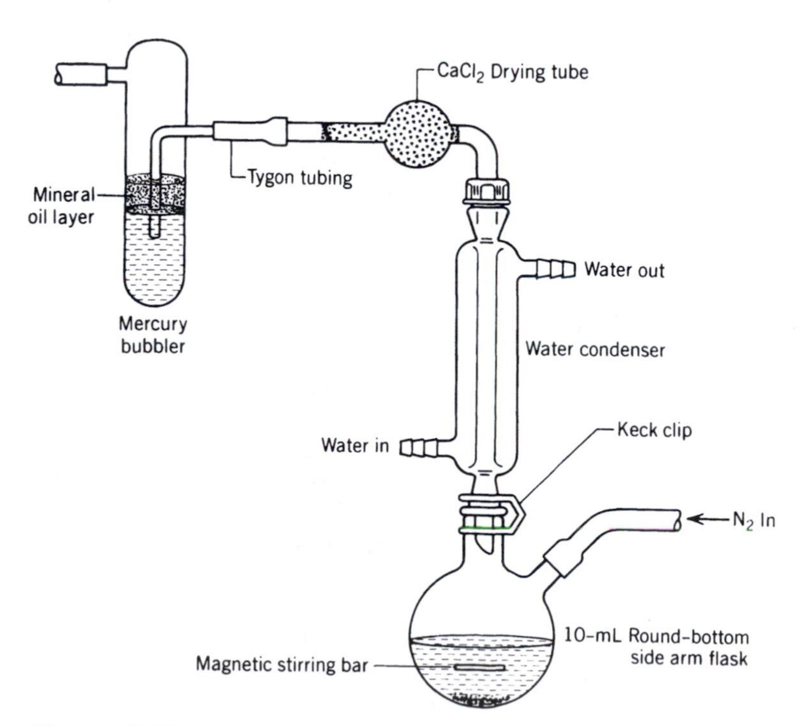

Figure 9.12. *Apparatus for Experiment 45.*

monitor the temperature. Flush the system with a rapid flow of nitrogen for 5 min.

While maintaining the flow of nitrogen, momentarily remove the condenser and syringe 1 mL (1.49 g, 7.65 mmol) of $Fe(CO)_5$ into the reaction flask, against the countercurrent of N_2.

> **NOTE: *$Fe(CO)_5$ is an extremely toxic liquid, having a characteristic musty smell. It has a high volatility, with a boiling point of 103* °C. Gloves must be worn when handling this compound.**

Add 6.4 mL (47.7 mmol) of dicyclopentadiene using a graduated cylinder. Replace the condenser and discontinue the flow of N_2 gas. Heat the reaction mixture, with stirring, to about 135 °C, and maintain this temperature overnight (or 8–10 h), making sure that cold water constantly circulates through the condenser during the heating period. The temperature must not exceed 140 °C; at higher temperatures the compound is unstable and forms highly pyrophoric metallic iron.

Isolation of Product

Restart the flow of N_2 gas and then allow the system to cool to room temperature. Remove the flask from the sand bath. Collect the violet crystals of product by suction filtration using a Hirsch filter. Wash the crystalls with several 100-μL portions of pentane. The resulting crystals are sufficiently pure for IR and NMR investigation. If desired, the compound may be recystallized from a minimum of 1:1 chloroform and hexane, concentrating the solution until crystallization begins.

Part B: Variable Temperature ^{13}C NMR Investigation of $[(\eta^5\text{-}C_5H_5)Fe(CO)_2]_2$

Dissolve 25 mg of sample in approximately 500 μL of CD_2Cl_2 solvent. Obtain the ^{13}C NMR spectrum at the following temperatures.

−75 °C −65 °C −35 °C −10 °C 25 °C (the signal may be difficult to observe) 55 °C

Signals for the carbonyl groups in the compound appear in the range 200–300 ppm downfield from TMS, while the C_5H_5 signal appears at 85 ppm.

The following features are important:

1. Below −60 °C, there is no free rotation about the Fe—Fe bond. Rotation is not rapid on the NMR timescale until 0 °C.
2. There are two isomers present at low temperature, the cis and the trans, as shown in Figure 9.11.
3. The bridging carbonyl groups open in a *trans* manner, resulting in two carbonyls and one cyclopentadienyl group on each iron when the bridges open. The opening and closing of the bridges is rapid even at −75 °C.

On the basis of these statements, explain the changes in the spectra of the complex at various temperatures, based upon the ^{13}C NMR spectra obtained. The following questions may help you formulate your explanations.

QUESTIONS

1. Draw Newman projections of the cis and trans isomers, once the carbonyl bridges have opened. Identify the formerly bridging carbonyls with a star.
2. From the projections in Question 1, recalling that there is no free rotation about the Fe—Fe bond, how many chemically different carbonyl groups are there in each isomer? How many signals would you expect to see for the carbonyls in the trans isomer? In the cis isomer?
3. Referring once again to the Newman projections, since the carbonyl bridges open and close only by a trans mechanism (i.e., the CO groups must be trans to each other), in which isomers can the bridge and terminal carbonyls interconvert?
4. What effect does being able to rotate about the Fe—Fe bond have on this interconversion at higher temperature?
5. Iron, in its organic chemistry, forms more dinuclear species than almost any other metal. Many of these are quite complex and interesting. From the literature, describe a representative variety of these compounds, their preparations, and spectral characteristics.

REFERENCES

1. Gansow, O. A.; Burke, A. R.; Vernon, W. D. *J. Am. Chem. Soc.* **1972,** *94,* 2550.
2. Adams, R. D.; Cotton, F. A. *J. Am. Chem. Soc.* **1973,** *95,* 6589.
3. Mann, B. E.; Taylor, B. F., *^{13}C NMR Data for Organometallic Compounds,* Academic Press: London, 1981.

GENERAL REFERENCE

Fehlhammer, W. P.; Stolzenberg, H., "Dinuclear Iron Compounds with Hydrocarbon Ligands" in *Comprehensive Organometallic Chemistry,* G. Wilkinson, Ed., Pergamon: Oxford, 1982, Vol. 4, Chapter 31.4, p. 513.

Chapter 10
Bioinorganic Chemistry

Also see the following experiments of bioinorganic interest:

Experiment 46 Synthesis of Palladium Nucleosides

Part A: Preparation of *cis*-[Dichlorobis(inosine)palladium(II)]

Part B: Preparation of *cis*-[Bis(inosinato)palladium(II)]

Part C: Preparation of *trans*-[Bis(inosinato)palladium(II)]

INTRODUCTION

The interaction of metals with the organic base purine and its derivatives is of great biological significance. Both the metal ions and the nucleoside serve as cofactors in various enzymatic reactions. Of particular interest is the fact that antitumor activity, for example, of the platinum complexes (see Experiment 48), may also hold true for those of palladium. In this series of experiments we examine the reaction of inosine (hypoxanthine riboside), a biochemical containing the purine nucleus, with the palladium metal ion. Inosine is found in many plants and is a close relative of guanosine, which is a building block of RNA and DNA.

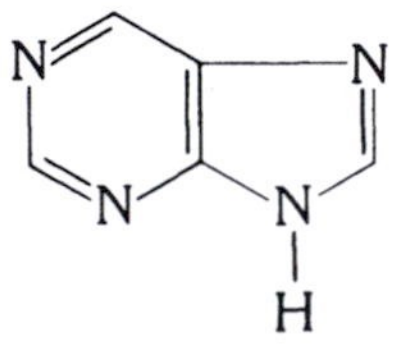

Purine

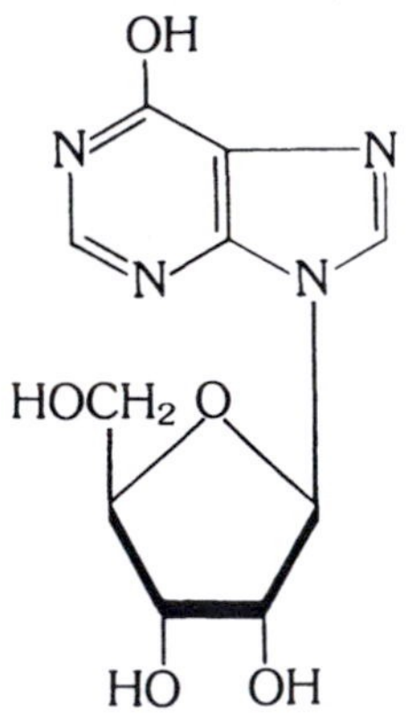

Inosine

The following experiments demonstrate the ease with which representative complexes of this class containing palladium can be prepared. The proposed structure of *cis*- and *trans*-[Pd(inosine-H^+)$_2Cl_2$] are shown in Figure 10.1.

Infrared data sheds light on the bonding of inosine to the metal atom. It should be noted that the basic purine nucleus in the inosine molecule is hypoxanthine and that this entity can exist in tautomeric equilibrium as depicted in Figure 10.2.

The carbonyl stretching frequency of the inosine molecule is shifted to a lower value by ~60 cm^{-1} when Pd—O interaction is observed. A similar shift has

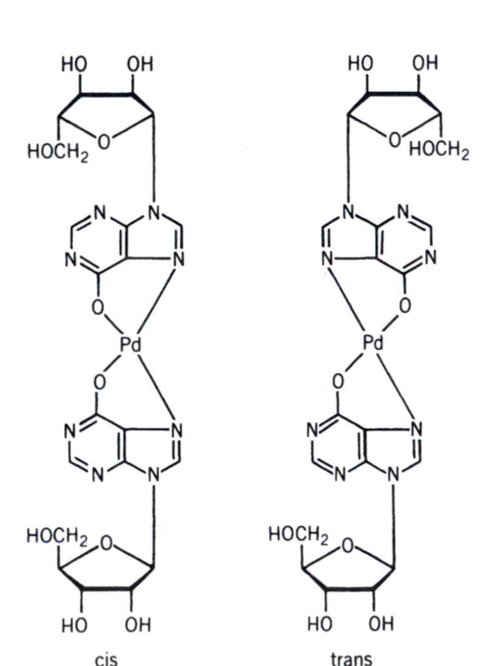

Figure 10.1. *Isomers of [Pd(inosine-H^+)$_2$].*

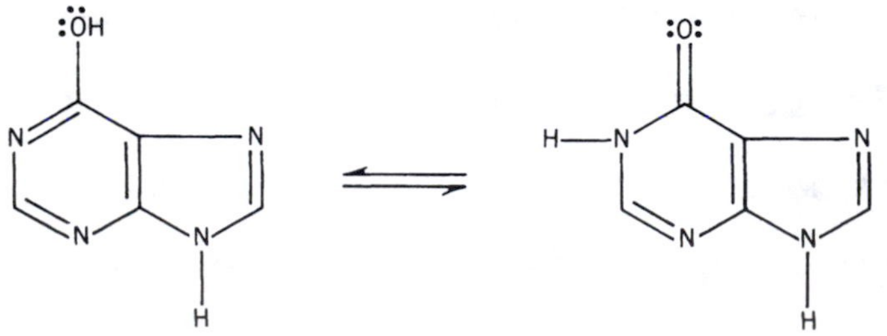

Figure 10.2. *Tautomers of hypoxanthine.*

always been observed in the IR spectra of metal complexes of inosine.[1] Table 10.1 summarizes the data taken in our laboratories for the Pd–inosine complexes prepared in this experiment. Nuclear magnetic resonance data may also be used to substantiate the formation of the complexes.[2]

Table 10.1 Infrared Data of the Pd–Inosine Complexes

Compound	ν(C=O) (cm^{-1})	Comments
Inosine	1697	Normal stretching frequency
cis-[Pd(inosine)$_2Cl_2$]	1697	No Pd—O interaction indicated
cis-[Pd(inosine-H^+)$_2$]	1636	Pd—O bonding; shift observed
trans-[Pd(inosine-H^+)$_2$]	1648	Pd—O bonding; shift observed

Prior Reading and Techniques

Section 5.D.3: Isolation of Crystalline Products (Suction Filtration)

Section 6.C: Infrared Spectroscopy

Section 6.D: Nuclear Magnetic Resonance Spectroscopy

Related Experiments

Palladium Chemistry: Experiments 20B, 39, and 41

Isomerism: 26, 27, 37, and 49

EXPERIMENTAL SECTION

Safety Recommendations

Palladium(II) chloride (CAS No. 7647-10-1): This compound may be fatal if swallowed, inhaled, or absorbed through the skin. It may be carcinogenic. ORL-RAT LD50: 2704 mg/kg.

Inosine (CAS No. 58-63-9): Toxicity data for this compound is not available. MTDS: dnd-mam: lyn 60 mmol/L.

Potassium tetrachloropalladate(II) (CAS No. 10025-98-6): This compound may be harmful if inhaled, ingested, or absorbed through the skin. IVN-RBT LD50: 6400 μg/kg.

Part A: Preparation of *cis*-[Dichlorobis(inosine)palladium(II)]

$$PdCl_2 + 2HCl \rightarrow H_2[PdCl_4]$$

$$H_2[PdCl_4] + 2\text{ inosine} \rightarrow cis\text{-}[Pd(inosine)_2Cl_2] + 2HCl$$

CHEMICAL DATA

Compound	FW	Amount	mmol	mp (°C)	Density
$PdCl_2$	177.31	86 mg	0.50		1.060
Inosine	268.23	268 mg	1.0	212[a]	

[a] Decomposes.

Required Equipment

Magnetic stirring hot plate, two 10-mL beakers, magnetic stirring bar, 10-mL graduated cylinder, Pasteur pipet, glass funnel, Hirsch funnel.

Time Required for Experiment: 3 h.

EXPERIMENTAL PROCEDURE[3]

Place 86 mg (0.5 mmol) of palladium chloride and 4 mL of 0.5*M* HCl (graduated cylinder) in a 10-mL beaker containing a magnetic stirring bar. The resulting suspension is heated to boiling, with stirring, on a magnetic stirring hot plate until complete dissolution is effected. Remove the beaker from the heat and allow the solution to cool to room temperature. If necessary, filter the solution by gravity through a small cotton plug placed in a small glass funnel.

In a second 10-mL beaker, place 268 mg (1 mmol) of inosine and 4 mL of 0.5*N* HCl solution. Swirl the solution to obtain dissolution (a Vortex mixer is helpful in this step). A magnetic stirring bar may also be used.

Transfer the palladium chloride solution using a Pasteur pipet into the inosine solution and also transfer the stirring bar using a pair of forceps. Stir the resulting solution at room temperature for approximately 2 h. A yellow precipitate forms during this period.

Isolation of Product

Collect the yellow complex by suction filtration using a Hirsch funnel. Wash the filter cake with two 1-mL portions of acetone followed by two 1-mL portions of ether. To further dry the material, place it in a vacuum desiccator containing t.h.e. SiO_2 desiccant for several hours. If desired, it may be further dried at 110 °C in a vacuum oven over potassium hydroxide pellets. Calculate the percentage yield.

Characterization of Product

Obtain the melting point of the product. Obtain the IR spectrum as a KBr pellet. The NMR spectrum in 3*N* DCl may be obtained for characterization purposes.[2] Compare the spectra with those obtained in subsequent parts of this experiment.

Part B: Preparation of *cis*-[Bis(inosinato)palladium(II)]

$$cis\text{-}[Pd(inosine)_2Cl_2] + 2KOH \rightarrow cis\text{-}[Pd(inosine\text{-}H^+)_2] + 2KCl + 2H_2O$$

Required Equipment

Magnetic stirring hot plate, 10-mL beaker, magnetic stirring bar, 10-mL graduated cylinder, Pasteur pipet, and 2-mL sintered glass filter funnel.

Time Required for Experiment: 3 h.

EXPERIMENTAL PROCEDURE[3]

Place 50 mg (0.11 mmol) of *cis*-[bis(inosine)dichloropalladium(II)], prepared in Part 46.A, in a 10-mL beaker containing 5 mL of distilled water and a magnetic stirring bar. Stir the suspension at room temperature until dissolution is effected.

Adjust the resulting acidic solution to a pH of ~6 by the dropwise addition (Pasteur pipet) of 0.1*M* KOH solution. This requires ~1 mL of the alkaline solution.

Isolation of Product

A light yellow suspension is formed by continuing stirring over a 1.5-h period. Collect the precipitate by suction filtration through a sintered glass filter (2-mL size). This step is very slow since the material tends to clog the filter. Wash the yellow solid with two 0.5-mL portions of methanol followed by two 0.5-mL portions of diethyl ether. Dry the yellow product in a vacuum desiccator oven t.h.e. SiO_2 drying agent. It may be further dried in a vacuum over at 110 °C.

Characterization of Product

Obtain the melting point of the product. Obtain the IR spectrum as a KBr pellet. The ^1H NMR spectrum in 3*N* DCl may be obtained for characterization purposes.[2] Compare the spectra with those obtained in the other parts of this experiment.

Part C: Preparation of *trans*-[Bis(inosinato)palladium(II)]

$$K_2[PdCl_4] + 2\ \text{inosine} \xrightarrow{\text{KOH}} trans\text{-}[Pd(\text{inosine-}H^+)_2] + 2KCl + 2HCl$$

CHEMICAL DATA

Compound	FW	Amount (mg)	mmol	mp (°C)	Density
K_2PdCl_4	326.42	33	0.10	105[a]	2.670
Inosine	268.23	54	0.20	212[a]	

[a] Decomposes.

Required Equipment: Same as Part 46.B.

Time Required for Experiment: 2 h.

EXPERIMENTAL PROCEDURE[3]

Place 54 mg (0.17 mmol) of inosine in a 10-mL beaker containing 5 mL of water and a magnetic stirring bar. Add (Pasteur pipet) a solution previously prepared by dissolving 33 mg (0.1 mmol) of potassium tetrachloropalladate(II) in 1 mL of distilled water.

Mixing of the two solutions creates an immediate light yellow precipitate. The pH of the supernatant liquid should be ~2. While stirring the mixture at room temperature, add 0.1*N* KOH solution dropwise over the period of 30 min. A total of 1.5 mL will produce a pH of ~6. Stir this mixture for an additional 30 min.

Collect the resulting light yellow precipitate by suction filtration using a sintered glass filter funnel. This is a very slow filtration. Wash the product with two 1-mL portions of cold methanol followed by two 1-mL portions of diethyl ether. Dry the material under vacuum at 110 °C.

Characterization of Product

Obtain the melting point of the product. Obtain the IR spectrum as a KBr pellet. The ^1H NMR spectrum in 3*N* DCl may be obtained for characterization purposes.[2] Compare the spectra with those obtained in the previous parts of this experiment.

QUESTIONS

1. Compare and contrast the structures of the isomers prepared in this experiment with the copper glycine complexes prepared in Experiment 49.
2. How can the IR spectrum of the isomers prepared in this experiment elucidate the nature of bonding to the oxygen and nitrogen atoms of the inosine? For some related work using IR frequency shifts, see Experiment 20.
3. The interaction of metals with the organic base purine and its derivatives is of great biological significance. Give a brief overview of this statement.
4. Based upon the literature, determine the function of guanosine in DNA and RNA. (This experiment could have been carried out equally effectively substituting guanosine for inosine.)

REFERENCES

1. Hadjiliadis, N.; Theophanides, T. *Inorg. Chim. Acta* **1976,** *16,* 77 and references cited therein.

2. Pneumatikakis, G.; Hadjiliadis, N.; Theophanides, T. *Inorg. Chem.* **1978,** *17,* 915.
3. Hadjiliadis, H.; Mascharak, P. K.; Lippard, S. J. *Inorg. Syn.* **1985,** *23,* 51.

GENERAL REFERENCES

Hughes, M. N., "Coordination Compounds in Biology" in *Comprehensive Coordination Chemistry,* G. Wilkinson, Ed., Pergamon: Oxford, 1987, Vol. 6, Chapter 62.1, p. 541.

Hodgson, D. J., "The Stereochemistry of Metal Complexes of Nucleic Acid Constituents" in *Progress in Inorganic Chemistry,* S. J. Lippard, Ed., Interscience: New York, 1977, Vol. 23, p. 211.

Marzilli, L. G. "Metal-ion Interactions with Nuclei Acids and Nucleic Acid Derivatives" in *Progress in Inorganic Chemistry,* S. J. Lippard, Ed., Interscience: New York, 1977, Vol. 23, p. 255.

Experiment 47 Metal Complexes of Saccharin

Part A: Preparation of Tetraaqua-bis(*o*-sulfobenzoimido)copper(II)

Part B: Preparation of Tetraaqua-bis(*o*-sulfobenzoimido)cobalt(II)

INTRODUCTION

Synthetic sweetening agents such as saccharin (**I,** 1,2-benzisothiazol-3(2*H*)-one,1,1-dioxide) and aspartame were developed over the years to eliminate the caloric intake in the diet associated with carbohydrate sugars. Saccharin, discovered in 1879, was used extensively as a sweetening agent until studies indicated that this substance might produce urinary bladder carcinomas when implanted in the bladders of mice. Since that time (1957), extensive work was carried out to investigate the effect of saccharin on human metabolism.

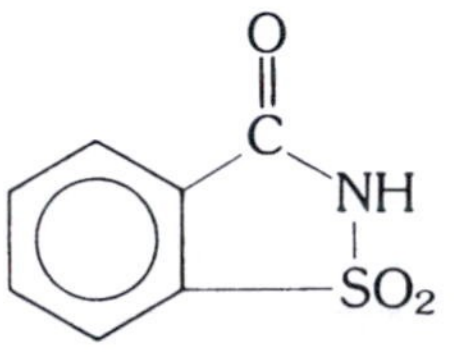

O
C
NH
SO_2
I

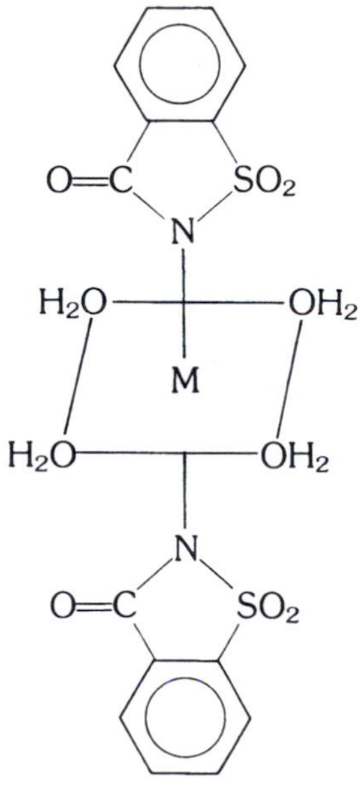

O=C SO_2
N
H_2O OH_2
M
H_2O OH_2
N
O=C SO_2
II

The compound itself does not have a high solubility, so when used as a sweetening agent, the sodium or calcium salt is generally used. It is not metabolized and has no food value. Whether saccharin poses a serious threat to humans through its consumption in soft drinks, and so on, is still open to debate. Nevertheless, its use was sharply curtailed over the past several years.

This experiment demonstrates a simple, direct approach to the synthesis of copper(II) and cobalt(II) complexes of saccharin(**II**). These substances were shown to play a role in human metabolism. Other metal complexes, such as the iron(II), nickel(II), and zinc(II) may be prepared in a like manner.

The general reaction for the preparation of the metal complexes of saccharin, using copper(II) as an example, is (Sac = $NSO_3H_4C_7$):

$$CuSO_4 \cdot 5H_2O + 2NaSac + H_2O = [Cu(Sac)_2(H_2O)_4] \cdot 2H_2O + Na_2SO_4$$

As indicated, the soluble sodium salt of the saccharin species is used in these preparations. In saccharin, the lone pair of electrons on the nitrogen *p* orbital is extensively conjugated with the *d* orbitals of the sulfur. The electrons are not, therefore, available for donation to the metal, and it is difficult to prepare metal saccharinates by direct reaction with saccharin itself. In the saccharide salt, however, the second nitrogen lone pair is readily available for donation, and the metal saccharinates form readily.

Prior Reading and Techniques

Section 5.B: Thermal Analysis

Section 5.D.3: Isolation of Crystalline Products (Suction Filtration)

Section 5.F.2: Evaporation Techniques

Section 6.B: Visible Spectroscopy

Section 6.C: Infrared Spectroscopy

Related Experiments

Cobalt Chemistry: Experiments 7B, 17, 26, 27, 30, and 35

Copper Chemistry: Experiments 20A, 24B, and 49

EXPERIMENTAL SECTION

Safety Recommendations

Saccharin, sodium salt (CAS No. 82385-42-0): This compound has been shown to have some positive activity in the EPA Genetox program. It is harmful if swallowed, inhaled, or absorbed through the skin. ORL-RAT LD50: 14,200 mg/kg.

Copper(II) sulfate pentahydrate (CAS No. 20919-8): This compound is not normally considered dangerous, but the usual precautions should be taken. ORL-RAT LD50: 300 mg/kg, ORL-HMN LDLo: 1088 mg/kg.

Cobalt(II) chloride hexahydrate (CAS No. 7791-13-1): The compound is harmful if swallowed, inhaled, or absorbed through the skin. ORL-RAT LD50: 766 mg/kg.

CHEMICAL DATA

Compound	FW	Amount (mg)	mmol	mp (°C)	Density
$NaNSO_3H_4C_7 \cdot H_2O$	205.17	100	0.49		
$CuSO_4 \cdot 5H_2O$	249.65	52	0.21		
$CoCl_2 \cdot 6H_2O$	237.93	48	0.20	86	1.920

Required Equipment
Magnetic stirring hot plate, 10-mL beaker, magnetic stirring bar, sand bath, Hirsch funnel, ice–water bath, clay tile or filter paper.

Time Required for Experiment: 1.5 h.

Part A: Preparation of Tetraaqua-bis(*o*-sulfobenzoimido)copper(II)

EXPERIMENTAL PROCEDURE[1,2]

In a 10-mL beaker containing a magnetic stirring bar, place 52 mg (0.2 mmol) of copper(II)sulfate pentahydrate, 100 mg (0.49 mmol) of sodium saccharinate hydrate and 6 mL of water. The mixture is stirred until dissolution occurs. Slight warming on a magnetic stirring hot plate hastens this operation.

Place the light blue solution on a warm sand bath (~140 °C) and, with stirring, concentrate the solution to a volume of ~2.5–3 mL. A slow stream of nitrogen impinging on the surface of the solution hastens this process.

Isolation of Product
Remove the beaker and contents from the sand bath and allow them to cool slowly to room temperature. Light blue crystals form during this time. Cool the beaker further in an ice bath (30 min) and collect the resulting crystals by suction filtration using a Hirsch funnel. Wash the robin's-egg blue crystals with two 1-mL portions of ice-cold water, and dry them on a clay tile or on filter paper. Further drying may be accomplished by placing the crystals in a small vial over silica gel in a desiccator.

Characterization of Product

Spectroscopic Characterization Obtain an IR spectrum of the complex as a KBr pellet. Visible spectra of the complexes can be measured using DMF as the solvent. The literature reports the following data.

Visible Data

Complex	Absorption maxima (nm)
$[Co(C_7H_4SO_3N)_2(H_2O)_4]\cdot 2H_2O$	525; 280
$[Cu(C_7H_4SO_3N)_2(H_2O)_4]\cdot 2H_2O$	784; 355

Thermal Characterization Obtain the TGA thermogram of the product over the range 25–450 °C. Dehydration and oxide formation occur over this temperature range.

Part B: Preparation of Tetraaqua-bis(*o*-sulfobenzoimido)cobalt(II)

EXPERIMENTAL PROCEDURE

This material is prepared using the same procedure given above for the copper complex. Use 48 mg (0.2 mmol) of cobalt(II) chloride hexahydrate, 100 mg (0.49 mmol) of sodium saccharinate hydrate dissolved in 6 mL of water. Characterize the product as above.

QUESTIONS

1. Give a brief historical overview of the use and prohibition of saccharine as a sweetening agent.

2. Formose sugars were also proposed as possible sweetening agents. Why did they not cause caloric intake? Why are they not in current use?
3. Aspartame is the current sugar substitute in commercial use. By what trade name is it known? What type of organic compound is it?
4. Zinc plays an important role in bioinorganic chemistry. Based on the literature, discuss its role in catalyzing enzymatic reactions.

REFERENCES

1. Haider, S. Z.; Malik, K. M. A.; Ahmed, K. J. *Inorg. Syn.* **1985,** *23,* 47.
2. Kirk–Othmer *Encyclopedia of Chemical Technology,* 3rd ed., Wiley; New York, 1979, Vol. 2, p. 448.

GENERAL REFERENCE

Hughes, M. N., "Coordination Compounds in Biology" in *Comprehensive Coordination Chemistry,* G. Wilkinson, Ed., Pergamon: Oxford, 1987, Vol. 6, Chapter 62.1, p. 541.

Experiment 48 Synthesis of *cis*-Diamminedihaloplatinum(II) Compounds

Part A: Preparation of *cis*-Diamminediiodoplatinum(II)

Part B: Preparation of *cis*-Diamminedichloroplatinum(II), Cisplatin

INTRODUCTION

Platinum(II) complexes have been extensively studied as anticancer chemotherapeutic agents. One particularly effective anticancer drug is *cis*-$[Pt(NH_3)_2Cl_2]$, *cis*-diaminedichloroplatinum(II). The commercial name of this drug is cisplatin. In this experiment, a number of Pt(II) complexes are synthesized, resulting in the synthesis of cisplatin.

In the first step of the synthesis, potassium tetraiodoplatinate(II) is prepared from potassium tetrachloroplatinate(II) by a metathesis reaction.

$$K_2PtCl_4 + 4KI \rightarrow K_2PtI_4 + 4KCl$$

Two of the iodide ligands are replaced with ammonia or some other ammine (am) ligand, forming *cis*-diamminediiodoplatinum(II), and a byproduct of potassium iodide. The ammines add stepwise. No isomerism is possible when the first ammine substitutes into the complex [resulting in the formation of the intermediate monoamminetriiodoplatinate(II)]. The iodide ligands are the stronger trans directors, so that the second ammine will add trans to one of the remaining iodides, resulting in the cis complex (see Experiment 33 for a discussion of the trans effect).

$$K_2PtI_4 + 2\ am \rightarrow cis\text{-}[Pt(am)_2I_2] + 2KI$$

The platinum complex is then reacted with silver ion, which precipitates the remaining iodide ligands, which are replaced by water.

$$cis\text{-}[Pt(am)_2I_2] + Ag_2SO_4(aq) \rightarrow cis\text{-}[Pt(am)_2(H_2O)_2]SO_4 + 2AgI$$

Silver sulfate is used to accomplish the precipitation instead of the more obvious choice of silver nitrate, as formation of the chloroplatinum complex from the sulfate complex is more favorable than from the analogous nitrate complex. This results in a higher yield of the final product.

Finally, the water ligands are easily replaced with an alkali halide, specifically KCl.

$$cis\text{-}[Pt(am)_2(H_2O)_2]^{2+} + 2MX \rightarrow cis\text{-}[Pt(am)_2X_2] + 2M^+$$

Alternatively, appropriate soluble barium salts can be used to isolate various *cis*-diamminedianionicplatinum(II) complexes. This is not desirable in the case of the synthesis of cisplatin, as insoluble $BaSO_4$ will precipitate, necessitating an additional filtration step in the synthesis.

Prior Reading and Techniques

Section 5.D.3: Isolation of Crystalline Products (Suction Filtration)

Related Experiments

Platinum Chemistry: Experiments 36–38

Trans Effect: Experiment 37

EXPERIMENTAL SECTION

Part A: Preparation of *cis*-Diamminediiodoplatinum(II)

Safety Recommendations

Potassium tetrachloroplatinate(IV) (CAS No. 10025-99-7): This compound is harmful if swallowed, inhaled, or absorbed through the skin. It is classified as an anticancer agent. IPR-MUS LD50: 45 mg/kg.

Potassium iodide (CAS No. 7681-11-0): This compound is harmful if swallowed, inhaled, or absorbed through the skin. No toxicity data is available. It has been shown to have deleterious effects on newborns and on pregnancy.

CHEMICAL DATA

Compound	FW	Amount	mmol	mp (°C)	Density
K_2PtCl_4	415.26	125 mg	0.30		
KI	166.01	300 mg	1.81	681	3.130
NH_3, 2*M*	17.03	500 μL	1.00		

Required Equipment

Magnetic stirring hot plate, 10-mL beaker, magnetic stirring bar, automatic delivery pipet, sand bath, Hirsch funnel.

Time Required for Experiment: 1 h.

EXPERIMENTAL PROCEDURE

> **NOTE:** ***Bright light should be avoided in this experiment. This will minimize the formation of iodoplatinum precipitates.***

Place 125 mg (0.300 mmol) of potassium tetrachloroplatinate in a 10-mL beaker containing a stirring bar.

> **NOTE:** ***If potassium tetrachloroplatinate is not available, it can be prepared from chloroplatinic acid by reduction with a stoichiometric amount of hydrazine sulfate in aqueous solution, in the presence of KCl.[1]***

Add 200 μL of water with an automatic delivery pipet and heat the solution with stirring in a sand bath to 40 °C. Add a solution of 300 mg (1.81 mmol) of KI dissolved in 500 μL of warm water. Upon the addition of KI, the solution changes from red-brown to dark brown in color.

Heat the mixture to 70 °C with continuous stirring. **Do not overheat the solution!** As soon as this temperature is reached, cool the mixture to room temperature.

Isolation of Product

Filter the solution using a Hirsch funnel to remove any solid impurities. Use a few drops of water to make the transfer as quantitative as possible. Add 400–500 μL (1 mmol) of ~2.0*M* NH_3 solution (automatic delivery pipet) dropwise to the filtrate. Stir the solution. As soon as the ammonia is added fine yellow crystals of *cis*-diamminediiodoplatinum(II) should precipitate. If the supernatent liquid is still dark yellow in color, add a few more drops of ammonia to complete the reaction. Allow the beaker to stand for an additional 20 min at room temperature. Filter the yellow crystalline compound using a Hirsch funnel. Wash the product with ice-cold ethanol (500 μL) followed by ether (1.0 mL). Use these wash liquids to transfer as much solid as possible from the beaker to the filter. Air-dry the compound and determine the percentage yield.

Characterization of Product

Obtain the IR spectrum of the product as a Nujol mull. If a far-IR spectrometer is available, obtain the spectrum in the range 50–100 cm^{-1} (Pt—I stretch).

Part B: Preparation of *cis*-Diamminedichloroplatinum(II), Cisplatin

Additional Safety Recommendations

Potassium chloride (CAS No. 7447-40-7): This compound is not normally considered dangerous. ORL-RAT LD50: 2600 mg/kg.

Silver sulfate (CAS No. 10294-26-5): No toxicity data is available for this compound. It would be prudent to follow the normal precuations (Section 1.A.3), as silver salts have been found to act as heavy metal poisons.

CHEMICAL DATA

Compound	FW	Amount	mmol	mp (°C)	Density
$K_2Pt(NH_3)_2I_2$	482.94	100 mg	0.207		
KCl	74.56	330 mg	4.43	770	1.984
Ag_2SO_4	311.80	63 mg	0.202	652	5.450

Required Equipment

Magnetic stirring hot plate, 25-mL beaker, magnetic stirring bar, spatula, sand bath, ice–water bath, Hirsch funnel.

Time Required for Experiment: 1 h.

EXPERIMENTAL PROCEDURE

NOTE: *Bright light should be avoided in this experiment. This will minimize the formation of iodoplatinum precipitates.*

Prepare a solution of 63 mg (0.202 mmol) of silver sulfate in 10 mL of water in a 25-mL beaker containing a magnetic stirring bar. Add 100 mg (0.207 mmol) of the *cis*-diiodo derivative prepared in Part 48.A, in small portions, to this Ag^+ solution.

NOTE: *The diiodo derivative might remain suspended at the surface of the solution. If this occurs, stir the solution vigorously with a spatula, making sure that all the compound is well wetted.*

Heat the suspension, with stirring, on a sand bath (70–80 °C) for 10–12 min. Filter the mixture to separate the precipitate of AgI.

Isolation of Product

Concentrate the filtrate to a volume of about 2.0 mL. Treat this solution with 330 mg (4.43 mmol, a large excess) of KCl. Heat the mixture on a sand bath at 70–80 °C for 2–3 min. Bright yellow crystals of *cis*-diamminedichloroplatinum(II) should precipitate out. The heating is continued for an additional 5–8 min. Cool the mixture to 0 °C in an ice–water bath. Filter the product using a Hirsch funnel. Wash the crystals with 500 μL of ethanol followed by 1 mL of ether and dry them under suction in air. Determine the percentage yield.

Characterization of Product

Obtain the mid- and far-IR spectra of the compound (400–4000 cm^{-1}, 150–400 cm^{-1}). Assign the bands for the Pt—Cl stretches. Compare the spectra with those obtained in Part 48.A.

QUESTIONS

1. Draw a mechanism showing the substitution of two ammonia ligands onto potassium tetraiodoplatinate(II), keeping the trans effect in mind.
2. Cis geometry is maintained in the reaction step

$$cis\text{-}[Pt(am)_2I_2] + Ag_2SO_4(aq) \rightarrow cis\text{-}[Pt(am)_2(H_2O)_2]SO_4 + 2AgI$$

   Explain why.
3. All complexes prepared in this experiment are square planar in geometry. Why is this a favorable geometry for Pt(II)?
4. Provide a brief discussion of the anticancer role of cisplatin.
5. Many organic drugs have silicon analogs, which usually show reduced physiological activity. One exception is silatrane, which has no organic analog, and also shows anticancer activity. Why is there no organic analog to this drug? Discuss its anticancer activity.

REFERENCES

1. Livingstone, S. E. *Syn. Inorg. Metorg. Chem.* **1971,** *1,* 1.
2. Dhara, S. C. *Indian J. Chem.* **1970,** *8,* 193.
3. Harrison, R. C.; McAuliffe, C. A.; Zaki, A. M. *Inorg. Chim. Acta* **1980,** *46,* L15.

GENERAL REFERENCE Howard-Lock, H. E.; Lock, C. J. L., "Uses and Therapy" in *Comprehensive Coordination Chemistry*, G. Wilkinson, Ed., Pergamon: Oxford, 1987, Vol. 6, Chapter 62.2, p. 755.

Experiment 49 Preparation of Copper Glycine Complexes

Part A: Preparation of *cis*-Bis(glycinato)copper(II) Monohydrate

Part B: Preparation of *trans*-Bis(glycinato)copper(II)

INTRODUCTION Like the more familiar acetylacetone (see Experiment 22), the amino acid glycine **(I)** dissociates to form an anion (gly), which can coordinate to a wide variety of metal complexes.

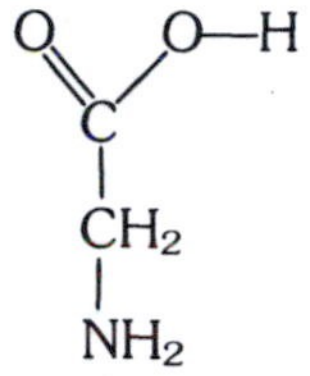

One major difference is that the glycine anion is not symmetric and structural isomers can arise depending on the relative orientation of the ligands.

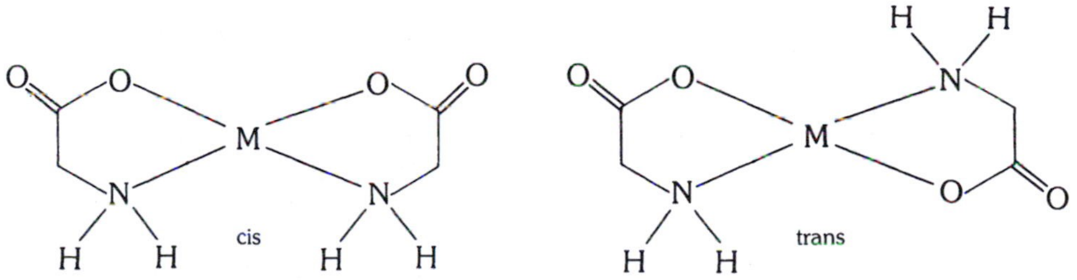

cis trans

In this experiment, the cis and trans copper glycinates are prepared. The direct reaction of copper(II) acetate monohydrate (see Experiment 24B for the preparation of this compound) and glycine results in an equilibrium mixture of the two isomers.

$$[(CH_3CO_2)_2Cu{\cdot}H_2O]_2 + H_2NCH_2CO_2H \rightarrow \textit{cis}\text{-}Cu(gly)_2{\cdot}H_2O + \textit{trans}\text{-}Cu(gly)_2{\cdot}H_2O$$

The cis isomer precipitates much more quicklly than the trans, leading to a shift in the equilibrium away from the trans, producing only the cis product. Interestingly, even though the cis isomer is the kinetically favored product, the trans isomer is thermodynamically favored. The cis isomer may be converted to the trans simply by heating it at 180 °C for 15 min.

Prior Reading and Techniques

Section 5.D.3: Isolation of Crystalline Products (Suction Filtration)

Section 6.C: Infrared Spectroscopy

Related Experiments

Copper Chemistry: Experiments 20A, 24B, and 47A
Isomerism: Experiments 26, 27, 37, and 46

EXPERIMENTAL SECTION

Safety Recommendations

Copper(II) acetate monohydrate (CAS No. 66923-66-8): This compound is harmful if swallowed, inhaled, or absorbed through the skin. ACGIH TLV-TWA: 1 mg/m^3.

Glycine (CAS No. 56-40-6): No toxicity data is available for this compound, but it would be prudent to follow the normal precautions (Section 1.A.3).

Part A: Preparation of *cis*-Bis(glycinato)copper(II) Monohydrate

CHEMICAL DATA

Compound	FW	Amount	mmol	mp (°C)	Density
$(CH_3CO_2)_2Cu \cdot H_2O$	199.65	100 (mg)	0.5		1.882
Glycine	75.07	75 (mg)	1.0	245[a]	

[a] Decomposes

Required Equipment

Two 10-mL Erlenmeyer flasks, magnetic stirring hot plate, magnetic stirring bar, calibrated Pasteur pipet, Pasteur pipet, ice–water bath, Hirsch funnel, clay tile or filter paper, side arm test tube, aluminum block.

Time Required for Experiment: 3 h.

EXPERIMENTAL PROCEDURE[1,2]

In a 10-mL Erlenmeyer flask containing a magnetic stirring bar, dissolve 100 mg (0.5 mmol) of copper(II) acetate monohydrate in 1.5 mL of hot deionized water. Add 1.0 mL (calibrated Pasteur pipet) of hot 95% ethanol to the solution. Maintain the temperature of the solution at 70 °C.

In a separate 10-mL Erlenmeyer flask, dissolve 75 mg (1 mmol) of glycine in 1 mL of hot deionized water by swirling the flask. Transfer the glycine solution (Pasteur pipet) into the copper(II) acetate solution and stir briefly. Discontinue the stirring and allow the solution to cool to room temperature.

Isolation of Product

Complete the precipitation of product by transferring the Erlenmeyer flask into a ice–water bath for 10 min. Collect the product by suction filtration using a Hirsch funnel. Wash the product with a 100 µL portion of ice-cold ethanol and dry the crystals on a clay tile or on filter paper.

Part B: Preparation of *trans*-Bis(glycinato)copper(II)

EXPERIMENTAL PROCEDURE

Place ~35 mg of the cis product produced in Part 49.A into a stoppered side arm test tube. Place the test tube on an aluminum block set on a magnetic stirring hot plate and heat the block to approximately 220 °C for 15 min.

> **NOTE:** ***If a muffle furnace is available, it may be used instead of the aluminum block.***

Remove the test tube from the block and allow it to cool to room temperature. The cis product from Part 48.A has been converted to the trans product.

Characterization of Products

Obtain an IR spectrum (KBr pellet) of the cis and trans products. If a far-IR spectrometer is available, prepare the material as a Nujol mull. The Cu—N stretches may be observed from 450–500 cm^{-1}, and the Cu—O stretches from 250–350 cm^{-1}. Differential scanning calorimetry is also informative, as the temperature for cis to trans conversion may be easily determined.

QUESTIONS

1. Assign the heavy atom framework for the cis and trans isomer to its proper point group.
2. Why is the IR spectrum of the trans compound much simpler than that of the cis, especially in the fingerprint region (800–1200 cm^{-1})?
3. When a carboxylic acid ligand is monodentate, the separation of the symmetric and antisymmetric C═O stretching frequencies increases, compared to the free acid. Why?
4. Which of the other amino acids would you expect to give rise to similar complexes having geometric isomers? Search the literature to determine if any were prepared.

REFERENCES

1. Delf, B. W.; Gillard, R. D.; O'Brien, P. *J. Chem. Soc. Dalton Trans.* **1979,** 1901.
2. O'Brien, P. *J. Chem. Educ.* **1982,** *59,* 1052.

GENERAL REFERENCES

Laurie, S. H., "Amino Acids, Peptides and Proteins" in *Comprehensive Coordination Chemistry,* G. Wilkinson, Ed., Pergamon: Oxford, 1987, Vol. 2, Chapter 20.2, p. 740.

Hughes, M. N., "Coordination Compounds in Biology" in *Comprehensive Coordination Chemistry,* G. Wilkinson, Ed., Pergamon: Oxford, 1987, Vol. 6, Chapter 62.1, p. 541.

Hathaway, B. J., "Copper" in *Comprehensive Coordination Chemistry,* G. Wilkinson, Ed., Pergamon: Oxford, 1987, Vol. 5, Chapter 53.4, p. 720.

Appendix A
Safety Data for Common Solvents*

Acetone (CAS No. 67-64-1): Acetone is an extremely flammable liquid. It is not normally considered dangerous, but the normal precautions should be employed (Section 1.A.3). ORL-RAT LD50: 5800 mg/kg.

Acetonitrile (CAS No. 75-05-8): Acetonitrile is harmful if swallowed, inhaled, or absorbed through the skin. Overexposure has caused reproductive disorders in laboratory animals. ORL-RAT LD50: 2730 mg/kg.

Benzene (CAS No. 71-43-2): Benzene is harmful if swallowed, inhaled, or absorbed through the skin. It is classified as a carcinogen. IHL-HMN LCLo: 2 pph/5M. ORL-RAT LD50: 3360 mg/kg. It is extremely flammable. Toluene should, in general, replace benzene in all preparations.

Carbon tetrachloride (CAS No. 56-23-5): Carbon tetrachloride is harmful if inhaled, swallowed, or absorbed through the skin. It is classified as a carcinogen. IHL-HMN LCLo: 5 pph/5M. ORL-RAT LD50: 2350 mg/kg.

Chloroform (CAS No. 67-66-3): Chloroform is a potent narcotic agent. It may be fatal if inhaled, swallowed, or absorbed through the skin. It is classified as a carcinogen. IHL-HMN LCLo: 25,000 ppm/5M. ORL-RAT LD50: 908 mg/kg.

Cyclohexane (CAS No. 110-82-7): Cyclohexane is harmful if inhaled or swallowed. It is extremely flammable. ORL-RAT LD50: 12,705 mg/kg.

Diethyl ether (CAS No. 60-29-7): Diethyl ether is an extremely flammable solvent. Exposure to moisture tends to form peroxides, which may be explosive. The solvent is a potent narcotic. ORL-MAN LDLo: 260 mg/kg. ORL-RAT LD50: 1215 mg/kg.

***N,N*-Dimethylformamide** (CAS No. 68-12-2): DMF is harmful if swallowed, inhaled, or absorbed through the skin. ORL-RAT LD50: 2800 mg/kg.

Dimethyl sulfoxide (CAS No. 67-68-5): DMSO is harmful if swallowed, inhaled, or absorbed through the skin. Overexposure has been found to have effects on fertility. ORL-RAT LD50: 14,500 mg/kg.

* All safety data in this table and elsewhere in the text is derived from the Sigma–Aldrich Material Safety Data Sheets on CD-ROM, Aldrich Chemical Co., Inc., Milwaukee, WI, July 1989 version.

Ethanol (CAS No. 64-17-5): Ethanol may be fatal if inhaled, swallowed, or absorbed through the skin in large amounts. It has been shown to have effects on fertility and on embryo development. ORL-HMN LDLo: 1400 mg/kg. ORL-RAT LD50: 7060 mg/kg. The vapor may travel considerable distances to the source of ignition and flash back.

Hexane (CAS No. 110-54-3): Hexane is harmful if inhaled, swallowed, or absorbed through the skin. It is a flammable liquid. ORL-RAT LD50: 28,710 mg/kg.

2-Propanol (CAS No. 67-63-0): 2-Propanol (commercial name: rubbing alcohol) is not normally considered dangerous, but the usual precautions (Section 1.A.3) should be followed. ORL-HMN LDLo: 3570 mg/kg. ORL-RAT LD50: 5045 mg/kg.

Methanol (CAS No. 67-56-1): Methanol may be fatal if swallowed. It is harmful if inhaled or absorbed through the skin. It is a flammable liquid. ORL-HMN LDLo: 143 mg/kg. ORL-RAT LD50: 5628 mg/kg.

Methylene chloride (CAS No. 75-09-2): Methylene chloride is harmful if swallowed, inhaled, or absorbed through the skin. ORL-HMN LDLo: 357 mg/kg. ORL-RAT LD50: 1600 mg/kg. It is a possible carcinogen.

Pentane (CAS No. 109-66-0): Pentane is harmful if inhaled or swallowed. The compound is extremely flammable. IVN-MUS LD50: 446 mg/kg.

Tetrahydrofuran (CAS No. 109-99-9): THF may cause severe damage to the liver. The liquid is extremely flammable. ORL-RAT LD50: 2816 mg/kg. On exposure to air, THF forms peroxides that can explode on contact with strong bases.

Toluene (CAS No. 108-88-3): Toluene is a flammable liquid. ORL-HMN LDLo: 50 mg/kg. ORL-RAT LD50: 5000 mg/kg.

Appendix B
List of Common Acids and Bases

Acetic acid (CAS No. 64-19-7): Glacial acetic acid is available in up to 100% purity. FW = 60.05. mp = 16.2 °C. bp = 116–118 °C. density = 1.049. Concentration is 17.5*M*. The acid is corrosive and toxic and has a pungent odor.

Ammonium hydroxide (CAS No. 1336-21-6): Ammonium hydroxide is available as a 28–30% solution. FW = 17 (as NH_3). density = 0.900. Concentration is ~15*M*. The base is corrosive and toxic, and has a pungent odor.

Hydrochloric acid (CAS No. 7647-01-0): Hydrochloric acid is available as a 37% solution. FW = 36.46. density = 1.200. Concentration is about 12*M*. The acid is extremely corrosive and toxic.

Nitric acid (CAS No. 7697-37-2): Nitric acid is available as a 69–71% solution. FW = 63.01. density = 1.400. Concentration is about 15.6*M*. The acid is extremely corrosive and toxic. Toxic fumes of NO_2 may be given off. Strong oxidizing agent.

Phosphoric acid (CAS No. 7664-38-2): Phosphoric acid is available as an 85% solution. FW = 98.00. density = 1.685. Concentration is about 14.6*M*. The acid is corrosive.

Sulfuric acid (CAS No. 7664-93-9): Sulfuric acid is available as a 95–98% solution. FW = 98.08. density = 1.840. Concentration is about 18*M*. The acid is extremely corrosive. Strong oxidizing and dehydrating agent.

Appendix C
Table of Reagents and Selected Solvents Used in Experiments

Reagent or Solvent	Experiment Number
Acetic acid	13, 24, 39
Acetic anhydride	40
Acetylacetone	22
Allyl bromide	39
Ammonia	1
Ammonium bifluoride	5
Ammonium chloride	10
Ammonium fluoride	5
Ammonium hydroxide	14
Ammonium metavanadate	28
Ammonium oxalate monohydrate	2
Ammonium tetrafluoroborate	5
Antimony pentachloride	15
Barium carbonate	2
Benzonitrile	41
Benzoyl chloride	25
Beryllium hydroxide	5
Bis-diphenylphosphinopropane	31
Boric acid	4, 5, 8
Boron trichloride	6
Bromine	35
Bromobenzene	32
N-Bromosuccinimide	23
Calcium carbonate	2, 3
Carbon tetrachloride	23, 42
o-Carborane	7
Chlorine	10
Chloroform	42
Chloroplatinic acid	38
Chromium(III) acetylacetonate	23
Chromium(III) chloride hexahydrate	22, 29

Reagent or Solvent	Experiment Number
Chromium(III) nitrate nonahydrate	29, 33
Cobalt(II) acetate tetrahydrate	30
Cobalt(II) chloride hexahydrate	7, 26, 47
Cobalt(II) nitrate hexahydrate	17
Cobalt(II) sulfate heptahydrate	35
Copper(II) acetate monohydrate	49
Copper(II) chloride	20
Copper(II) sulfate pentahydrate	24, 47
Cyclohexene	34
1,5-Cyclooctadiene	38
Cyclopentadienylirondicarbonyl dimer	45
Diazald®	44
Dichlorodimethylsilane	8
Dicyclopentadiene	7, 40, 45
Dimethyldichlorosilane	8
N,N-Dimethylformamide	42
Dimethyl sulfoxide	20, 30, 40
1,3-Bis(diphenylphosphino)propane	31
EDTA, disodium salt	33, 36
Ethyl acetate	17
Ethylene	34, 41
Ethylenediamine	26, 29, 30
Ethylene glycol	40
Ferric chloride (anhydrous)	32
Fluoroboric acid	35
Glycine	49
n-Heptanal	34
Hexachlorocyclotriphosphazene	12
Hydrogen	34
Inosine	46
Iodine	9, 14, 16, 18, 19, 40
Iron(II) chloride tetrahydrate	40
Iron(III) chloride (anhydrous)	32
Iron(III) nitrate nonahydrate	1
Iron pentacarbonyl	45
Lead(II) dichloride	10
Lithium bromide	34
Magnesium	3
Magnesium oxide	2
Manganese(II) chloride hexahydrate	22
Manganese dioxide	16
2-Mercapto-1-methylimidazole	17
Mercury(II) chloride	28
Methanol	7, 29
Methylcyclopentadienylmanganese tricarbonyl	43
Methylene chloride	9, 15
Nickel(II) chloride hexahydrate	31
N-Bromosuccinimide	23
N-Methyl, *N*-nitroso-*p*-toluenesulfonamide	44
4-Nitrophenol	12

Reagent or Solvent	Experiment Number
Palladium(II) chloride	20, 39, 41, 46
2,4-Pentanedione	22
1,10-Phenanthroline hydrate	35
4-Phenyl-3-thiosemicarbazide	11
Phosphonitrilic chloride trimer	12
Phosphoryl chloride	13
Poly(vinylalcohol)	4
Potassium antimonyl-*d*-tartrate hydrate	27, 35
Potassium bromide	35
Potassium chlorate	19
Potassium chloride	48
Potassium hexafluorophosphate	35
Potassium hydroxide	7, 12, 36, 40
Potassium iodide	16, 48
Potassium 4-nitrophenoxide	12
Potassium permanganate	22
Potassium tetrachloropalladate(II)	46
Potassium tetrachloroplatinate(IV)	36, 37, 48
n-Propanol	4
Pyridine	18, 21, 37
Rhodium(III) chloride hydrate	21, 24, 34, 42
Ruthenium(III) chloride trihydrate	20, 44
Saccharine, sodium salt	47
Salicylaldehyde	30
Silicone oil	7, 40
Silver nitrate	18
Silver sulfate	48
Sodium	1
Sodium acetate trihydrate	22, 24
Sodium bicarbonate	40
Sodium bisulfite	42
Sodium bromide	35
Sodium hydroxide	24
Sodium hypophosphite hydrate	21
Sodium nitrite	11
Sodium thiosulfate pentahydrate	16
Strontium carbonate	2
Tetra-*n*-butylammonium bromide	12, 25
Tetrabutylammonium perrhenate	25
Tetracosane	22, 46
Tetraphenyltin	6, 15
Thiosemicarbazide	11
Tin	9
Tin(IV) chloride (anhydrous)	10
Toluene	4
Triethylamine	44
Triethyl orthoformate	17
Tris(2,4-pentanedionato)chromium(III)	23
Triphenylphosphine	34, 36, 42, 43, 44
Urea	2, 22
Zinc	9, 28, 29

Appendix D

Table of Techniques Employed in Experiments

Exp. Number	Gas Cyl.	Inert Atmosphere	Reflux	Filtr.	Recryst.	Melting Point	Magnetic Susceptibility	Thermal	Spectroscopy	Chromatography	Other
1	X	X	—	—	—	—	—	—	—	—	NH_3 (ℓ)
2	—	—	—	X	—	—	—	TGA	—	—	Homo prep
3	X	—	—	—	—	—	—	—	AA	—	Ser dilution
4	—	—	X	—	—	—	—	—	IR, NMR	—	—
5	—	—	—	X	X	—	—	—	NMR	—	—
6	X	X	—	—	—	—	—	—	IR, NMR	—	—
7	—	X	X	X	—	X	X	—	IR, NMR, UV, Vis	—	—
8	—	—	—	—	—	—	—	—	—	FL	—
9	—	—	X	X	X	X	—	—	—	—	—
10	X	—	—	X	—	X	—	—	—	—	—
11	—	—	—	X	X	X	—	—	IR, UV	—	—
12	—	—	X	X	X	X	—	X	IR	—	—
13	—	—	—	X	—	X	—	—	IR	—	—
14	—	—	—	—	—	—	—	—	—	—	—
15	—	—	X	X	—	X	—	—	IR	—	—
16	—	—	—	X	—	—	—	TGA	IR	—	—
17	—	—	X	X	—	X	X	TGA	IR	—	—
18	—	—	—	X	X	X	—	—	IR	—	—
19	—	—	—	X	X	—	—	—	—	—	—
20	—	—	X	X	—	X	—	—	IR	—	—
21	—	—	X	X	X	—	—	—	IR	—	—
22	—	—	—	X	X	X	X	—	IR	—	Homo prep
23	X	—	—	—	—	—	—	—	—	GC	—
24	X	X	X	X	X	—	X	—	IR	—	—
25	X	X	X	X	—	—	X	—	IR	—	—
26	—	—	—	X	—	—	—	—	IR, Vis	—	—
27	—	—	—	X	—	—	—	—	—	—	Spec Rot
28	—	—	—	—	—	—	—	—	Vis	ION	—
29	—	—	X	X	—	—	—	—	Vis	—	—
30	X	X	—	X	—	X	—	—	—	—	Abs Oxy
31	—	—	—	X	—	X	X	—	IR, NMR	—	—
32	—	—	X	X	—	—	—	—	—	HPLC	—
33	—	—	—	—	—	—	—	—	Vis	—	pH
34	X	—	X	X	—	X	—	—	IR, NMR	GC	Abs Hyd
35	—	—	X	X	X	X	—	—	IR, NMR	—	Spec Rot
36	—	—	—	X	X	—	—	—	IR	—	—
37	—	—	—	X	X	X	—	—	IR	—	—
38	—	—	—	X	—	—	—	—	IR, NMR	—	—
39	—	—	X	X	X	—	—	—	IR, NMR	—	Rot Evap
40	X	X	X	X	Sub	X	—	—	IR, NMR	TLC, CC, HPLC	—
41	X	—	—	X	—	X	—	—	IR	—	—
42	X	—	X	X	—	—	—	—	IR	—	Rot Evap
43	X	X	—	X	X	X	—	—	IR	—	Photochem
44	—	—	X	X	X	X	—	—	IR	—	—
45	X	X	X	X	X	—	—	—	NMR	—	—
46	—	—	—	X	—	X	—	—	IR, NMR	—	—
47	X	—	—	X	—	—	—	TGA	Vis, IR	—	—
48	—	—	—	X	—	—	—	—	IR	—	—
49	—	—	—	X	—	—	—	DSC	IR	—	—

Appendix E
Companies and Addresses

Ace Glass, Inc.
PO Box 688, Vineland, NJ 08360

Aldrich Chemical Company, Inc.
PO Box 355, Milwaukee, WI 53201

American Chemical Society
1155 16th St. NW, Washington, DC 20036

Brinkman Instruments
Cantiaque Road, Westbury, NY 11590

Cricket Software
30 Valley Stream Pkwy.
Malvern, PA 19355

CAS Customer Service
P.O. Box 3012, Columbus, OH 43210

DuPont Analytical Instruments
Quillen Building, Concord Plaza, Wilmington, DE 19888

Fisher Scientific
50 Fadem Rd., Springfield, NJ 07081

Fluka Chemical Corp.
980 South 2nd St., Ronkonkoma, NY 11779-7238

D.C. Heath Co.
125 Spring St., Lexington, MA 02173

ICN K&K Laboratories
4911 Commerce Parkway, Cleveland, OH 44128

Institute for Scientific Information
3501 Market St., Philadelphia, PA 19104

Instruments for Research and Industry (I^2R), Inc.
108 Franklin Ave., Cheltenham, PA 19012

Johnson Matthey, Catalytic Systems Division
Wayne, PA 19087

Macmillan Publishing Co.
866 Third Ave., New York, NY 10022

Matheson Gas Products, Inc.
PO Box 1587, Secaucus, NJ 07094

McGraw–Hill Book Co.
1221 Avenue of the Americas, New York, NY 10020

Merck and Company, Inc.
Rahway, NJ 07065

Nicolet Instruments
5225 Verona Road, Madison, WI 53711

Perkin–Elmer Corp.
761 Main Ave., Norwalk, CT 06859-0219

Pergamon Press
Headington Hill Hall, Oxford OX3 OBW, England

Pike, Szafran, and Singh, Inc.
Dept. of Chemistry, Merrimack College, N. Andover, MA 01845

Rainin Instrument Co.
Mack Road, Woburn, MA 01801

Sadtler Research Laboratories, Division of Bio-Rad Laboratories, Inc.
3316 Spring Garden St., Philadelphia, PA 19104

Sargent Welch Scientific Co., a VWR Company
7350 North Linder Ave., PO Box 1026, Skokie, IL 60077-1026

Saunders College Publishing Co., A Division of Harcourt,
Brace, Jovanovich, Inc.,
Independence Square, Philadelphia, PA 19106

Sigma Chemical Company
PO Box 14508, St. Louis, MO 63178-9974

STN^{SM} International
2540 Olen Tangy River Road, PO Box 02228, Columbus, OH 43202

Thomas Scientific
99 High Hill Rd. at I-295, PO Box 99, Swedesboro, NJ 08085-0099

Varian Techtron Pty Limited
Mulgrave, Australia

Wheaton
1000 North Tenth St., Millville, NJ 08332

John Wiley & Sons, Inc.
605 Third Ave., New York, NY 10158

Index

H

I

K

L

M

S

T

V

W

CPSIA information can be obtained at www.ICGtesting.com
Printed in the USA
BVOW051507080112

279999BV00001B/7/A